T0282503

LONDON MATHEMATICAL SOCIETY LECTURE NOTE SE

Managing Editor: Professor J.W.S. Cassels, Department of Pure Mathemati
University of Cambridge, 16 Mill Lane, Cambridge CB2 1SB, England.

The titles below are available from booksellers, or, in case of difficulty, from Cambridge University Press.

34 Representation theory of Lie groups, M.F. ATIYAH *et al*
46 p-adic analysis: a short course on recent work, N. KOBLITZ
50 Commutator calculus and groups of homotopy classes, H.J. BAUES
59 Applicable differential geometry, M. CRAMPIN & F.A.E. PIRANI
66 Several complex variables and complex manifolds II, M.J. FIELD
69 Representation theory, I.M. GELFAND *et al*
76 Spectral theory of linear differential operators and comparison algebras, H.O. CORDES
77 Isolated singular points on complete intersections, E.J.N. LOOIJENGA
83 Homogeneous structures on Riemannian manifolds, F. TRICERRI & L. VANHECKE
86 Topological topics, I.M. JAMES (ed)
87 Surveys in set theory, A.R.D. MATHIAS (ed)
88 FPF ring theory, C. FAITH & S. PAGE
89 An F-space sampler, N.J. KALTON, N.T. PECK & J.W. ROBERTS
90 Polytopes and symmetry, S.A. ROBERTSON
92 Representation of rings over skew fields, A.H. SCHOFIELD
93 Aspects of topology, I.M. JAMES & E.H. KRONHEIMER (eds)
94 Representations of general linear groups, G.D. JAMES
95 Low-dimensional topology 1982, R.A. FENN (ed)
96 Diophantine equations over function fields, R.C. MASON
97 Varieties of constructive mathematics, D.S. BRIDGES & F. RICHMAN
98 Localization in Noetherian rings, A.V. JATEGAONKAR
99 Methods of differential geometry in algebraic topology, M. KAROUBI & C. LERUSTE
100 Stopping time techniques for analysts and probabilists, L. EGGHE
104 Elliptic structures on 3-manifolds, C.B. THOMAS
105 A local spectral theory for closed operators, I. ERDELYI & WANG SHENGWANG
107 Compactification of Siegel moduli schemes, C-L. CHAI
108 Some topics in graph theory, H.P. YAP
109 Diophantine analysis, J. LOXTON & A. VAN DER POORTEN (eds)
110 An introduction to surreal numbers, H. GONSHOR
113 Lectures on the asymptotic theory of ideals, D. REES
114 Lectures on Bochner-Riesz means, K.M. DAVIS & Y-C. CHANG
115 An introduction to independence for analysts, H.G. DALES & W.H. WOODIN
116 Representations of algebras, P.J. WEBB (ed)
118 Skew linear groups, M. SHIRVANI & B. WEHRFRITZ
119 Triangulated categories in the representation theory of finite-dimensional algebras, D. HAPPEL
121 Proceedings of *Groups - St Andrews 1985*, E. ROBERTSON & C. CAMPBELL (eds)
122 Non-classical continuum mechanics, R.J. KNOPS & A.A. LACEY (eds)
125 Commutator theory for congruence modular varieties, R. FREESE & R. MCKENZIE
126 Van der Corput's method of exponential sums, S.W. GRAHAM & G. KOLESNIK
128 Descriptive set theory and the structure of sets of uniqueness, A.S. KECHRIS & A. LOUVEAU
129 The subgroup structure of the finite classical groups, P.B. KLEIDMAN & M.W. LIEBECK
130 Model theory and modules, M. PREST
131 Algebraic, extremal & metric combinatorics, M-M. DEZA, P. FRANKL & I.G. ROSENBERG (eds)
132 Whitehead groups of finite groups, ROBERT OLIVER
133 Linear algebraic monoids, MOHAN S. PUTCHA
134 Number theory and dynamical systems, M. DODSON & J. VICKERS (eds)
135 Operator algebras and applications, 1, D. EVANS & M. TAKESAKI (eds)
136 Operator algebras and applications, 2, D. EVANS & M. TAKESAKI (eds)
137 Analysis at Urbana, I, E. BERKSON, T. PECK, & J. UHL (eds)
138 Analysis at Urbana, II, E. BERKSON, T. PECK, & J. UHL (eds)
139 Advances in homotopy theory, S. SALAMON, B. STEER & W. SUTHERLAND (eds)
140 Geometric aspects of Banach spaces, E.M. PEINADOR & A. RODES (eds)
141 Surveys in combinatorics 1989, J. SIEMONS (ed)
144 Introduction to uniform spaces, I.M. JAMES
145 Homological questions in local algebra, JAN R. STROOKER
146 Cohen-Macaulay modules over Cohen-Macaulay rings, Y. YOSHINO
147 Continuous and discrete modules, S.H. MOHAMED & B.J. MÜLLER
148 Helices and vector bundles, A.N. RUDAKOV *et al*
149 Solitons, nonlinear evolution equations and inverse scattering, M. ABLOWITZ & P. CLARKSON
150 Geometry of low-dimensional manifolds 1, S. DONALDSON & C.B. THOMAS (eds)

151 Geometry of low-dimensional manifolds 2, S. DONALDSON & C.B. THOMAS (eds)
152 Oligomorphic permutation groups, P. CAMERON
153 L-functions and arithmetic, J. COATES & M.J. TAYLOR (eds)
154 Number theory and cryptography, J. LOXTON (ed)
155 Classification theories of polarized varieties, TAKAO FUJITA
156 Twistors in mathematics and physics, T.N. BAILEY & R.J. BASTON (eds)
157 Analytic pro-*p* groups, J.D. DIXON, M.P.F. DU SAUTOY, A. MANN & D. SEGAL
158 Geometry of Banach spaces, P.F.X. MÜLLER & W. SCHACHERMAYER (eds)
159 Groups St Andrews 1989 volume 1, C.M. CAMPBELL & E.F. ROBERTSON (eds)
160 Groups St Andrews 1989 volume 2, C.M. CAMPBELL & E.F. ROBERTSON (eds)
161 Lectures on block theory, BURKHARD KÜLSHAMMER
162 Harmonic analysis and representation theory, A. FIGA-TALAMANCA & C. NEBBIA
163 Topics in varieties of group representations, S.M. VOVSI
164 Quasi-symmetric designs, M.S. SHRIKANDE & S.S. SANE
165 Groups, combinatorics & geometry, M.W. LIEBECK & J. SAXL (eds)
166 Surveys in combinatorics, 1991, A.D. KEEDWELL (ed)
167 Stochastic analysis, M.T. BARLOW & N.H. BINGHAM (eds)
168 Representations of algebras, H. TACHIKAWA & S. BRENNER (eds)
169 Boolean function complexity, M.S. PATERSON (ed)
170 Manifolds with singularities and the Adams-Novikov spectral sequence, B. BOTVINNIK
171 Squares, A.R. RAJWADE
172 Algebraic varieties, GEORGE R. KEMPF
173 Discrete groups and geometry, W.J. HARVEY & C. MACLACHLAN (eds)
174 Lectures on mechanics, J.E. MARSDEN
175 Adams memorial symposium on algebraic topology 1, N. RAY & G. WALKER (eds)
176 Adams memorial symposium on algebraic topology 2, N. RAY & G. WALKER (eds)
177 Applications of categories in computer science, M. FOURMAN, P. JOHNSTONE, & A. PITTS (eds)
178 Lower K- and L-theory, A. RANICKI
179 Complex projective geometry, G. ELLINGSRUD *et al*
180 Lectures on ergodic theory and Pesin theory on compact manifolds, M. POLLICOTT
181 Geometric group theory I, G.A. NIBLO & M.A. ROLLER (eds)
182 Geometric group theory II, G.A. NIBLO & M.A. ROLLER (eds)
183 Shintani zeta functions, A. YUKIE
184 Arithmetical functions, W. SCHWARZ & J. SPILKER
185 Representations of solvable groups, O. MANZ & T.R. WOLF
186 Complexity: knots, colourings and counting, D.J.A. WELSH
187 Surveys in combinatorics, 1993, K. WALKER (ed)
188 Local analysis for the odd order theorem, H. BENDER & G. GLAUBERMAN
189 Locally presentable and accessible categories, J. ADAMEK & J. ROSICKY
190 Polynomial invariants of finite groups, D.J. BENSON
191 Finite geometry and combinatorics, F. DE CLERCK *et al*
192 Symplectic geometry, D. SALAMON (ed)
193 Computer algebra and differential equations, E. TOURNIER (ed)
194 Independent random variables and rearrangement invariant spaces, M. BRAVERMAN
195 Arithmetic of blowup algebras, WOLMER VASCONCELOS
196 Microlocal analysis for differential operators, A. GRIGIS & J. SJÖSTRAND
197 Two-dimensional homotopy and combinatorial group theory, C. HOG-ANGELONI, W. METZLER & A.J. SIERADSKI (eds)
198 The algebraic characterization of geometric 4-manifolds, J.A. HILLMAN
199 Invariant potential theory in the unit ball of $\mathbf{C}^n$, MANFRED STOLL
200 The Grothendieck theory of dessins d'enfant, L. SCHNEPS (ed)
201 Singularities, JEAN-PAUL BRASSELET (ed)
202 The technique of pseudodifferential operators, H.O. CORDES
203 Hochschild cohomology of von Neumann algebras, A. SINCLAIR & R. SMITH
204 Combinatorial and geometric group theory, A.J. DUNCAN, N.D. GILBERT & J. HOWIE (eds)
205 Ergodic theory and its connections with harmonic analysis, K. PETERSEN & I. SALAMA (eds)
206 Lectures on noncommutative geometry, J. MADORE
207 Groups of Lie type and their geometries, W.M. KANTOR & L. DI MARTINO (eds)
208 Vector bundles in algebraic geometry, N.J. HITCHIN, P. NEWSTEAD & W.M. OXBURY (eds)
209 Arithmetic of diagonal hypersurfaces over finite fields, F.Q. GOUVÊA & N. YUI
210 Hilbert C*-modules, E.C. LANCE
211 Groups 93 Galway / St Andrews I, C.R. CAMPBELL *et al*
212 Groups 93 Galway / St Andrews II, C.R. CAMPBELL *et al*
215 Number theory, S. DAVID (ed)
216 Stochastic partial differential equations, A. ETHERIDGE (ed)
221 Harmonic approximation, S. GARDINER

London Mathematical Society Lecture Note Series. 215

Number Theory

Séminaire de Théorie des Nombres de Paris 1992-3

Edited by

Sinnou David
Université Pierre et Marie Curie, Paris

Number Theory
Paris 1992–93

Liste des conférenciers

5 octobre : H. CARAYOL. — *Représentations galoisiennes et congruences*

12 octobre : E. PEYRE. — *Points de hauteur donnée sur une surface de Del Pezzo*

D. BROWNAWELL. — *Transcendance on Drinfeld modules*

19 octobre : I. PAYS. — *Algèbres de quaternions et formes quadratiques entières*

2 novembre : G. HARMAN. — *Types of fractions near irrationals*

9 novembre : P. RAMBOUR. — *Propriétés galoisiennes d'un anneau d'entiers en caractéristique p*

16 novembre : T. HALES. — *Sphere packings*

23 novembre : E. FOUVRY. — *Nombres premiers supersinguliers*

30 novembre : V. GRITSENKO. — *Maass wave functions on four dimensional hyperbolic space*

7 décembre : T.N. SHOREY. — *Squares in products from a block of consecutive integers*

14 décembre : R. TIJDEMAN. — *Complementing sets of integers*

4 janvier : M. HINDRY. — *Hauteurs de Néron sur les variétés abéliennes*

11 janvier : Q. LIU. — *Modèles stables des courbes de genre deux*

18 janvier : G. CHRISTOL. — *Indice d'opérateurs différentiels*

25 janvier : R. COULANGEON. — *Réseaux unimodulaires et quaternioniens*

1er février : N. BOSTON. — *Some applications of families of Galois representations*

8 février : H. OUKHABA. — *Unités elliptiques et unités cyclotomiques dans les corps de fonctions*

1er mars : J. BOXALL. — *Sous-variétés algébriques de variétés abéliennes sur un corps fini*

8 mars : J. MARTINET. — *Structures algébriques sur les réseaux*

15 mars : E. ULLMO. — *Géométrie d'Arakelov des courbes elliptiques*

22 mars : J.-M. COUVEIGNES. — *Calcul et rationalité des fonctions de Belyi*

29 mars : R. MURTY. — *The supersolvable reciprocity law*

5 avril : M.L. BROWN. — *Modules singuliers et modules supersinguliers de modules de Drinfeld*

26 avril : L. SCHNEPS. — *Groupe de Grothendieck-Teichmüller et automorphismes de groupes de tresses*

3 mai : J.-P. WINTENBERGER. — *Relèvements de représentations adéliques associées aux motifs de modules de Drinfeld*

10 mai : D. BLASIUS. — *Good reduction of Shimura varieties*

17 mai : S. SAITO. — *Cohomological Hasse principle for a surface over a number field*

24 mai : H. HIDA. — *Λ-adic forms of half-integral weight*

7 juin : W. RASKIND. — *Applications d'Abel-Jacobi ℓ-adiques supérieures*

Y. ANDRÉ. — *Pour une théorie inconditionnelle des motifs*

14 juin : K. ALLADI. — *The combinatorics of words and applications to partitions*

21 juin : P. COHEN. — *Propriétés transcendantes des fonctions automorphes*

28 juin : M. KUWATA. — *Points rationnels sur les tordues de courbes elliptiques*

Y. ZARHIN. — *Hodge and Tate cycles on simple abelian fourfolds*

Les textes qui suivent sont pour la plupart des versions écrites de conférences données pendant l'année 1992–93 au Séminaire de Théorie des Nombres de Paris. Ce séminaire est financièrement soutenu par le C.N.R.S. et regroupe des arithméticiens de plusieurs universités et est dotée d'un conseil scientifique et éditorial. Ont été aussi adjoints certains textes dont la mise à la disposition d'un large public nous a paru intéressante. Les articles présentés ici exposent soit des résultats nouveaux, soit des synthèses originales de questions récentes; ils ont en particulier tous fait l'objet d'un rapport.

Ce recueil doit bien sûr beaucoup à tous les participants du séminaire et à ceux qui ont accepté d'en réviser les textes. Il doit surtout à Monique Le Bronnec qui s'est chargée du secrétariat et de la mise au point définitive du manuscrit; son efficacité et sa très agréable collaboration ont été cruciales dans l'élaboration de ce livre.

Pour le Conseil éditorial et scientifique

S. DAVID

Number Theory
Paris 1992–93

The method of weighted words and applications to partitions

Krishnaswami Alladi

ABSTRACT. The study of identities of Rogers-Ramanujan type forms an important part of the theory of partitions and q–series. These identities relate partitions whose parts satisfy certain difference conditions to partitions whose parts satisfy congruence conditions. Lie Algebras have provided a natural setting in which many such identities have arisen. In this paper a new technique called "the method of weighted words" is discussed and various applications illustrated. The method is particularly useful in obtaining generalisations and refinements of various Rogers-Ramanujan type identities. In doing so, new companions to familiar identities emerge. Gordon and I introduced the method a few years ago to obtain generalisations and refinements of the celebrated 1926 partition theorem of Schur. The method has now been improved in collaboration with Andrews and Gordon thereby increasing its applicability. The improved method yielded a generalisation and a strong refinement of a recent partition conjecture of Capparelli which arose in a study of Lie Algebras. Another application is a refinement and generalisation of a deep partition theorem of Göllnitz. A unified approach to these partition identities is presented here by blending the ideas in four of my recent papers with Andrews and Gordon. Proofs of many of the results are given, but for those where the details are complicated, only the main ideas are sketched.

1. — Introduction

Identities of Rogers-Ramanujan type form an important part of the theory of partitions and q–series. Generally, one side of these identities is in the form of an infinite series, while the other side is an infinite product. Usually, the series is the generating function for partitions whose parts satisfy certain difference conditions whereas the product is the generating function for partitions whose parts satisfy congruence conditions. The literature on such identities is vast (see for instance Andrews [10], [11]).

Andrews' monograph [11] gives a quick overview of several recent major advances and discusses applications to many areas within mathematics, and even to Physics.

The generic name "Identities of Rogers-Ramanujan Type", stems from the two celebrated identities due independently to L. J. Rogers and S. Ramanujan which are the prototype, namely,

$$(1.1)\quad \sum_{n=0}^{\infty} \frac{q^{n^2}}{(1-q)(1-q^2)\dots(1-q^n)} = \prod_{m=0}^{\infty} \frac{1}{(1-q^{5m+1})(1-q^{5m+4})}$$

and

$$(1.2)\quad \sum_{n=0}^{\infty} \frac{q^{n^2+n}}{(1-q)(1-q^2)\dots(1-q^n)} = \prod_{m=0}^{\infty} \frac{1}{(1-q^{5m+2})(1-q^{5m+3})}.$$

Indeed, even today, (1.1) and (1.2) are unmatched in simplicity, elegance and depth. These two identities have nice combinatorial interpretations as observed by MacMahon and Schur :

THEOREM R. — *For $i = 1, 2$, the number of partitions of an integer n into parts with minimal difference 2 and each part $\geq i$, is equal to the number of partitions of n into parts $\equiv \pm i \pmod 5$.*

However, no simple combinatorial proof of (1.1) and (1.2) is known. One reason for the difficulty is that we do not know how to refine (1.1) and (1.2) by introducing a free parameter whose power would represent an important statistic in the partitions being counted. The term refinement is explained below.

There are several examples of Rogers-Ramanujan type identities for which refinements are known, the most famous being the 1926 partition theorem of Schur [22] :

SCHUR'S THEOREM. — *Let $S(n)$ denote the number of partitions of n into distinct parts $\equiv \pm 1 \pmod 3$.*

Let $S_1(n)$ denote the number of partitions of n into parts with minimal difference 3, and such that no two consecutive multiples of 3 can occur as parts. Then

$$S(n) = S_1(n).$$

The equality in Schur's theorem can be refined to

$$(1.3)\quad S(n;k) = S_1(n;k),$$

for any positive integer k, where $S(n;k)$ and $S_1(n;k)$ count partitions of the type counted by $S(n)$ and $S_1(n)$, but with the added restriction that there are precisely k parts, and with the convention that parts $\equiv 0 \pmod 3$ are counted twice. Combinatorial proofs can usually be given for partition identities which permit refinements. This is also the case with Schur's theorem; for combinatorial proofs of Schur's theorem, see Bressoud [14] or Alladi-Gordon [1].

Schur had originally stated the equality of three partition functions $T(n) = S(n) = S_1(n)$, where $T(n)$ is the number of partitions of n into parts $\equiv \pm 1 \pmod 6$. We do not discuss $T(n)$ in this paper because refinements of the type (1.3) are not possible with $T(n)$. The equality $T(n) = S_1(n)$ can be considered as the next case beyond the Rogers–Ramanujan identities because the minimal difference 2 is replaced by 3 and the modulus 5 is replaced by 6. But in doing so, Schur realised that he needed the extra condition that consecutive multiples of 3 should not occur as parts.

The purpose of this paper is to discuss a new technique called *the method of weighted words* originally due to Alladi and Gordon [1], [2], and which has recently been improved by Alladi, Andrews and Gordon [3], [4]. This method is particularly useful in the study of Rogers-Ramanujan identities which permit refinements. Indeed, the method provides substantial refinements and generalisations often involving several free parameters which keep track of the number of parts in various residue classes. In many instances these refinements lead to bijective proofs of the partition theorem in question, in a natural way.

The basic idea of the method is to consider positive integers which occur in various colours denoted by $a, b, c, \ldots$. We then form words whose letters are symbols $a, b, c, \ldots$ with subscripts, where the subscript denotes the integer or part of the partition, and the symbols $a, b, \ldots$ will denote the colour of that part. Special types of words are considered by impossing certain order rules on the symbols and certain gap conditions on the subscripts. These will correspond to the gap conditions on the parts in the partition theorem being discussed. The usefulness in the method lies in the fact that the symbols $a, b, c, \ldots$ play a dual role. On the one hand they represent colours, and on the other, they are free parameters when computing generating functions. The upshot of all this is that the partition theorem in question becomes a special case under certain dilations and translations, and the free parameters $a, b, c, \ldots$ provide the necessary refinements of the partition theorem. By changing the order rules on the symbols $a, b, c, \ldots$, several new companion partition identities are generated.

The method was introduced in [1] to obtain refinements and generalisations of Schur's theorem. Subsequently [2], it was noticed that by changing the order rules for the symbols several companions to Schur's theorem would be generated. More precisely, the function $S_1(n)$ is only one of six partition functions $S_\mu(n), \mu = 1, 2, \ldots, 6$, all equal. Generalisations, refinements and companions to Schur's theorem are discussed in §§4, 5.

In recent years, Lie Algebras have provided a general setting where Rogers-Ramanujan type identities have been discovered (see Lepowsky and Wilson [19], [20], [21]). Motivated by a study of Vertex Operators in Lie Algebras, Capparelli [15] [16] made the following conjecture :

CAPPARELLI'S CONJECTURE. — *Let $C^*(n)$ denote the number of partitions of into parts* $\equiv \pm 2, \pm 3 \pmod{12}$.

Let $D(n)$ denote the number of partitions of n into parts > 1 with minimal difference 2, where the difference is ≥ 4 unless consecutive parts are multiples of 3 or add up to a multiple of 6.

Then $C^*(n) = D(n)$.

Capparelli's conjecture was proved by Andrews in 1992 [12] using generating functions. Since there are similarities between Capparelli's conjecture and Schur's theorem, it was natural to see whether the method of weighted words would apply. And indeed it did, but the key idea was to replace $C^*(n)$ by the equivalent function $C(n)$ which denotes the number of partitions of n into distinct parts $\equiv 2, 3, 4$ or $6 \pmod 6$. This is for the purpose of refinements. In otherwords, $C^*(n)$ is like $T(n)$ in Schur's theorem for which refinements are not possible and $C(n)$ is like $S(n)$ in Schur's theorem which permits refinements. Andrews, Gordon and I [3] obtained a three parameter refinement and generalisation (see Theorem 14 in §7) from which Capparelli's conjecture followed as a special case under suitable dilations and translations.

The deepest application of the method of weighted words so far has been a proof of a substantial generalisation and refinement due to Andrews, Gordon and myself [4], of the following formidable theorem of Göllnitz [18] :

GÖLLNITZ'S THEOREM. — *Let $A(n)$ denote the number of partitions of n into parts* $\equiv 2,\ 5,\ 11 \pmod{12}$.

Let $B(n)$ denote the number of partitions of n into distinct parts $\equiv 2,\ 4,$ *or* $5 \pmod 6$.

Let $C(n)$ denote the number of partitions of n in the form $m_1 + m_2 + \cdots + m_\nu$, no part equal to 1 or 3, and such that $m_l - m_{l+1} \geq 6$ with strict inequality if $m_l \equiv 6,\ 7$ or $9 \pmod 6$.

Then $A(n) = B(n) = C(n)$.

The equality $A(n) = B(n)$ is trivial, and so the real challenge is the equality $B(n) = C(n)$. Moreover, the function $A(n)$ is like $C^*(n)$ in Capparelli's conjecture and $T(n)$ in Schur's Theorem, permitting no refinements. So we will not discuss $A(n)$ in this paper.

Göllnitz [18] actually established the refinement

$$B(n;\nu) = C(n;\nu), \tag{1.4}$$

where $B(n;\nu)$ and $C(n;\nu)$ denote the number of partitions counted by $B(n)$ and $C(n)$ respectively, with the additional restriction that there are precisely ν parts, and the convention that parts $\equiv 6,\ 7$ or $9 (\mathrm{mod}\, 6)$ are counted twice. His proof is complicated and the details are forbidding. Andrews [8] gave a proof using generating functions, and subsequently gave a second proof ([11], § 10) where he used computer algebra to simplify the calculations. He then asked for a proof which lends more insight into the equality (1.4). We believe that our proof by the method of weighted words does provide this insight (see § 9 for a sketch of the main ideas behind this proof).

The main idea is to consider integers in six colours, of which three colours a, b and c are primary, and three colours ab, ac and bc are secondary. We then impose certain order rules on the coloured integers. Göllnitz' theorem is viewed as emerging from an incredible key identity (see (9.11) in § 9). The proof of this key identity is deep and difficult and may be found in [4]. It requires not only Watson's q-analog of Whipple's theorem but also the ${}_6\Psi_6$ summation of Bailey. This is substantially more than what is required to prove either Schur's theorem or the Rogers-Ramanujan identities.

One advantage in using primary and secondary colours is that this explains the requirement for strict inequalities in Göllnitz' theorem when $m_l \equiv 6,\ 7$ or $9 (\mathrm{mod}\, 6)$. As will be seen in § 9, the residues $2, 4$ and $5 (\mathrm{mod}\, 6)$ correspond to the primary colours a, b and c whereas the residues 6, 7 and $9 (\mathrm{mod}\, 6)$ correspond to the secondary colours ab, ac and bc.

Another advantage with our approach is that Schur's theorem falls out as a special case by setting $c = 0$. For then, colours c, ac and bc disappear and we are left only with three colours a, b and ab, which corresponds to Schur's theorem.

The paper concludes with § 10 where some further problems going beyond Göllnitz' theorem are described.

2. — Notations

We adopt the standard notation

$$(a)_n = (a;q)_n = \prod_{j-0}^{n-1} (1 - aq^j)$$

for any complex number a, and

$$(a)_\infty = \lim_{n\to\infty} (a)_n = \prod_{j=0}^{\infty} (1 - aq^j), \text{ for } |q| < 1. \tag{2.1}$$

In fact (2.1) can be used to define $(a)_n$ for all real n by means of the relation

$$(a)_n = \frac{(a)_\infty}{(aq^n)_\infty}. \tag{2.2}$$

For a positive integer t, the q–multinomial coefficient of order t is defined by

$$\begin{bmatrix} n_1+n_2+\cdots+n_t \\ n_1, n_2, \ldots, n_t \end{bmatrix} = \begin{bmatrix} n_1+n_2+\cdots+n_t \\ n_1, n_2, \ldots, n_t \end{bmatrix}_q = \frac{(q)_{n_1+n_2+\cdots+n_t}}{(q)_{n_1}(q)_{n_2}\cdots(q)_{n_t}}. \tag{2.3}$$

If the base for the multinomial coefficient is q, then as in (2.3) we sometimes do not write it; if the base is anything other than q, it will be displayed. Multinomial coefficients of order $t = 3$ will be used in the proof of Schur's theorem while those of order $t = 6$ will appear in the discussion on Gollnitz's theorem. When $t = 2$, the expression $\begin{bmatrix} n_1+n_2 \\ n_1, n_2 \end{bmatrix}$ is the q–binomial coefficient. This is often denoted by $\begin{bmatrix} n_1+n_2 \\ n_1 \end{bmatrix}$. Binomial coefficients to bases q and q^2 will appear in the discussion of Capparelli's conjecture.

Given a partition π, we let $\sigma(\pi)$ denote the sum of its parts, and $\nu(\pi)$, the number of parts of π. Also $\lambda(\pi)$ will denote the largest part of π. Sometimes the parts will be arranged according to a specific lexicographic ordering (not necessarily the standard ordering of the positive integers) and in such cases $\lambda(\pi)$ will denote the largest part of π with respect to this ordering. Quite often we will be counting the number of parts of π of a specific type, and this will be denoted by a subscript, eg : $\nu_a(\pi)$ will be the number of a–parts of π (precise definition will be given when necessary).

If $\pi = b_1+b_2+b_3+\cdots$ and $\pi' = b_1'+b_2'+b_3'+\cdots$ are two partitions whose parts, as is customary, are written in decreasing order, then $\pi'' = \pi + \pi'$ is defined to be the partition $b_1'' + b_2'' + b_3'' + \ldots$, where $b_j'' = b_j + b_j'$ for each j. If $\nu(\pi) \neq \nu(\pi')$, then in place of b_j or b_j' (whichever is missing), the number 0 is substituted while computing b_j''. Thus $\nu(\pi + \pi') = \max(\nu(\pi), \nu(\pi'))$.

We also adopt the convention that $j \equiv b (\bmod m)$ means $j = b + lm$, with $j > 0$ and $l \geq 0$.

Finally, whenever we refer to a minimal partition, we mean a partition for which $\sigma(\pi)$ is minimal subject to the given conditions.

3. — Lower bound gap conditions

For partitions π which are defined by imposing lower bound conditions on the gaps or differences between consecutive parts, their generating function can be computed by considering first the minimal partitions of the type under discussion.

For instance, consider the generating function for partitions π into n distinct parts. The smallest partition into n distinct parts is $n(n+1)/2 = n + (n-1) + \cdots + 1$. Hence the generating function is

$$q^{n(n+1)/2}/(q)_n.$$

Similarly, in connection with the Rogers-Ramanujan identities, consider the generating function for partitions into n parts with minimal difference 2. The minimal partition in this case is $n^2 = (2n-1)+(2n-3)+\cdots+3+1$. Hence the generating function is

$$q^{n^2}/(q)_n,$$

which is the n–th term in (1.1).

More generally, consider the generating function

$$G(q) = \sum_{\pi} q^{\sigma(\pi)} \tag{3.1}$$

for partitions π which are given by specifying $\nu(\pi) = m$, the number of parts of π, and by imposing certain lower bound gap conditions on the parts. Let

$$H(q) = \sum_{\pi^*} q^{\sigma(\pi^*)} \tag{3.2}$$

be the generating function for all minimal partitions π^* that can be constructed subject to these conditions. We then have

LEMMA 1. — $G(q) = H(q)/(q)_m$.

Proof : lemma 1 is an immediate consequence of the observation that the partitions π counted in (3.1) can all be realised in the form

$$\pi = \pi^* + \pi', \tag{3.3}$$

where π' is a partition which satisfies $\nu(\pi') \le \nu(\pi^*) = m$. The best way to see (3.3) is to draw the Ferrers graphs of π, π^* and π'. Since the generating function of the graphs π' is $1/(q)_m$, Lemma 1 follows from (3.3).

The lower bound gap conditions defining π can be quite complicated. Lemma 1 is extremely useful in such situations.

4. — The method of weighted words

Let each integer $j \ge 2$ occur in three colours, red, blue and purple, and let the integer 1 occur only in two colours, red and blue. We introduce the symbols a_j, b_j and c_j to represent the integer (part) j in colours red, blue and purple respectively. Sometimes we refer to a_j as an a–part, b_j as a b–part and c_j as a c–part. We think of the letters a, b, and c as free parameters. Under the transformation

$$\text{(4.1)} \quad \text{(dilation) } q \mapsto q^3 \quad \text{and (translations) } a \mapsto aq^{-2}, b \mapsto bq^{-1}, c \mapsto cq^{-3},$$

the powers of the parameters a, b and c will represent the number of parts in residue classes 1, 2 and 3(mod 3) in partitions counted by $S_1(n)$. In order to make the transition from $S_1(n)$ to $S(n)$ we will need to choose $c = ab$. That is why we think of c as having colour purple=red plus blue.

The gap between any two symbols representing coloured integers is defined to be the absolute value of the difference between their subscripts. The gap is a non-negative integer without colour. For example, the gap between a_5 and c_5 is 0, and between a_5 and b_7 is 2. The weight of a symbol is its subscript. For example the weight of b_6 is 6, and b_6 represents the integer 6 coloured blue. In order to discuss partitions using these symbols, we need a lexiographic ordering of the symbols.

First we consider the following lexiographic ordering :

$$\text{(4.2)} \qquad a_1 \prec b_1 \prec c_2 \prec a_2 \prec b_2 \prec c_3 \prec a_3 \prec b_3 \prec \cdots$$

We shall refer to this as Scheme 1. Now by a word π we mean a collection of these symbols arranged in non-increasing order according to the specified scheme (lexicographic ordering), in the present case, Scheme 1. By $\sigma(\pi) = n$ we mean that the sum of the weights of π is n. In this sense we may think of π as a partition of n into coloured integers and we use the terms word and partition interchangeably as convenient. For example, $\pi = a_7c_7b_5b_5a_4c_4c_2b_1a_1a_1$ is a word with $\sigma(\pi) = 37$. To specify the partition of 37 into coloured integers we sometimes write $a_7 + c_7 + b_5 + b_5 + a_4 + c_4 + c_2 + b_1 + a_1 + a_1$. Let $\nu_a(\pi), \nu_b(\pi), \nu_c(\pi)$ denote the number of a–parts, b–parts and c–parts of π respectively. In this example $\nu_a(\pi) = 4$, $\nu_b(\pi) = 3$, and $\nu_c(\pi) = 3$.

We now consider words $e_1e_2 \ldots e_\nu$, where the e_l are symbols from (4.2), such that the gap between the symbols is ≥ 1, with the added restriction that the gap between consecutive symbols e_l and e_{l+1} is > 1 if

$$\text{(4.3)} \qquad \begin{cases} e_l \text{ is red, } e_{l+1} \text{ is blue} \\ \qquad \text{or } e_l \text{ is purple.} \end{cases}$$

We refer to this as a Type 1 word or a Type 1 partition. At first glance condition (4.3) might seem artificial but it is a natural generalisation of the gap conditions defining $S_1(n)$. Also, there are interesting generating functions attached to Type 1 partitions. More precisely we have the following lemma which was established in [1] :

LEMMA 2. — *Let* $G = G(i, j, k; q)$ *be the generating function for all Type* 1 *partitions* π *having* $\nu_a(\pi) = i, \nu_b(\pi) = j$ *and* $\nu_c(\pi) = k$. *Then*

$$G = \frac{q^{T_{i+j+k}+T_k}}{(q)_i (q)_j (q)_k},$$

where $T_m = \frac{m(m+1)}{2}$ *is the* m*–th triangular number.*

Since all partitions π counted by G have $\nu(\pi) = i+j+k$, it follows from Lemma 1 that Lemma 2 is equivalent to the statment that

$$H(i, j, k) = G(i, j, k)(q)_{i+j+k} = \begin{bmatrix} i+j+k \\ i, j, k \end{bmatrix} q^{T_{i+j+k}+T_k}. \tag{4.4}$$

Here $H(i, j, k)$ is the generating function for all minimal Type 1 partitions π^* with $\nu_a(\pi^*) = i$, $\nu_b(\pi^*) = j$ and $\nu_c(\pi^*) = k$. Since the multinomial coefficients satisfy a recurrence relation, it turns out that (4.4) can be proved by induction on $i + j + k$, and indeed this was how Lemma 2 was proved in [1].

For the purpose of the induction, it is convenient to consider the decomposition

$$H = H_a + H_b + H_c, \tag{4.5}$$

where H_a, H_b and H_c are the generating functions for partitions counted by H, with the additional restriction that the smallest part is an a–part, b–part and c–part respectively. It can be showed by induction on $i + j + k$ (see [1]) that :

$$\begin{aligned} H_a(i, j, k) &= q^{T_{i+j+k}+T_k} \begin{bmatrix} i+j+k-1 \\ i-1, j, k \end{bmatrix} \\ H_b(i, j, k) &= q^{T_{i+j+k}+T_k+i} \begin{bmatrix} i+j+k-1 \\ i, j-1, k \end{bmatrix} \\ H_c(i, j, k) &= q^{T_{i+j+k}+T_k+i+j} \begin{bmatrix} i+j+k-1 \\ i, j, k-1 \end{bmatrix}. \end{aligned} \tag{4.6}$$

Once this is done, (4.4) follows from (4.5) and (4.6) because of the recurrence

$$\begin{aligned} \begin{bmatrix} i+j+k \\ i, j, k \end{bmatrix} = \begin{bmatrix} i+j+k-1 \\ i-1, j, k \end{bmatrix} + q^i \begin{bmatrix} i+j+k-1 \\ i, j-1, k \end{bmatrix} + \\ + q^{i+j} \begin{bmatrix} i+j+k-1 \\ i, j, k-1 \end{bmatrix} \end{aligned} \tag{4.7}$$

satisfied by the multinomial coefficients.

Next, let $B_1(n; i, j, k)$ denote the number of Type 1 partitions π of having $\nu_a(\pi) = i$, $\nu_b(\pi) = j$ and $\nu_c(\pi) = k$. Then from Lemma 2 we see that

$$\sum_{i,j,k,n} B_1(n; i, j, k) a^i b^j c^k q^n = \sum_{i,j,k} a^i b^j c^k \, G(i, j, k) = \tag{4.8}$$
$$\sum_{i,j,k} a^i b^j c^k \, \frac{q^{T_{i+j+k}+T_k}}{(q)_i (q)_j (q)_k}.$$

The main result which leads to a refinement and generalisation of Schur's theorem is

LEMMA 3. — *Let* $r, s \geq 0$ *be given integers. Then*

$$\sum_{0 \leq m \leq \min(r,s)} G(r-m, s-m, m) = \sum_{0 \leq m \leq \min(r,s)} \frac{q^{T_{r+s-m}+T_m}}{(q)_{r-m}(q)_{s-m}(q)_m} = \frac{q^{T_r+T_s}}{(q)_r (q)_s}.$$

In [1] we give two proofs of Lemma 3 one of which is combinatorial. Lemma 2 has a nice combinatorial interpretation from which Schur's theorem follows and we describe this now.

Note that

$$\sum_{r,s} \frac{a^r b^s q^{T_r+T_s}}{(q)_r (q)_s} = (-aq)_\infty (-bq)_\infty = \sum_{n,r,s} V(n; r, s) a^r b^s q^n, \tag{4.9}$$

where $V(n; r, s)$ is the number of (vector) bi-partitions $(\pi_1; \pi_2)$ of n such that π_1 has r distinct red parts and π_2 has s distinct blue parts. So Lemma 3 is the assertion that the generating functions in (4.8) and (4.9) are equal when $c = ab$. In this case with

$$i = r - m, j = s - m, k = m,$$

we have

$$i + j + 2k = r + s. \tag{4.10}$$

So we get the following combinatorial result :

THEOREM 1. — *Let* $t \geq 0$ *be given and* $V(n;t) = \sum_{r+s=t} V(n;r,s)$. *Then*

$$V(n;t) = \sum_{i+j+2k=t} B_1(n;i,j,k).$$

Theorem 1 gives a refinement of Schur's theorem under the transformations (4.1). More generally, under the transformations

$$q \mapsto q^M, \quad a \mapsto aq^{\alpha - M}, \quad b \mapsto bq^{\beta - M}$$

applied to (4.7) and (4.8) we get the following generalization and refinement of Schur's theorem :

THEOREM 2. — *Let* $M \geq 3$ *and* $0 < \alpha < \beta < M \leq \alpha + \beta$.

Let $A(n;k)$ *denote the number of partitions of* n *into* k *distinct parts* $\equiv \alpha$ *or* $\beta (\bmod M)$.

Let $B(n;k)$ *denote the number of partitions of* n *into* k *parts* $\equiv \alpha, \beta$ *or* $\alpha + \beta (\bmod M)$ *such that*

(i) *the difference between any two parts is* $\geq M$.

(ii) *the difference between parts* $\equiv (\alpha + \beta)(\bmod M)$ *is* $> M$.

(iii) *the parts* $\equiv \alpha + \beta (\bmod M)$ *are counted twice.*

Then

$$A(n;k) = B(n;k).$$

In [1] a more general form of Theorem 2 is stated by relaxing the condition $0 < \alpha < \beta < M \leq \alpha + \beta$ to the simpler condition that α, β and $\alpha + \beta$ are incongruent $(\bmod M)$. The reason condition (4.9) enters into Theorem 1 is because, given a bi-partition of n into r red and s blue parts, one takes m of the red parts and m of the blue parts to form the purple parts, leaving behind $r - m$ red parts and $s - m$ blue parts. Hence the purple parts have to be counted twice.

5. — Six companions to Schur

In 1971, using a computer search, Andrews [9] found the following companion to Schur's theorem :

THEOREM A. — *Let* $S_2(n)$ *denote the number of partitions of* n *into parts* $e_1 + e_2 + \cdots + e_\nu$ *such that* $e_l - e_{l+1} \geq 3, 2$ *or* 5 *according as* $e_l \equiv 1, 2$ *or* 3 $(\bmod 3)$. *Then* $S(n) = S_2(n)$.

Andrews gave a proof of Theorem A using generating functions in a manner similar to his proof of Schur's theorem [5]. But the exact connections between Theorem A and Schur's theorem remained unclear. These connections will now become clear by means of the method of weighted words. What is more, this approach will show that $S_1(n)$ and $S_2(n)$ are only two of six partition functions $S_\mu(n), \mu = 1, 2, \ldots, 6$, all equal to $S(n)$.

Under the transformations (4.1), the symbols in Scheme 1 become

$$a_m = 3m - 2, b_m = 3m - 1 \text{ and } c_{m+1} = (ab)_{m+1} = 3m, \tag{5.1}$$

where $c = ab$. In this case, the lexicographic ordering (4.2) for Scheme 1 yields the natural ordering

$$1 < 2 < 3 < 4 < \cdots$$

for the positive integers. From now on we will refer to the transformations in (4.1) as standard transformations.

Let $x_m = x_m^{(1)}$ denote the symbol occupying the m–th position in (4.2). That is $x_1 = a_1, x_2 = b_1, x_3 = c_2, \ldots$, and so on. Instead of defining Type 1 partitions by means of the gap conditions (4.3), we may define them as follows : Type 1 partitions are those of the form $x_{m_1} + x_{m_2} + \cdots + x_{m_\nu}$, where

$$m_l - m_{l+1} \geq 3 \text{ with strict inequality if } x_{m_l} \text{ is a } c\text{–part.} \tag{5.2}$$

This is a direct translation of the difference conditions defining $S_1(n)$ to the more general situation involving weighted symbols. From now on we will refer to the inequalities (5.2) as standard gap conditions.

In order to understand the difference conditions defining Andrews' $S_2(n)$, consider another lexicographic ordering of the symbols, namely, Scheme 2, given by

$$a_1 \prec c_2 \prec b_1 \prec a_2 \prec c_3 \prec b_2 \prec a_3 \prec c_4 \prec b_3 \prec \cdots . \tag{5.3}$$

Under the standard transformations and with $c = ab$, Scheme 2 in (5.3) yields the following different ordering of the positive integers :

$$1 \prec 3 \prec 2 \prec 4 \prec 6 \prec 5 \prec 7 \prec 9 \prec 8 \prec \cdots . \tag{5.4}$$

Next let $x_m^{(2)}$ denote the symbol occupying the m–th position in (5.4). Then the difference conditions defining Andrews' function $S_2(n)$ are equivalent to the statement that we consider partitions of the form $x_{m_1}^{(2)} + x_{m_2}^{(2)} + \cdots + x_{m_\nu}^{(2)}$, where $x_{m_l} = x_{m_l}^{(2)}$ satisfy the standard gap conditions. More generally, if $x_m^{(2)}$ denotes the symbol occupying the m–th position in (5.3), let a Type 2 partition be one of the form $x_{m_1}^{(2)} + x_{m_2}^{(2)} + \cdots + x_{m_\nu}^{(2)}$, with $x_{m_l} = x_{m_l}^{(2)}$ satisfying (5.2). Also, let $B_2(n; i, j, k)$ denote the number of Type 2 partitions π of n having $\nu_a(\pi) = i$, $\nu_b(\pi) = j$, $\nu_c(\pi) = k$. We then have :

THEOREM 3. — *Let $n > 0$ and $i, j, k \geq 0$ be integers. Then*

$$B_1(n; i, j, k) = B_2(n; i, j, k).$$

Theorem A is a consequence of Theorem 3 and Theorem 1, under the standard transformations. Indeed, Theorem 3 yields a refinement of Theorem A.

One way to prove Theorem 3 is to show that the generating functions of $B_1(n; i, j, k)$ and $B_2(n; i, j, k)$ are the same, that is show that

$$\sum_{i,j,k,n} B_2(n; i, j, k) a^i b^j c^k q^n = \sum_{i,j,k} a^i b^j c^k \frac{q^{T_{i+j+k}+T_k}}{(q)_i (q)_j (q)_k} \tag{5.5}$$

and compare with (4.8).

The proof of (5.5) proceeds in exactly the same way as that of (4.8). The only difference is that we now use another recurrence for the multinomial coefficients instead of (4.7), namely,

$$\begin{bmatrix} i+j+k \\ i,j,k \end{bmatrix} = \begin{bmatrix} i+j+k-1 \\ i-1,j,k \end{bmatrix} + q^i \begin{bmatrix} i+j+k-1 \\ i,j,k-1 \end{bmatrix} + q^{i+k} \begin{bmatrix} i+j+k-1 \\ i,j-1,k \end{bmatrix} \tag{5.6}$$

to establish the corresponding generating function formulae for minimal Type 2 partitions.

Scheme 1 was generated by the standard ordering $a_1 \prec b_1 \prec c_2$ while Scheme 2 was generated by a different ordering $a_1 \prec c_2 \prec b_1$. More generally, there are six schemes given by the six permutations of the symbols a_1, b_1 and c_2. They are

Scheme 1 : $a_1 \prec b_1 \prec c_2 \prec a_2 \prec b_2 \prec c_3 \prec \cdots$
Scheme 2 : $a_1 \prec c_2 \prec b_1 \prec a_2 \prec c_3 \prec b_2 \prec \cdots$
Scheme 3 : $c_2 \prec a_1 \prec b_1 \prec c_3 \prec a_2 \prec b_2 \prec \cdots$
Scheme 4 : $b_1 \prec a_1 \prec c_2 \prec b_2 \prec a_2 \prec c_3 \prec \cdots$
Scheme 5 : $b_1 \prec c_2 \prec a_1 \prec b_2 \prec c_3 \prec a_2 \prec \cdots$
Scheme 6 : $c_2 \prec b_1 \prec a_1 \prec c_3 \prec b_2 \prec a_2 \prec \cdots$

Actually, only three of these schemes are essentially different, because Schemes 4, 5 and 6 are obtained from Schemes 1, 2 and 3 by interchanging the roles of a and b. But there are certain advantages in discussing all six schemes as will be seen soon.

Let $x_m^{(\mu)}$ denote the symbol occupying position m in Scheme $\mu, \mu = 1, 2, \ldots, 6$. By a partition π of Type μ we mean an expression $x_{m_1}^{(\mu)} + x_{m_2}^{(\mu)} + \cdots + x_{m_\nu}^{(\mu)}$, where the $x_{m_l} = x_{m_l}^{(\mu)}$ satisfy the standard gap conditions (5.2). Let $B_\mu(n; i, j, k)$ denote the number of partitions π of n of Type μ such that $\nu_a(\pi) = i$, $\nu_b(\pi) = j$ and $\nu_c(\pi) = k$. Then, extending Theorems 3 and 1 we have

THEOREM 4. — *Let $n \geq 1$ and $i, j, k \geq 0$ be given integers. Then*

$$B_1(n; i, j, k) = B_2(n; i, j, k) = \cdots = B_6(n; i, j, k).$$

THEOREM 5. — *Let $r, s \geq 0$ be given integers. Then*

$$V(n; r, s) = \sum_{0 \leq m \leq \min(r,s)} B_\nu(n; r-m, s-m, m).$$

for $\nu = 1, 2, \ldots, 6$. Consequently, if $r + s = t$, then

$$V(n; t) = \sum_{i+j+2k=t} B_\mu(n; i, j, k).$$

Under the standard transformations, the gap conditions (5.2) defining Type μ partitions, $\mu = 1, 2, \ldots, 6$, become the following : we are now counting partitions of the form $e_1 + e_2 + \cdots + e_\nu$, where the e_l are (ordinary) positive integers satisfying the difference conditions given by :

$$\text{Type 1}: \ e_l - e_{l+1} \geq 3, 3 \text{ or } 4, \ \text{if } e_l \equiv 1, 2 \text{ or } 3 \pmod 3. \tag{5.7}$$

$$\text{Type 2}: \ e_l - e_{l+1} \geq 3, 2 \text{ or } 5, \ \text{if } e_l \equiv 1, 2 \text{ or } 3 \pmod 3. \tag{5.8}$$

$$\text{Type 3}: e_l - e_{l+1} \begin{cases} = 1 \text{ or } \geq 3, \ \text{if } e_l \equiv 1 \pmod 3, \\ \geq 2 \text{ or } 6, \ \text{if } e_l \equiv 2 \text{ or } 3 \pmod 3. \end{cases} \tag{5.9}$$

$$\text{Type 4}: e_l - e_{l+1} \begin{cases} \geq 2 \text{ or } 4, \ \text{if } e_l \equiv 1 \text{ or } 3 \pmod 3, \\ = 3 \text{ or } \geq 5, \ \text{if } e_l \equiv 2 \pmod 3. \end{cases} \tag{5.10}$$

$$\text{Type 5}: e_l - e_{l+1} \begin{cases} \geq 1, \ \text{if } e_l \equiv 1 \pmod 3, \\ = 3 \text{ or } \geq 5, \ \text{if } e_l \equiv 2 \pmod 3, \\ = 4 \text{ or } \geq 6, \ \text{if } e_l \equiv 3 \pmod 3. \end{cases} \tag{5.11}$$

$$\text{Type 6}: e_l - e_{l+1} \begin{cases} \geq 1 \text{ or } 6, \ \text{if } e_l \equiv 1 \text{ or } 3 \pmod 3, \\ = 2, 3 \text{ or } \geq 5, \ \text{if } e_l \equiv 2 \pmod 3. \end{cases} \tag{5.12}$$

Note that the difference conditions in (5.7) are precisely those defining $S_1(n)$ in Schur's theorem, whereas the conditions in (5.8) are the same as those given by Andrews for $S_2(n)$ in Theorem A.

Next, let $S_\mu(n; i, j, k)$ denote the number of Type μ partitions (those given by conditions (5.7)–(5.12) corresponding to Type μ), having i parts $\equiv 1$ (mod 3), j parts $\equiv 2$ (mod 3) and k parts $\equiv 3$ (mod 3). Then using Theorems 4 and 5 and the standard transformations, Schur's theorem and Theorem A can be improved to :

THEOREM 6. — *Given integers* $n \geq 1$ *and* $i, j, k \geq 0$,

$$S_1(n; i, j, k) = S_2(n; i, j, k) = \cdots = S_6(n; i, j, k).$$

THEOREM 7. — *Given integers* $n \geq 1$ *and* $r, s \geq 0$, *let* $S(n; r, s)$ *denote the number of partitions of* n *into* r *distinct parts* $\equiv 1 \pmod 3$ *and* s *distinct parts* $\equiv 2 \pmod 3$. *Then*

$$S(n; r, s) = \sum_{0 \leq m \leq \min(r,s)} S_\mu(n; r-m, s-m, m), \mu = 1, 2, \ldots, 6.$$

The difference conditions (5.9)–(5.12) defining $S_\mu(n)$ for $\mu = 3$, 4, 5 and 6 are more complicated than those defining $S_1(n)$ and $S_2(n)$. Hence the partition functions $S_\mu(n)$, $\mu \geq 3$, did not show up in Andrews' computer search [9]. This illustrates the usefulness of the weighted words approach. The partition functions $B_\mu(n; i, j, k)$ deal with the base case without any dilations and translations and so the gap condition defining them, namely (5.2), is uniform and elegant for all six functions. It is the lexicographic orderings that distinguish these six functions.

Since Theorems 5, 6, and 7 are consequences of Theorem 4 (and Theorem 1), we will now describe how to prove Theorem 4 utilising recurrence relations for the q–multinomial coefficients.

The q–multinomial coefficients $\begin{bmatrix} i+j+k \\ i, j, k \end{bmatrix}$ satisfy a total of six basic recurrences depending on the order in which the letters i, j and k are reduced by one. Two recurrences have already been given, namely, (4.7) and (5.6). Yet another recurrence is

$$\begin{bmatrix} i+j+k \\ i, j, k \end{bmatrix} = \begin{bmatrix} i+j+k-1 \\ i, j, k-1 \end{bmatrix} + q^k \begin{bmatrix} i+j+k-1 \\ i-1, j, k \end{bmatrix} + q^{k+i} \begin{bmatrix} i+j+k-1 \\ i, j-1, k \end{bmatrix}. \tag{5.13}$$

There are three more recurrences which we will not write down.

Next, let $H_a^{(\mu)}(i, j, k)$, $H_b^{(\mu)}(i, j, k)$, and $H_c^{(\mu)}(i, j, k)$ denote the generating functions for all minimal Type μ partitions using i a–parts, j b–parts and k c–parts respectively, such that the smallest part is a_1, b_1 and c_2. Then these generating functions can be computed by induction on $i + j + k$ in a manner similar to (4.6) as will be described presently.

For Type 1 partitions, the starting triple for Scheme 1 ordering is

$$\text{Scheme } 1: \ a_1 \prec b_1 \prec c_2.$$

Hence the generating functions $H_a = H_a^{(1)}$, $H_b = H_b^{(1)}$ and $H_c = H_c^{(1)}$ in (4.6) are given in terms of q–multinomial coefficients by replacing i by $i-1$, j by $j-1$, and k by $k-1$, in that order.

For Type 2 partitions, the starting triple in Scheme 2 ordering is

$$\text{Scheme 2}: \ a_1 \prec c_2 \prec b_1.$$

So in this case we have

$$H_a^{(2)}(i,j,k) = q^{T_{i+j+k}+T_k} \begin{bmatrix} i+j+k-1 \\ i-1, j, k \end{bmatrix}$$

$$\text{(5.14)} \qquad H_c^{(2)}(i,j,k) = q^{T_{i+j+k}+T_k+i} \begin{bmatrix} i+j+k-1 \\ i, j, k-1 \end{bmatrix}$$

$$H_c(i,j,k) = q^{T_{i+j+k}+T_k+i+k} \begin{bmatrix} i+j+k-1 \\ i, j-1, k \end{bmatrix}.$$

Equations (5.14) can be established by induction on $i+j+k$. In Scheme 2, since a_1 occurs first followed by c_2 and then by b_1, the letters i, k and j are reduced by 1 and in that order in (5.14). From (5.14) and (5.6) it follows that

$$\text{(5.15)} \qquad H_a^{(2)} + H_b^{(2)} + H_c^{(2)} = H^{(2)} = H = G \cdot (q)_{i+j+k}$$

with H and G as in (4.4). This yields Theorem 3, which is the first equality in Theorem 4.

Similarly, since the starting triple for Scheme 3 ordering is

$$\text{Scheme 3}: \ c_2 \prec a_1 \prec b_1,$$

we have

$$H_c^{(3)}(i,j,k) = q^{T_{i+j+k}+T_k} \begin{bmatrix} i+j+k-1 \\ i, j, k-1 \end{bmatrix}$$

$$\text{(5.16)} \qquad H_a^{(3)}(i,j,k) = q^{T_{i+j+k}+T_k+k} \begin{bmatrix} i+j+k-1 \\ i-1, j, k \end{bmatrix}$$

$$H_b^{(3)}(i,j,k) = q^{T_{i+j+k}+T_k+k+i} \begin{bmatrix} i+j+k-1 \\ i, j-1, k \end{bmatrix},$$

where in (5.16) the letters k, i and j are decreased by 1 in that order to correspond to the starting triple in Scheme 3. Formula (5.16) is easily established by induction on $i+j+k$. From (5.16) and (5.13) it follows that

$$H_a^{(3)} + H_b^{(3)} + H_c^{(3)} = H^{(3)} = H,$$

and this yields the equality $B_1 = B_2 = B_3$ in Theorem 4.

The treatment of the generating functions $H_a^{(\mu)}, H_b^{(\mu)}$ and $H_c^{(\mu)}$, for $\mu = 4$, 5 and 6 is similar. We summarize the ideas of this section in the form of

THEOREM 8. — *The six schemes correspond to the six ways in which the q–multinomial coefficient* $\begin{bmatrix} i+j+k \\ i,j,k \end{bmatrix}$ *can be expanded in terms of* $\begin{bmatrix} i+j+k-1 \\ i-1,j,k \end{bmatrix}$, $\begin{bmatrix} i+j+k-1 \\ i,j-1,k \end{bmatrix}$ *and* $\begin{bmatrix} i+j+k-1 \\ i,j,k-1 \end{bmatrix}$. *In particular, under the standard transformations, the six companion functions* $S_\mu(n) = 1, 2, \ldots, 6$, *to Schur's partition function* $S(n)$, *correspond to these six recurrences for the q–multinomial coefficients.*

6. — Combinatorial proof of Theorems 4 and 1

The combinatorial proof given here is an extension of the method in [2] which itself is based on certain ideas due to Bressoud [13].

Let r, s be given. We start with a bi–partition $(\pi_1; \pi_2)$ counted by $V(n; r, s)$. So $\nu(\pi_1) = r$ and $\nu(\pi_2) = s$. In what follows several steps will be given and illustrated with

$$\pi_1 = a_7 + a_6 + a_3 + a_2 + a_1 \text{ and } \pi_2 = b_{13} + b_{12} + b_8 + b_7 + b_5 + b_4 + b_2.$$

Step 1 : decompose π_2 into two partitions π_4 and π_5, where π_4 has the parts of π_2 which are $\leq r$ and π_5 has the remaining parts. Let $\nu(\pi_4) = m$. So $\nu(\pi_5) = s - m$.

$$\pi_4 = b_5 + b_4 + b_2\,, \quad \pi_5 = b_{13} + b_{12} + b_8 + b_7\,.$$

Step 2 : consider the conjugate of the Ferrers graph of π_4 and circle the bottom node in each column. Denote this graph by π_4^*. Construct the Ferrers graph π_6, where the number of nodes in each row of π_6 is the sum of the number of nodes in the corresponding rows of π_1 and π_4^*. The parts of π_6 ending in the circled nodes are coloured purple and the remaining parts of π_6 are red. Thus π_6 has $r - m$ red parts and m purple parts.

$$\pi_6 = \pi_1 + \pi_4^* = a_{10} + c_9 + a_5 + c_4 + c_2.$$

Step 3 : write the parts of π_5 in a column in descending order and below them write the parts of π_6 in descending order. Draw a line to separate the parts of π_5 and π_6.

Step 4 : substract 0 from the bottom element, 1 from the element above that, 2 from the one above that etc..., and display the new values and the subtracted values in two adjacent columns $C_1|C_2$. The elements of C_2 have no colour while those of C_1 retain the colour of the parts from which they were derived.

Step 5 : (penultimate Step)

Rearrange the entries of C_1 in descending order according to Scheme μ, to form a column C_1^R.

Step 6 : (final Step)

Add the corresponding elements of C_1^R and C_2 to get a partition π_3 counted by $B_\mu(n; r-m, s-m, m)$.

Each of these steps is a one–to–one correspondence and so this combinatorial procedure has provided a proof of Theorem 1 with any B_μ in place of B_1. Since any function $B_\mu, \mu = 1, 2, \ldots, 6$ can be used in Theorem 1, the statement of Theorem 4 is an immediate consequence with $i = r - m$, $j = s - m$, $k = m$.

To simply get a bijection between $B_\mu(n; i, j, k)$ and $B_\omega(n; i, j, k)$, proceed from Step 6 to Step 5, and replace Scheme μ ordering by Scheme ω ordering and return to Step 6. We illustrate the above steps for Schemes 1 and 2 below.

Step 3
π_5/π_6
b_{13}
b_{12}
b_8
b_7
a_{10}
c_9
a_5
c_4
c_2

Step 4	
C_1	C_2
b_5	8
b_5	7
b_2	6
b_2	5
a_6	4
c_6	3
a_3	2
c_3	1
c_2	0

Step 5 (Scheme 1)		Step 6	Step 5 (Scheme 2)		Step 6
C_1^R	C_2	π_3	C_1^R	C_2	π_3
a_6	8	a_{14}	a_6	8	a_{14}
c_6	7	c_{13}	b_5	7	b_{12}
b_5	6	b_{11}	b_5	6	b_{11}
b_5	5	b_{10}	c_6	5	c_{11}
a_3	4	a_7	a_3	4	a_7
c_3	3	c_6	b_2	3	b_5
b_2	2	b_4	b_2	2	b_4
b_2	1	b_3	c_3	1	c_4
c_2	0	c_2	c_2	0	c_2

The combinatorial proof given above leads to several improvements of Theorems 1 and 4, a few of which will be described below. For a full description of these improvements and for details of proofs see [2].

Let $\mathcal{F}_{\mu,\omega}$ denote the bijection which is described above converting a Type μ partition π_μ to a partition π_ω of Type ω. More precisely, this is the bijection which is the result of starting with a Type μ partition in Step 6, then going back to Step 5 with Scheme μ ordering, then rearranging the elements of column C_1^R in Step 5 according to Scheme ω ordering, and finally perform Step 6 to get the partition π_ω or Type ω. So,

$$\mathcal{F}_{\mu,\omega}(\pi_\mu) = \pi_\omega. \tag{6.1}$$

The mappings $\mathcal{F}_{\mu,\omega}$ tell us a lot about the generating functions

(6.2)
$$F_\mu(x_m^{(\mu)}) = F_\mu(x_m^{(\mu)}; a, b, c, q) = \sum_{\substack{\pi_\mu \text{ of Type } \mu \\ \lambda(\pi_\mu) \preceq x_m^{(\mu)}}} a^{\nu_a(\pi_\mu)} b^{\nu_b(\pi_\mu)} c^{\nu_c(\pi_\mu)} q^{\sigma(\pi_\mu)}.$$

The most striking result concerning these generating function is

THEOREM 9. — *For all postive integers m*

$$F_1(x_{3m}^{(1)}) = F_2(x_{3m}^{(2)}) = \cdots = F_6(x_{3m}^{(6)}).$$

The proof of Theorem 9 (see [2]) makes use of several identities connecting $F_\mu(x_m^{(\mu)})$ for different values of m and μ, these identities being direct consequences of the bijections $\mathcal{F}_{\mu,\omega}$. The principal reason for the equality of the six generating functions at all positions $3m$ is because the alphabet lists in all six schemes are identical when truncated at $3m$. That is, for all m, the set of values

$$\begin{aligned}(6.3)\quad &\left\{x_1^{(\mu)}, x_2^{(\mu)}, \ldots, x_{3m}^{(\mu)}\right\} \\ &= \left\{a_1, a_2, \ldots, a_m, b_1, b_2, \ldots, b_m, c_2, c_3, \ldots, c_{m+1}\right\}, \mu = 1, 2, \ldots, 6.\end{aligned}$$

In particular, under the standard transformations, the set of symbols $\preceq x_{3m}^{(\mu)}$ considered in (6.3) for each of the Schemes become the set of positive integers $\leq 3m$. That is, although the natural ordering of the positive integers is altered in Schemes μ, for $\mu \geq 2$, the set of positive integers considered is the same in all Schemes when truncated at position $3m$. So, if $K_\mu = K_\mu(m; a, b, c, q)$ denotes the $a - b - c - q$ generating function for all Type μ partitions (under standard transformations) such that all parts are $\leq 3m$, then Theorem 9 yields

THEOREM 10. — $K_1(3m) = K_2(3m) = \ldots = K_6(3m)$.

By comparing the coefficient of $a^i b^j c^k q^n$ in the equalities of Theorem 10 we get the following improvement of Theorem 6 :

THEOREM 11. — *Let $S_\mu(n; i, j, k, 3m)$ denote the number of partitions counted by $S_\mu(n; i, j, k)$ with the additional restriction that all parts are $\leq 3m$. Then, for each integer $m \geq 1$,*

$$S_1(n; i, j, k, 3m) = S_2(n; i, j, k, 3m) = \cdots = S_6(n; i, j, k, 3m).$$

7. — Refinements of Capparelli's conjecture

In recent years, a number of classical partition identities have been found to lie at the heart of the interaction of vertex operators and representations of affine Lie Algebras (see Lepowsky and Wilson [19], [20], [21]). In the course of a study of standard modules of level 3 for $A_2^{(2)}$, Capparelli [15] was led to the partition conjecture stated in § 1. This conjecture is stated after his Theorem 21 in [16]. Andrews [12] recently proved Capparelli's conjecture using generating functions.

We now describe how to obtain substantial refinements as well as generalisations of this partition result by the method of weighted words.

From the point of view of refinements it is preferable to replace $C^*(n)$ by the function $C(n)$ which denotes the number of partitions of n into distinct parts $\equiv 2, 3, 4$ or $6 \pmod 6$. Clearly $C(n) = C^*(n)$ because

$$\begin{aligned}\sum_{n=0}^{\infty} C^*(n)q^n &= \frac{1}{(q^2;q^{12})_\infty(q^3;q^{12})_\infty(q^9;q^{12})_\infty(q^{10};q^{12})_\infty} \\ &= (-q^2;q^6)_\infty(-q^4;q^6)_\infty(-q^3;q^3)_\infty = \sum_{n=0}^{\infty} C(n)q^n .\end{aligned} \tag{7.1}$$

With the function $C(n)$ the following three parameter refinement of Capparelli's conjecture can be proved using the method of weighted words :

THEOREM 12. — *Let $C(n; i, j, k)$ denote the number of partitions counted by $C(n)$ with the additional restriction that there are precisely i parts $\equiv 4 \pmod 6$, j parts $\equiv 2 \pmod 6$, and of those $\equiv 0 \pmod 3$, exactly k are $> 3(i + j)$.*

Let $D(n; i, j, k)$ denote the number of partitions counted by $D(n)$ with the additional restriction that there are precisely i parts $\equiv 1 \pmod 3$ and j parts $\equiv 2 \pmod 3$ and k parts $\equiv 0 \pmod 3$. Then

$$C(n; i, j, k) = D(n; i, j, k).$$

It is possible to prove Theorem 12 combinatorially. Although the combinatorial proof is similar to the proof of Schur's theorem given in the previous section, there are some important differences, and so we give this combinatorial proof in the next section. As a consequence of that proof we get a further improvement of Theorem 12, namely,

THEOREM 13. — *Let $C(n; i, j, k, N)$ denote the number of partitions π of n counted by $C(n; i, j, k)$ such that $L + 3k \le 3N - 2$, where L is the largest part $\not\equiv 0 \pmod 3$ among the parts of π.*

Let $D(n; i, j, k, N)$ denote the number of partitions of n counted by $D(n; i, j, k)$ such that the largest part is $\le 3N - 2$. Then

$$C(n; i, j, k, N) = D(n; i, j, k, N) .$$

The method of weighted words yields the more general result stated as Theorem 14 below from which Theorem 12 follows as a special case under

$$\text{(dilation) } q \mapsto q^3 \qquad \text{and (translations) } a \mapsto aq^{-2}, b \mapsto bq^{-4}. \tag{7.2}$$

In a certain sense Capparelli's conjecture represents the most interesting dilations and translations in Theorem 14, but there are other nice special cases. We do not discuss them here and refer the reader to [3].

For Capparelli's problem, we assume that the integer 1 occurs in two colours a and c, and that integers ≥ 2 occur in three colours a, b and c. As before, the symbols a_j, b_j and c_j represent the integer j in colours a, b and c respectively. The lexicographic ordering that we choose is

$$a_1 < b_2 < c_1 < a_2 < b_3 < c_2 < a_3 < b_4 < c_3 < \cdots . \tag{7.3}$$

The Capparelli problem corresponds to the transformations

$$a_j \mapsto 3j-2, \qquad b_j \mapsto 3j-4, \qquad c_j \mapsto 3j \tag{7.4}$$

in which case the inequalities (7.3) become

$$1 < 2 < 3 < 4 < \cdots ,$$

the natural ordering among the positive integers.

Let $K(n; i, j, k)$ denote the number of vector partitions of n in the form (π_1, π_2, π_3) such that π_1 has distinct even a–parts, π_2 has distinct even b–parts and π_3 has distinct c–parts such that $\nu(\pi_1) = i$, $\nu(\pi_2) = j$ and the number of parts of π_3 which are $> (i + j)$ is k. Let $G(n; i, j, k)$ denote the number of partitions (words) of n into symbols a_j, b_j, c_j such that each part is $> a_1$ and the gap between consecutive symbols is given by the matrix below :

	a	b	c
a	2	2	1
b	0	2	0
c	2	3	1 .

We then have

THEOREM 14. — $K(n; i, j, k) = G(n; i, j, k)$.

Note : this matrix is to be read row-wise. For instance, if an a–part has weight j, then the next larger part, if it is a b–part, must have weight $\geq j+2$; if the next larger part is a c–part, its weight is $\geq j + 2$.

Theorem 14 is a generalisation of Theorem 12 which itself is a refinement of Capparelli's conjecture. From the point of view of generating functions, Theorem 14 is a consequence of

LEMMA 4. — *Let* $T_i = \frac{i(i+1)}{2}$ *denote the i-th Triangular number. Then*

$$\text{(a)} \quad \sum_{i,j,k,n} K(n;i,j,k)a^i b^j c^k q^n = \sum_{i,j} \frac{a^i b^j q^{2T_i+2T_j}(-q)_{i+j}(-cq^{i+j+1})_\infty}{(q^2;q^2)_i(q^2;q^2)_j}$$

$$\text{(b)} \quad \sum_{i,j,k,n} G(n;i,j,k)a^i b^j c^k q^n = \sum_{i,j,k} \frac{a^i b^j c^k q^{2T_i+2T_j+T_k+(i+j)k}}{(q)_{i+j+k}} \begin{bmatrix} i+j+k \\ i+j,k \end{bmatrix}_q \begin{bmatrix} i+j \\ i,j \end{bmatrix}_{q^2}.$$

Remark : theorem 14 is the statement that the generating functions in (a) and (b) of Lemma 4 are equal. To see this, all one needs to do is to sum the function in (b) over k to get the function in (a). More precisely

$$\begin{aligned} &\sum_{i,j,k} \frac{a^i b^j c^k q^{2T_i+2T_j+T_k+(i+j)k}}{(q)_{i+j+k}} \cdot \frac{(q)_{i+j+k}}{(q)_{i+j}(q)_k} \frac{(q^2;q^2)_{i+j}}{(q^2;q^2)_i(q^2;q^2)_j} \\ &= \sum_{i,j} \frac{a^i b^j q^{2T_i+2T_j}(-q)_{i+j}}{(q^2;q^2)_i(q^2;q^2)_j} \sum_k \frac{c^k q^{T_k+(i+j)k}}{(q)_k} \\ &= \sum_{i,j} \frac{a^i b^j q^{2T_i+2T_j}(-q)_{i+j}(-cq^{i+j+1})_\infty}{(q^2;q^2)_i(q^2;q^2)_j}, \end{aligned}$$

which is the generating function in Lemma 4 (a).

If we take $c = 1$, then the generating function in Lemma 4 (a) becomes a product, because

$$(7.5) \qquad (-q)_\infty \sum_{i,j} \frac{a^i b^j q^{2T_i+2T_j}}{(q^2;q^2)_i(q^2;q^2)_j} = (-q)_\infty(-aq^2;q^2)_\infty(-bq^2;q^2)_\infty.$$

In (7.5) replace $q \mapsto q^3$, $a \mapsto q^{-2}$, $b \mapsto q^{-4}$ to get

$$\prod_{m=1}^{\infty}(1+q^{6m-2})(1+q^{6m-4})(1+q^{3m}) = \sum_{n=0}^{\infty} C(n)q^n,$$

the generating function in (7.1).

So, in order to prove Theorem 14, we need to prove Lemma 4. Part (a) of the lemma is clear from the definition of $K(n;i,j,k)$. It is the proof of part (b) which is deeper and we sketch the main ideas below.

Given i, j and k, we need to show that

$$G(i,j,k) = \sum_n G(n;i,j,k)q^n = \frac{q^{2T_i+2T_j+T_k+(i+j)k}}{(q)_{i+j+k}} \begin{bmatrix} i+j+k \\ i+j, k \end{bmatrix}_q \begin{bmatrix} i+j \\ i, j \end{bmatrix}_{q^2}. \tag{7.6}$$

Since the partitions counted by $G(i,j,k)$ have $i+j+k$ parts and since these partitions are given by lower bound gap conditions as in the matrix, the generating function for all minimal partitions $H(i,j,k)$, using i a–parts, j b–parts and k c–parts is given by Lemma 1 and (7.6) to be
(7.7)

$$H(i,j,k) = G(i,j,k)(q)_{i+j+k} = q^{2T_i+2T_j+T_k+(i+j)k} \begin{bmatrix} i+j+k \\ i+j, k \end{bmatrix}_q \begin{bmatrix} i+j \\ i, j \end{bmatrix}_{q^2}.$$

The proof of (7.7) is by induction on $i+j+k$, the length of the word, and makes use of the functions $H_a(i,j,k)$, $H_b(i,j,k)$ and $H_c(i,j,k)$. These are the generating functions for partitions counted by $H(i,j,k)$ but with additional restriction that the smallest part is a_2, b_2 and c_1 respectively. From the definition of H_a, H_b and H_c it is clear that

$$H_a + H_b + H_c = H\,. \tag{7.8}$$

The q–binomial coefficients in (7.6) satisfy two recurences each, namely,

$$\begin{bmatrix} i+j+k \\ i+j, k \end{bmatrix} = \begin{bmatrix} i+j+k-1 \\ i+j, k-1 \end{bmatrix} + q^k \begin{bmatrix} i+j+k-1 \\ i+j-1, k \end{bmatrix} = \begin{bmatrix} i+j+k-1 \\ i+j-1, k \end{bmatrix} + q^{i+j} \begin{bmatrix} i+j+k-1 \\ i+j, k-1 \end{bmatrix} \tag{7.9}$$

and

$$\begin{bmatrix} i+j \\ i, j \end{bmatrix}_{q^2} = \begin{bmatrix} i+j-1 \\ i-1, j \end{bmatrix}_{q^2} + q^{2i} \begin{bmatrix} i+j-1 \\ i, j-1 \end{bmatrix}_{q^2} = \begin{bmatrix} i+j-1 \\ i, j-1 \end{bmatrix}_{q^2} + q^{2j} \begin{bmatrix} i+j-1 \\ i-1, j \end{bmatrix}_{q^2}. \tag{7.10}$$

All the recurences in (7.9) and (7.10) are necessary to establish the formulae for H_a, H_b and H_c. The details are a bit complicated and so we do not give them here; they may be found in [3]. Once these formulae for H_a, H_b and H_c are established, (7.8) will yield (7.7) which in turn will yield (7.6).

Remarks : in the discussion of Schur's theorem and companions in §§ 4,5, we made crucial use of the q–multinomial coefficient of order 3, namely

$$\begin{bmatrix} i+j+k \\ i, j, k \end{bmatrix} = \frac{(q)_{i+j+k}}{(q)_i (q)_j (q)_k}.$$

As is well known, the q–multinomial coefficient and the q–binomial coefficient are related by

$$\begin{bmatrix} i+j+k \\ i,j,k \end{bmatrix} = \begin{bmatrix} i+j+k \\ i+j,k \end{bmatrix}_q \begin{bmatrix} i+j \\ i,j \end{bmatrix}_q . \tag{7.11}$$

The main difference here is that instead of (7.11) we are making use of the product

$$\begin{bmatrix} i+j+k \\ i+j,k \end{bmatrix}_q \begin{bmatrix} i+j \\ i,j \end{bmatrix}_{q^2}$$

in Lemma 4(b).

8. — Combinatorial proof of Theorem 14

It is convenient to introduce the concept of level for the symbols a_j, b_j and c_j. More precisely, in the lexicographic ordering

$a_1 < b_2 < c_1$	$< a_2 < b_3 < c_2$	$< a_3 < b_4 < c_3$	$< a_4 < b_5 < c_4$	$< \cdots$
level 1	level 2	level 3	level 4	

we think of a_1, b_2, c_1 as at level 1, a_2, b_3, c_2 as at level 2, and so on. The symbol a_1 is necessary in this list even though it is never counted by the functions $K(n)$ and $G(n)$.

Combinatorial Proof : we begin with a partition $(\pi_1; \pi_2)$ counted by $K(n)$, where π_1 has distinct even a–parts and j distinct even b–parts, and π_2 has distinct c–parts. Several constructions will now be given and illustrated with

$$\pi_1 = a_{14} + b_{14} + b_{12} + b_8 + a_6 + a_4 + b_4 + a_2 + b_2$$

$$\pi_2 = c_{13} + c_{12} + c_{10} + c_5 + c_4 + c_2 + c_1 .$$

In this example $i = 4$, $j = 5$.

Step 1 : split π_2 into $\pi_4 \cup \pi_5$ where π_4 has the parts of π_2 with weights $\leq i+j$ and π_5 has the remaining parts.

$$\pi_4 = c_5 + c_4 + c_2 + c_1, \quad \pi_5 = c_{13} + c_{12} + c_{10}, \quad i+j = 9, k = 3.$$

Step 2 : consider the Ferrers graph of π_1 and to its side place the graph of the conjugate of π_4 – call this π_4^*.

$$\pi_4^* = 4 + 3 + 2 + 2 + 1 \text{ (uncoloured).}$$

Step 3 : consider the partition π_6 obtained by adding the number of nodes in the corresponding rows of π_1 and π_4^*. That is $\pi_6 = \pi_1 + \pi_4^*$. The parts of π_4^* have no colour, while the parts of π_6 retain the colour of the parts of π_1 from which they were derived.

$$\pi_6 = a_{18} + b_{17} + b_{14} + b_{10} + a_7 + a_4 + b_4 + a_2 + b_2.$$

Important Observation : the correspondence $(\pi_1; \pi_4^*) \leftrightarrow \pi_6$ is one–to–one. Observe that π_6 could have both odd and even parts. To extract π_4^* out of π_6, start from the lowest part of π_6, move upward and note the position of the first odd weight. This corresponds to the length of the first column of π_4^*. Next, note the position of the first even weight beyond this point. This corresponds to the length of the second column of π_4^*. Proceeding beyond this point note the position of the next odd weight, and so on, at each stage keeping track of the positions where there is a change in parity of the weights. These positions will give the columns of π_4^*.

Step 4 : write the parts of π_5 in descending order and below them write the parts of π_6 in descending order.

Step 5 : subtract 0 from the bottom element of this column, 1 from the element above that, 2 from the next one above, and so on, and display the new values and the amounts subtracted in two adjacent columns $C_1|C_2$. The elements of C_2 have no colour while those of C_1 retain the colour of the part from which they were derived.

Step 6 : rearrange the elements of C_1 to form a column C_1^R by inserting the c–parts at the appropriate levels. That is, if c_j is an element of C_1^R, then all elements below c_j are $\leq c_j$ and those above c_j are $\geq c_j$. Note that c_j may repeat in C_1 and C_1^R, as may also the parts a_j and b_j. In C_1^R the lexicographic ordering of a_j and b_j may not correspond to (7.3), but the worst that can happen is a switch within the same level. For example, although $a_1 < b_2$, we could have a_1 above b_2 in C_1 and C_1^R. But a symbol a_j or b_j which is a level higher than a_k or b_k will always occur above a_k or b_k in C_1 and C_1^R. Note also that a_1 can occur as an element of C_1 or C_1^R.

Step 7 : add the corresponding elements of C_1^R and C_2 to form a partition π_3 counted by $G(n; i, j, k)$. The parts of π_3 retain the colours of the elements of C_1^R from which they were derived. Note that all parts of π_3 are $> a_1$.

Step 4
π_5/π_6
c_{13}
c_{12}
c_{10}
a_{18}
b_{17}
b_{14}
b_{10}
a_7
a_4
b_4
a_2
b_2

Step 5	
C_1	C_2
c_2	11
c_2	10
c_1	9
a_{10}	8
b_{10}	7
b_8	6
b_5	5
a_3	4
a_1	3
b_2	2
a_1	1
b_2	0

Step 6	
C_1^R	C_2
a_{10}	11
b_{10}	10
b_8	9
b_5	8
a_3	7
c_2	6
c_2	5
c_1	4
a_1	3
b_2	2
a_1	1
b_2	0

Step 7
π_3
a_{21}
b_{20}
b_{17}
b_{13}
a_{10}
c_8
c_7
c_5
a_4
b_4
a_2
b_2

Each of these steps is a one–to–one correspondence and so this completes the proof of Theorem 14.

9. — Extensions of Göllnitz's theorem

Göllnitz's theorem stated in § 1 is one of the deepest in the theory of partitions. Our approach to this theorem via the method of weighted words not only lends insight into the structure of the partition functions concerned, but also provides substantial refinements and generalisations as can be seen from Theorems 15 and 16 below. Owing to the intricacy of the proofs of these theorems, we provide here only a description of the main ideas and refer to [4] for details.

We assume that the integer 1 occurs in three primary colours, red, blue and yellow, and that integers $j \geq 2$ occur in six colours, of which red, blue and yellow are primary and purple, orange and green are secondary. We use the symbols a_j, b_j, c_j, d_j, e_j and f_j to denote the integer j in colours red, blue, yellow, purple, orange and green respectively. As before, a, b, c, d, e, f are free parameters. However, in making the transition from $C(n)$ to $B(n)$ in Göllnitz' theorem, we need to take $d = ab$, $e = ac$, $f = bc$. That is why the subscripts of d, e and f are integers $j \geq 2$ since they are obtained by combining two parts in primary colours. That is also why we think of d, e and f as secondary colours.

The lexicographic ordering we use for Göllnitz's theorem is

$$a_1 \prec b_1 \prec c_1 \prec d_2 \prec e_2 \prec a_2 \prec f_2 \prec b_2 \prec c_2 \prec d_3 \prec e_3 \prec a_3 \prec \dots . \tag{9.1}$$

We will call this Scheme 1. Under the transformations

$$\begin{cases} \text{(dilation) } q \mapsto q^6, \text{ (translations) } a \mapsto aq^{-4}, b \mapsto bq^{-2}, c \mapsto cq^{-1}, \\ \text{and (combinations) } d = ab, e = ac, f = bc, \end{cases} \tag{9.2}$$

the symbols become

$$\begin{cases} a_j = 6j-4, b_j = 6j-2, c_j = 6j-1, \\ d_j = 6j-6, e_j = 6j-5, f_j = 6j-3, \end{cases} \tag{9.3}$$

and so the lexicographic ordering in (9.1) becomes

$$2 < 4 < 5 < 6 < 7 < 8 < 9 < 10 < 11 < 12 < 13 \dots , \tag{9.4}$$

the standard ordering of the positive integers. In view of this, the transformations in (9.2) will be called standard transformations. Observe that the integers 1 and 3 are absent in (9.4). This is because $e_1 = (ac)_1$ and $f_1 = (bc)_1$ are absent in (9.1). For reasons which will become clear towards the end of this section, it is sometimes useful to consider the full list of symbols

$$\underline{d}_1 \prec \underline{e}_1 \prec a_1 \prec \underline{f}_1 \prec b_1 \prec c_1 \prec d_2 \prec e_2 \prec a_2 \prec f_2 \prec b_2 \prec c_2 \prec \cdots , \tag{9.5}$$

where d, e, and f in (9.5) are underlined to indicate that they do not occur in (9.1).

Observe that for symbols with the same weight, the colours occur in the following order :

$$d \prec e \prec a \prec f \prec b \prec c . \tag{9.6}$$

Given two colours, we use (9.6) to determine which of the two is of lower order. For example between e and b, e is of lower order and b of higher order.

Next, consider partitions $\pi = m_1 + m_2 + \cdots$, using symbols in Scheme 1 such that the gap between symbols is ≥ 1 with the added restriction that the gap between consecutive symbols m_l and m_{l+1} is ≥ 2 if

$$\begin{cases} m_l \text{ is of lower order and } m_{l+1} \text{ of higher order} \\ \text{or if } m_l \text{ and } m_{l+1} \text{ are of the same secondary colour.} \end{cases} \tag{9.7}$$

We will refer to such a π as a Type 1 partition. The principal partition result that we get by this approach is Theorem 15, from which Göllnitz' theorem falls out as a special case.

THEOREM 15. — *Let $B(n; i, j, k)$ denote the number of vector partitions $\pi' = (\pi_1; \pi_2; \pi_3)$ of n such that π_1, π_2 and π_3 have distinct a–parts, b–parts and c–parts respectively, and also $\nu(\pi_1) = i, \nu(\pi_2) = j$, and $\nu(\pi_3) = k$.*

Let $C(n; \alpha, \beta, \gamma, \delta, \epsilon, \phi)$ denote the number of Type 1 partitions π of n such that $\nu_a(\pi) = \alpha, \nu_b(\pi) = \beta, \ldots, \nu_f(\pi) = \phi$. Then

$$B(n; i, j, k) = \sum_{\alpha+\delta+\varepsilon=i,\ \beta+\delta+\phi=j,\ \gamma+\varepsilon+\phi=k} C(n; \alpha, \beta, \gamma, \delta, \epsilon, \phi).$$

Note that under the standard transformations, Theorem 15 yields a strong refinement of Göllnitz' theorem. Under the standard transformations the gap conditions (9.7) defining Type 1 partitions become the difference conditions defining the function $C(n)$ in Göllnitz' theorem. Also, (9.7) provides a natural explanation as to why there is strict inequality in Göllnitz' theorem when $m_l \equiv 6, 7$ or $9 \pmod 6$. This is because the residue classes 6, 7 and $9 \pmod 6$ correspond to the secondary colours ab, ac and bc respectively.

More generally, under the substitutions

$$\text{(dilation) } q \mapsto q^M \text{ and (translations) } a \mapsto aq^{r_1-M}, \quad b \mapsto bq^{r_2-M}, c \mapsto cq^{r_3-M}. \tag{9.8}$$

Theorem 15 yields the following result.

THEOREM 16. — *Let $M \geq 6$ and r_1, r_2, r_3 residues such that*

$$0 < r_1 < r_2 < r_3 < M \leq r_1 + r_2 \text{ and } r_1 + M < r_2 + r_3.$$

Let $B(n; \nu)$ denote the number of partitions of n into ν distinct parts $\equiv r_1, r_2$ or $r_3 \pmod M$.

Let $C(n; \nu)$ denote the number of partitions of n into ν distinct parts $m_1 > m_2 > \cdots$ such that

(i) *each part m_l is $\equiv r_1, r_2, r_3, r_1 + r_2, r_1 + r_3$ or $r_2 + r_3 \pmod M$.*

(ii) *$m_l - m_{l+1} \geq M$ with strict inequality if $m_l \equiv r_1 + r_2, r_1 + r_3$, or $r_2 + r_3$ $\pmod M$.*

(iii) *the parts $\equiv r_1 + r_2, r_1 + r_3$ and $r_2 + r_3 \pmod M$ are counted twice.*

Then

$$B(n; \nu) = C(n; \nu).$$

It is to be noted that in Theorem 16 we have almost total freedom in choosing the three residues $r_1, r_2, r_3 \pmod M$. (Göllnitz [18], Satz (4.8 and 4.10)) obtained two extensions of his basic theorem with the modulus 6 replaced by $M+4$ for $M \geq 2$, but in each theorem he prescribes a fixed set of residues $\bmod(M+4)$. Göllnitz' Satz (4.8) and (4.10) follow as special cases of Theorem 16.

The combinatorial explanation for counting the parts of secondary colour twice in Theorem 15 is as follows : Given a partition counted by $B(n; i, j, k)$, take δ of the a-parts and δ of the b-parts to form the ab-parts (d-parts). Similarly, ϵ of the a-parts and ϵ of the c-parts combine to yield the ac-parts. Finally ϕ of the b-parts and ϕ of the c-parts combine to form the bc-parts. Thus in partitions counted by the function C we have

$$\alpha = i - \delta - \epsilon,\ \beta = j - \delta - \phi,\ \gamma = k - \epsilon - \phi. \tag{9.9}$$

Therefore

$$\alpha + \beta + \gamma + 2\delta + 2\epsilon + 2\phi = i + j + k.$$

Finally the number of parts in partitions counted by $C(n; \alpha, \beta, \gamma, \delta, \epsilon, \phi)$ is

$$s = \alpha + \beta + \gamma + \delta + \epsilon + \phi. \tag{9.10}$$

In what follows we assume (9.9) and (9.10).

From the point of view of the method of weighted words, Theorem 15 is seen as emerging from the following incredible Key Identity :

$$\begin{aligned}
&\sum_{i,j,k} a^i b^j c^k \sum_{\substack{s=\alpha+\beta+\gamma+\delta+\epsilon+\phi \\ i=\alpha+\delta+\epsilon,\, j=\beta+\delta+\phi,\, k=\gamma+\epsilon+\phi}} \frac{q^{T_s+T_\delta+T_\epsilon+T_{\phi-1}}(1-q^\alpha(1-q^\phi))}{(q)_\alpha(q)_\beta(q)_\gamma(q)_\delta(q)_\epsilon(q)_\phi} \\
&\sum_{i,j,k} \frac{a^i b^j c^k q^{T_i+T_j+T_k}}{(q)_i(q)_j(q)_k} = (-aq)_\infty(-bq)_\infty(-cq)_\infty.
\end{aligned} \tag{9.11}$$

Clearly, the generating function of $B(n; i, j, k)$ is

$$\sum_n B(n; i, j, k) q^n = \frac{q^{T_i+T_j+T_k}}{(q)_i(q)_j(q)_k},$$

which is the summand on the right hand side of (9.11). In order to prove Theorem 15 we need to do two things. Firstly, we need to show

LEMMA 4. —

$$\sum_n C(n; \alpha, \beta, \gamma, \delta, \epsilon, \phi) q^n = \frac{q^{T_s+T_\delta+T_\epsilon+T_{\phi-1}}(1-q^\alpha(1-q^\phi))}{(q)_\alpha(q)_\beta(q)_\gamma(q)_\delta(q)_\epsilon(q)_\phi}.$$

Secondly we need to prove the key identity (9.11). Both of these are quite difficult.

The proof of Lemma 4 is along the lines of the proof of Lemma 2 given in §4, but the details here are more difficult. First, we rewrite the expression on the right in Lemma 4 in the form

$$\text{(9.12)} \qquad \frac{q^{T_s+T_\delta+T_\epsilon+T_{\phi-1}}}{(q)_s}\begin{bmatrix} s \\ \alpha,\beta,\gamma,\delta,\epsilon,\phi \end{bmatrix}(1-q^\alpha(1-q^\phi)).$$

From (9.12) and Lemma 1 it follows that the generating function for all minimal Type 1 partitions having $\nu_a(\pi)=\alpha$, $\nu_b(\pi)=\beta,\ldots,\nu_f(\pi)=\phi$ is

$$\text{(9.13)} \quad H=H(\alpha,\beta,\gamma,\delta,\epsilon,\phi)=q^{T_s+T_\delta+T_\epsilon+T_{\phi-1}}(1-q^\alpha(1-q^\phi))\begin{bmatrix} s \\ \alpha,\beta,\gamma,\delta,\epsilon,\phi \end{bmatrix}.$$

The proof of (9.13) is by induction on s and makes use of recurrences for the multinomial coefficients of order $t=6$. Details may be found in [4].

The proof of the key identity (9.1) is very difficult (see [4]) and so we do not give it here. The depth of this identity becomes plain when we see that its proof requires not only Watson's q-analog of Whipple's theorem but also the ${}_6\Psi_6$ summation of Bailey (see Gasper and Rahman [17]).

Note that if we set $c=0$ in (9.11) and compare the coefficients on $a^r b^s$ on both sides, we get Lemma 3. Thus one advantage in this approach to Göllnitz' theorem is that Schur's theorem falls out as a special case by setting $c=0$.

We now discuss companions to the function $C(n)$ in Theorem 15 (hereinafter denoted by $C_1(n)$) obtained by considering lexicographic orderings other than Scheme 1. It is for this that the full list of symbols in (9.5) is useful. Before introducing these orderings we make an observation about Type 1 partitions.

Let $x_m = x_m^{(1)}$ denote the symbol occupying position m in the complete list (9.5). That is $x_0^{(1)} = \underline{d}_1$, $x_1^{(1)} = \underline{e}_1$, $x_2^{(1)} = a_1$, and so on. We set $x_0^{(1)} = \underline{d}_1$ because under the standard transformations, d_1 becomes the integer 0. With this notation the gap conditions defining Type 1 partitions in (9.7) can be recast in the following equivalent form : Type 1 partitions are those of the form $x_{m_1} + x_{m_2} + \cdots$, where :

$$\text{(9.13)} \qquad \begin{cases} m_l - m_{l+1} \geq 6 \text{ with strict inequality if} \\ x_{m_l} \text{ is of secondary colour.} \end{cases}$$

From now on we will refer to (9.13) as standard gap conditions.

In considering other lexicographic orderings we omit the symbol $\underline{d}_1$ because $\underline{d}_1 = 0$ under the standard transformations. We therefore choose any ordering of the symbols

$$\underline{e}_1,\ a_1,\ \underline{f}_1,\ b_1,\ c_1,\ d_2. \tag{9.14}$$

For instance, Scheme 1 is generated by the basic ordering

$$\textbf{Scheme 1:} \quad \underline{e}_1 \prec a_1 \prec \underline{f}_1 \prec \underline{b}_1 \prec c_1 \prec d_2 \tag{9.15}$$

We get the full list of symbols in (9.5) by increasing the weights in (9.15) by one in succession. Similarly, we may consider another ordering of the symbols, namely,

$$\textbf{Scheme 2:} \quad \underline{e}_1 \prec a_1 \prec \underline{f}_1 \prec b_1 \prec d_2 \prec c_1 \tag{9.16}$$

Then the full list of symbols generated by Scheme 2 is

$$\underline{e}_1 \prec a_1 \prec \underline{f}_1 \prec b_1 \prec d_2 \prec c_1 \prec e_2 \prec a_2 \prec f_2 \prec b_2 \prec d_3 \prec c_2 \prec \cdots \tag{9.17}$$

As in (9.5), the symbols $\underline{e}_1$ and $\underline{f}_1$ are underlined in (9.17) to indicate that they will never appear in partitions we will be considering presently.

More generally one may consider all $6\,! = 720$ orderings of the symbols in (9.14) and the 720 Schemes thus generated. Let $x_m = x_m^{(\mu)}$ denote the symbol occupying position m in the full list of symbols in Scheme μ. We define a Type μ partition to be one of the form $x_{m_1} + x_{m_2} + \cdots$, where $x_{m_l} = x_{m_l}^{(\mu)}$ satisfy the standard gap conditions (9.13). Next let $C_\mu(n; \alpha, \beta, \gamma, \delta, \epsilon, \phi)$ denote the number of Type μ partitions π of n with $\nu_a(\pi) = \alpha$, $\nu_b(\pi) = \beta, \ldots, \nu_f(\pi) = \phi$. Then we have

THEOREM 17. — *For* $\mu = 2, 3, \ldots, 720$

$$C_1(n; \alpha, \beta, \gamma, \delta, \epsilon, \phi) = C_\mu(n; \alpha, \beta, \gamma, \delta, \epsilon, \phi).$$

A combinatorial proof of Theorem 17 can be given in a manner identical to the bijective proof of the equality of the Schur companion functions given in §6. So we do not repeat the ideas here. A combinatorial proof of Theorem 17 may be found in [4].

Under the standard transformations the full list of symbols in (9.17) for Scheme 2 yields the following ordering of the positive integers :

$$\underline{1} \prec 2 \prec \underline{3} \prec 4 \prec 6 \prec 5 \prec 7 \prec 8 \prec 9 \prec 10 \prec 12 \prec 11 \prec \cdots . \tag{9.18}$$

This ordering leads to the following result which may be considered as a companion to Göllnitz' theorem just as Andrews' theorem is a companion to Schur's theorem.

THEOREM 18. — *Let* $C_2(n)$ *denote the number of partitions of* n *in the form* $m_1 + m_2 + \cdots$, *such that no* $m_l = 1$ *or* 3 *and* $m_l - m_{l+1} \geq 7, 6, 7, 6, 5$ *or* 8 *if* $m_l \equiv 7, 2, 9, 4, 5$ *or* $6 \pmod 6$. *Then*

$$C_1(n) = C_2(n).$$

Remarks : the inequalities defining most of the other companions $C_\mu(n)$ (under standard transformations) are generally more complicated than the ones for $C_1(n)$ and $C_2(n)$, and will be like (5.9)–(5.12). But here we have more than 700 such sets of inequalities and so we will not write them down! There may be a few functions $C_\mu(n)$ which are as nice as $C_1(n)$ and $C_2(n)$ and it may be worthwhile to determine them all.

The combinatorial proof of Theorem 17 alluded to above involves bijections $\mathcal{F}_{\mu,\omega}$ between partitions of Type μ and those of Type ω, for $\mu, \omega = 1, 2, \ldots, 720$; that is, bijections between partitions counted by $C_\mu(n)$ and $C_\omega(n)$. But it is an entirely different story concerning bijections between partition counted by $C_\mu(n)$ and $B(n)$, and indeed no such bijection is known at present even though Theorem 15 is a refinement of the equality $C_\mu(n) = B(n)$ involving many parameters. This appears to be extremely difficult.

10. — Beyond Göllnitz' theorem

The approach to Schur's theorem via the method of weighted words involved two primary colours a, b and one secondary colour ab. For Göllnitz' theorem we need three primary colours a, b, c and three secondary colours ab, ac and bc. The principal reason for the substantial increase in the depth and difficulty when making the transition from Schur's theorem to Göllnitz's theorem is that the ternary colour abc is dropped; that is, we are not dealing with the full non-empty alphabet of colours that can be generated using a, b, c, but only with a proper subset.

Andrews [6], [7] has obtained general partition theorems which extend Schur's theorem by choosing a set $a_1, a_2, \ldots, a_r$ of distinct residues $(\mathrm{mod}\, M)$, with $M \geq 2^r - 1$ and by considering all possible residue classes which are given as non–empty sums $\Sigma \varepsilon_i a_i$, $\varepsilon_i = 0$ or 1. The main reason he was able to obtain such extension was because he was dealing with the complete set of $2^r - 1$ residues generated by $a_1, a_2, \ldots, a_r$.

One way of extending Göllnitz's theorem would be to consider four primary colours a, b, c, d. We then have a choice of dropping either $abcd$, or may be even some of the secondary and ternary colours. At the moment we do not know which of the choices (if any) would be connected with the expansion of

$$(-aq)_\infty(-bq)_\infty(-cq)_\infty(-dq)_\infty.$$

An attempt at this question might give some clues about the general situation with r primary colours $a_1, a_2, \ldots, a_r$. Once again the emphasis is that we should not deal with the full alphabet of colours generated by $a_1, a_2, \ldots, a_r$. Such a study was attempted computationally in 1971 by Andrews [9; p. 384–385] but his effort was limited by the amount of computer power and memory available at that time. With the availability of modern computer algebra systems, it may not be unreasonable to consider these questions now.

Manuscrit reçu le 10 janvier 1994

References

[1] K. ALLADI and B. GORDON. — *Generalizations of Schur's partition theorem*, Manuscripta Mathematica **79** (1993), 113–126.

[2] K. ALLADI and B. GORDON. — *Schur's partition theorem, companions, refinements and generalisations*, (to appear).

[3] K. ALLADI, G.E. ANDREWS and B. GORDON. — *Refinements and generalisations of Capparelli's conjecture on partitions*, J. Algebra (to appear).

[4] K. ALLADI, G.E. ANDREWS and B. GORDON. — *Generalisations and refinements of a partition theorem of Göllnitz*, J. Reine und Angew. Math. (to appear).

[5] G.E. ANDREWS. — *On Schur's second partition theorem*, Glasgow Math. J. **9** (1967), 127–132.

[6] G.E. ANDREWS. — *A new generalisation of Schur's second partition theorem*, Acta. Arith. **4** (1968), 429–434.

[7] G.E. ANDREWS. — *A general partition theorem with difference conditions*, Amer. J. Math. **191** (1969), 18–24.

[8] G.E. ANDREWS. — *On a partition theorem of Göllnitz and related formulae*, J. Reine Angew. Math. **236** (1969), 37–42.

[9] G.E. ANDREWS. — *The use of computers in search of identities of Rogers-Ramanujan type*, in Computers in Number Theory (A.O.L. Atkin and B.J. Birch, Eds.), Academic Press (1971), 377–387.

[10] G.E. ANDREWS. — *The theory of partitions*, Encyclopedia of Math., Vol. 2 Addison Wesley, Reading (1976).

[11] G.E. ANDREWS. — *q–series : their development and applications in Analysis, Number Theory, Combinatorics, Physics and Computer Algebra*, NSF–CBMS Lectures, Vol. **66**, Amer. Math. Soc., Providence (1985).

[12] G.E. ANDREWS. — *Schur's theorem, Capparelli's conjecture and q–trinomial coefficients*, in Proc. Rademacher Centenary Conf. (1992), Contemp. Math., Amer. Math. Soc. (to appear).

[13] D.M. BRESSOUD. — *On a partition theorem of Göllnitz*, J. Reine und Angew. Math. **305** (1979), 215–217.

[14] D.M. BRESSOUD. — *A combinatorial proof of Schur's 1926 partition theorem*, Proc. Amer. Math. Soc. **79** (1980), 338–340.

[15] S. CAPPARELLI. — *Vertex operators for affine algebras and combinatorial identities*, Ph.D. Thesis, Rutgers Univ. (1988).

[16] S. CAPPARELLI. — *On some representations of twisted affine Lie Algebras and Combinatorial identities*, J. Algebra **154**, (1993), 335–355.

[17] G. GASPER and M. RAHMAN. — *Basic hyper-geometric series*, Encyclopedia of Mathematics and its Applications, Vol. 35, Cambridge (1990).

[18] H. GÖLLNITZ. — *Partitionen mit Differenzenbedingungen*, J. Reine Angew. Math. **225** (1967), 154–190.

[19] J. LEPOWSKY and R.L. WILSON. — *A new family of algebras underlying the Rogers-Ramanujan identities and generalisations*, Proc. Nat. Acad. Sci. USA **78** (1981), 7254–7258.

[20] J. LEPOWSKY and R.L. WILSON. — *A Lie–theoretic interpretation and proof of the Rogers-Ramanujan identities*, Adv. in Math. **45** (1982), 21–72.

[21] J. LEPOWSKY and R.L. WILSON. — *The structure of standard modules, I. Universal algebras and the Rogers-Ramanujan identities*, Invent. Math. **77** (1984), 199–290.

[22] I. SCHUR. — *Zur Addiven Zahlentheorie*, Gessammelte Abhandlungen, Vol. **2**, Springer, Berlin (1973), 43–50.

Krishnaswami ALLADI
Department of Mathematics
University of Florida
Gainesville, Florida 32611
U.S.A.

Number Theory
Paris 1992–93

Théorie des motifs et interprétation géométrique des valeurs p-adiques de G-functions (une introduction)

Yves André

1. — Introduction

Partons d'une question concrète, dans l'esprit de Hensel.

Considérons une série de puissances à coefficients rationnels, la plus simple pour commencer :

$$1 + x + x^2 + \cdots = (1 - x)^{-1}.$$

Pour $x = 2/3 < 1$, cette série converge. Mais $2/3$ a deux façons d'être < 1, l'usuelle et la dyadique; dans les deux cas bien entendu, la somme vaut 3 (dans $\mathbb{R}$ et dans $\mathbb{Q}_2$ respectivement).

Prenons un exemple moins élémentaire :

$$1 + x^2 + 3/2x^4 + \cdots + \binom{2n}{n} 2^{-n} x^{2n} + \cdots = (1 - 2x^2)^{-1/2}.$$

Pour $x = 2/3$, on trouve 3 en sommant dans $\mathbb{R}$. Dans $\mathbb{Q}_2$, on devine quc la somme est ± 3 en notant que le carré de la série de puissances est une fonction rationnelle. Pour déterminer le signe, il suffit dès lors de remarquer que cette somme est congrue à 1 modulo 4 : c'est donc -3. Ainsi, aux places de $\mathbb{Q}$ pour lesquelles $2/3$ est dans le disque de convergence, les sommes sont des nombres rationnels, mais distincts.

Plus généralement, considérons une série de puissances $y(x) = a_0 + a_1 x + a_2 x^2 + \cdots \in \mathbb{Q}[[x]]$, éventuellement transcendante, et supposons que pour une valeur $x \in \mathbb{Q}$, cette série converge dans $\mathbb{R}$ vers un nombre rationnel ou algébrique. Soit p un nombre premier tel que la série converge aussi p-adiquement.

Converge-t-elle encore vers un nombre rationnel, resp. algébrique, dans $\mathbb{Q}_p$?

Comme on peut s'y attendre, la réponse est non en général : d'ailleurs, en s'inspirant de l'exemple précédent, on construit aussitôt le contre-exemple suivant :

$$y(x) = [(1 - 2x^2)^{-1/2} - 3] \cdot \exp(2x), \qquad x = 2/3, \qquad p = 2.$$

On notera toutefois ici que $y(x)$ est solution d'une équation différentielle d'ordre 2, et le phénomène remarquable est que les évaluations en $2/3$, aussi bien réelles que dyadiques, de $y(x)$ et de sa dérivée $y'(x)$ sont rationnellement proportionnelles.

Ceci nous conduit à reformuler légèrement la question en termes de dépendance linéaire ou algébrique sur $\mathbb{Q}$ d'évaluations de $y(x)$ et de ses dérivées – nous limitant aux solutions d'équations différentielles linéaires.

Voici un exemple dû à F. Beukers [Be93], mettant en jeu la série hypergéométrique de Gauss $y(x) = {}_2F_1(1/12, 5/12, 1/2; x)$. Pour $x = 1323/1331$, cette série converge dans $\mathbb{R}$ vers le nombre algébrique $\frac{3}{4}\sqrt[4]{11}$ (miracle !).

Comme $1323 = 3^3 \cdot 7^2$, elle converge aussi dans $\mathbb{Q}_7$. Beukers montre qu'elle converge en fait vers le nombre algébrique $\frac{1}{4}\sqrt[4]{11}$ (second miracle !).

L'idée que je vais défendre ici est que ce phénomène a lieu chaque fois qu'il est *motivé*; plus précisément, je me propose de montrer comment la philosophie des motifs de A. Grothendieck conduit au principe heuristique (ou *conjecture*) suivant :

PRINCIPE DES RELATIONS GLOBALES. — *Soit* $(y_1(x), \dots, y_n(x)) \in \overline{\mathbb{Q}}[[x]]^n$ *une base de solutions d'une équation différentielle linéaire à coefficients dans* $\overline{\mathbb{Q}}(x)$, *facteur d'une équation de Picard–Fuchs. Soient* v, w *deux places de* $\overline{\mathbb{Q}}$, *dont l'une* v, *est archimédienne, et soit* ξ *un élément de* $\overline{\mathbb{Q}}$ *situé dans le disque de convergence* v*– et* w*–adique de* $y_1(x), \dots, y_n(x)$.

S'il existe une relation de dépendance algébrique sur $\overline{\mathbb{Q}}$ *entre les évaluations* v*–adiques* $y_1(\xi)_v, y_1'(\xi)_v, \dots, y_n^{(n-1)}(\xi)_v$ *des dérivées de* $y_1(x), \dots, y_n(x)$ *en* ξ, *ne provenant pas par spécialisation* $x \to \xi$ *d'une relation de dépendance algébrique sur* $\overline{\mathbb{Q}}(x)$ *entre les* $y_i^{(j)}(x)$, *alors il en est de même pour les évaluations* w*–adiques.*

Nous notons bien entendu ici $\overline{\mathbb{Q}}$ une clôture algébrique de $\mathbb{Q}$, et disons qu'une équation différentielle $\Lambda y = 0$ est facteur d'une autre $\Lambda' y = 0$ pour exprimer que Λ divise Λ' dans $\overline{\mathbb{Q}}(x)[d/dx]$ (à droite ou à gauche).

Rappelons qu'une équation de Picard–Fuchs est une équation différentielle qui régit la variation de la cohomologie de De Rham d'une $\overline{\mathbb{Q}}(x)$–variété algébrique en fonction du paramètre x (connexion de Gauss–Manin).

Ainsi, dans l'exemple de Beukers[1], l'équation hypergéométrique satisfaite par ${}_2F_1\ (1/12, 5/12, 1/2; 27t^2/4)$ est aussi satisfaite par la classe de cohomologie de la différentielle dX/Y sur la courbe elliptique d'équation $Y^2 = X^3 - X - t$.

Rappelons aussi [A89] que toute solution $y(x) = a_0 + a_1x + a_2x^2 + \cdots \in \overline{\mathbb{Q}}[[x]]$ d'une équation de Picard–Fuchs est ce que Siegel a appelé, il y a 65 ans, une *G–fonction* : elle définit pour toute place de $\overline{\mathbb{Q}}$ une fonction analytique au voisinage de 0, et, de plus, le dénominateur commun à a_0, $a_1, \ldots, a_m$ croît au plus exponentiellement en m.

Le principe des relations globales peut être étendu heuristiquement à toutes les G–fonctions; on n'y gagnerait guère, attendu qu'une conjecture de Bombieri–Dwork prédit que toute G–fonction "provient de la géométrie", i.e. est solution d'une extension (multiple) de facteurs de connexions de Gauss–Manin (exemple : les *polylogarithmes*). D'autre part, lorsque la connexion est semi–simple (par exemple lorsqu'elle gouverne la variation de cohomologie d'une $\overline{\mathbb{Q}}(x)$–variété propre et lisse), on peut omettre du "principe" la condition que v est archimédienne.

Le principe des relations globales est particulièrement intéressant en liaison avec le *théorème* suivant (qui reprend les mêmes notations), dû à Bombieri :

PRINCIPE DE HASSE POUR LES VALEURS DE G–FONCTIONS. — *Toute relation de dépendance algébrique de degré donné δ à coefficients dans $\overline{\mathbb{Q}}$ entre les évaluations $y_1(\xi), y_1'(\xi), \ldots, y_n^{(n-1)}(\xi)$, valide en toute place de convergence* [2]*, provient nécessairement, par spécialiation, d'une relation de dépendance algébrique entre les $y_i^{(j)}(x)$ à coefficients dans $\overline{\mathbb{Q}}(x)$, sauf si ξ appartient à un certain sous–ensemble de $\overline{\mathbb{Q}}$ de hauteur bornée (polynômialement en δ).*

Voir [Bo81], et aussi [A89].

En résumé, l'objet principal de cet article de synthèse est de justifier le principe des relations globales (à partir de conjectures bien connues de Grothendieck dans le cadre de sa théorie des motifs), et d'en esquisser une application. Cette justification comporte deux volets :

a) interprétation cohomologique des valeurs (archimédiennes ou non) de G–fonctions; nous nous attarderons sur des exemples mettant en

---

[1] cet exemple a d'ailleurs été conçu pour tester le principe ci–dessus, déjà esquissé dans [A89], cf. l'introduction de [Be93].

[2] Le principe ci–dessus permet de construire de telles relations, en multipliant entre elles les relations "locales" (ce qui augmente δ bien entendu).

jeu les variétés abéliennes, avant de développer le cas général (en vue duquel nous présentons brièvement la théorie des motifs et certaine variante non-conjecturale).

b) Elucider comment l'existence de relations algébriques "exceptionnelles" entre valeurs de solutions d'une équation de Picard–Fuchs est liée à la présence de cycles algébriques "exceptionnels" sur les fibres correspondantes de la famille à un paramètre de variétés sous-jacente.

Quant à l'application, nous nous bornerons à indiquer succinctement comment, en prenant les deux principes ci-dessus comme guide, on parvient à démontrer, inconditionnellement, une série de résultats du type suivant [A] :

THÉORÈME 1. — *Soit $f\,:\,\underline{A} \to S$ une famille de variétés abéliennes paramétrée par une courbe affine S sur $\overline{\mathbb{Q}}$. Une fibre $\underline{A}_s$ sera dite exceptionnelle si l'une de ses puissances porte un cycle algébrique qui ne provient pas par spécialisation*[3] *d'un cycle de Hodge absolu sur la puissance correspondante de la fibre générique géométrique $\underline{A}_{\overline{\eta}}$.*

Supposons qu'il existe une fonction rationnelle x sur S telle que les fibres de f au-dessus de $x = \infty$ soient de type CM.

Alors pour tout $\delta > 0$, il n'y a qu'un nombre fini de nombres algébriques ξ de degré $< \delta$ et p-entiers en tous les nombres premiers sauf au plus δ, tels que l'une des fibres de f au-dessus de $x = \xi$ soit exceptionnelle.

A noter, en particulier, qu'est exceptionnelle toute fibre $\underline{A}_s$ telle que l'homomorphisme de spécialisation $\mathrm{End}\,\underline{A}_{\overline{\eta}} \to \mathrm{End}\,\underline{A}_s$ ne soit pas surjectif. On pourra comparer avec le premier énoncé de ce type, obtenu dans [A86]. En fait, on peut préciser le théorème 1 : la hauteur (logarithmique) des ξ exceptionnels est bornée polynômialement en δ. Dans le cas d'une famille modulaire de courbes elliptiques, les fibres exceptionnelles correspondent aux moduli singuliers j, on peut poser $x = j^{-1}$, et on obtient ainsi que la hauteur de l'invariant j d'une courbe elliptique à multiplication complexes par un ordre quadratique imaginaire $\mathcal{O}$ est bornée par un certain polynôme en le nombre de classes de $\mathcal{O}$ et le nombre de premiers divisant la norme de j.

Précisons toutefois que les méthodes mises en oeuvre ici débordent largement le cadre des variétés abéliennes (cf. § 8d).

Plan de l'article :

2. Périodes
3. Une $\overline{\mathbb{Q}}$-structure dans la cohomologie cristalline?

---

[3] il est commode de considérer ici les réalisations de De Rham ou l-adiques, de sorte que la spécialisation est immédiate à définir.

4. Le cas des variétés abéliennes dégénérantes
5. Le cas des courbes elliptiques avec bonne réduction
6. Motifs
7. Réalisations de Betti, périodes p-adiques
8. Comment éviter les conjectures standard?

1. — Périodes

Soit $f : X \to S = \mathbb{P}^1 \backslash \{\zeta_1, \ldots, \zeta_s\}$ une famille de variétés projectives lisses définie sur un corps de nombres K. Quitte à retrancher d'autres points ζ_i, on peut supposer que les $\mathcal{O}_S$-modules de cohomologie de De Rham $H^q_{dR}(X/S) := \mathbb{R}^q f_* \Omega^\cdot_{X/S}$ sont libres. A l'aide d'une base $\omega_1, \ldots, \omega_n$, et en termes d'une coordonnée globale x sur S, la connexion de Gauss–Manin ∇ donne naissance à un système différentiel

(*) $\dfrac{dY}{dx} = \Gamma Y$, où Γ est une matrice $n \times n$ à coefficients dans $K(x)$.

Supposons les ζ_i tous non nuls pour simplifier, et considérons la matrice $Y(x)$ solution de (*) à coefficients dans $K[[x]]$ normalisée par $Y(0) = Id$.

Choisissons un plongement $K \hookrightarrow \mathbb{C}$.

On a :

$$(R^q f^{an} * \mathbb{Q}) \otimes_{\mathbb{Q}} \mathbb{C} \cong H^q_{dR}(X^{an}/S^{an})^\nabla \hookrightarrow H^q_{dR}(X/S) \otimes_{\mathcal{O}_S} \mathcal{O}_{S^{an}}.$$

Le choix d'une trivialisation locale de $R^q f^{an} * \mathbb{Q}$ au voisinage de $x = 0$ permet d'exprimer l'isomorphisme $H^q_{dR}(X/S) \otimes_{\mathcal{O}_S} \mathcal{O}_{S^{an}} \cong (R^q f^{an} * \mathbb{Q}) \otimes_{\mathbb{Q}} \mathcal{O}_{S^{an}}$ sous forme d'une matrice de périodes $\Omega(x) = (\Omega_{ij}(x) = \oint_{\gamma_j} \omega_i(x))$[4], satisfaisant à l'équation (*). Cette matrice est donc liée à $Y(x)$ par la formule

$$Yx) = \Omega(x)\Omega(0)^{-1}, \tag{**}$$

ce qui fournit une interprétation géométrique des évaluations complexes de $Y(x)$.

Si les évaluations en $\xi \in K$ ($\xi \neq \zeta_i$) des coefficients de la matrice $Y(x)$ sont algébriquement dépendantes sur K, on en déduit que les coefficients des matrices $\Omega(x)$ et $\Omega(0)$ le sont aussi. Or unc conjecture célèbre de Grothendieck [G66] affirme :

(P) *toute relation polynômiale à coefficients dans K entre les périodes d'une K-variété projective (lisse) devrait "provenir" de l'existence d'un cycle algébrique sur une puissance de cette variété*[5].

---

[4] sous cette forme traditionnelle, c'est en fait la *transposée* de la matrice de cet isomorphisme.

[5] Un tel cycle algébrique donne en effet lieu à des relations polynômiales à coefficients algébriques entre les périodes, en écrivant la compatibilité de ses composantes Betti et De Rham, et en utilisant la formule de Künneth.

Remarque : cette conjecture n'est connue que pour un petit nombre de variétés "très simples"[6] : surface cubique (la conjecture équivaut alors à la transcendance de π), courbe elliptique à multiplication complexe (Chudnovsky)...

Mais si l'on borne *a priori* le degré des relations de dépendance entre périodes (il en sera ainsi dans la situation considérée ci-dessus), alors le fait que ces relations proviennent de cycles algébriques peut s'établir, à l'aide de la théorie diophantienne des G-fonctions, dans des situations beaucoup plus générales. En voici un exemple, formulé en termes de "cycles motivés", légère variante des cycles algébriques développée au § 8 :

THÉORÈME 2. — *Soit* $g : Z \to S = \mathbb{P}^1 \backslash \{\zeta_1, \ldots, \zeta_s\}$ *un* K*-morphisme projectif et plat de dimension relative* q*, lisse en dehors de* 0*. On suppose que la fibre* Z_0 *est un diviseur à croisements normaux simples dont toutes les strates d'intersection sont lisses, et tel que la cohomologie de chaque strate soit formée de cycles motivés. Soient* $\lambda_1, \lambda_2, \ldots$ *des éléments linéairement indépendants de* $H_q(Z^{an}_{\frac{1}{m}}, \mathbb{Q})$ *situés dans l'image de la puissance* q*-ième du logarithme de la monodromie locale en* 0*, et soient* $\eta_1, \eta_2, \ldots$ *des éléments linéairement indépendants de* $H^q_{dR}(Z_{\frac{1}{m}})$*. Alors toute relation polynômiale homogène de degré* δ *à coefficients dans* K *entre les périodes* $\oint_{\lambda_j} \eta_i$ *provient d'un cycle motivé sur une puissance de la fibre* $Z_{\frac{1}{m}}$*, pourvu que* m *soit suffisamment grand (par rapport à* δ*).*

C'est un cas particulier du théorème principal de [A89] IX, sauf que dans l'hypothèse et la conclusion, cycles motivés remplacent cycles de Hodge; la preuve est identique. La borne pour m est en principe effective, du type exponentielle d'une puissance de δ (une particularité de la théorie diophantienne des G-fonctions fait qu'on ne sait pas remplacer dans l'énoncé $1/m$ par "un rationnel proche de 0").

3. — Une $\overline{\mathbb{Q}}$-structure dans la cohomologie cristalline ?

Examinons maintenant la possibilité d'une interprétation géométrique similaire des évaluations p-adiques de $Y(x)$.

Choisissons donc un plongement de K dans $\mathbb{C}_p$, le complété p-adique d'une clôture algébrique de $\mathbb{Q}_p$. Considérons le morphisme $X^{an} \to S^{an}$ de $\mathbb{C}_p$-variétés analytiques (au sens de Bourbaki, que nous qualifierons de "mou", par opposition à la théorie analytique rigide qui interviendra plus loin) associé à f. Le faisceau de germes horizontaux $H^q_{dR}(X^{an}/S^{an})^{\nabla} \hookrightarrow H^q_{dR}(X/S) \otimes_{\mathcal{O}_S} \mathcal{O}_{S^{an}}$ est localement constant (pour la topologie usuelle, "molle").

[6] du point de vue motivique, cf. infra

Supposons d'abord que f ait bonne réduction en p, i.e. se prolonge en un morphisme projectif lisse au-dessus du localisé en p de l'anneau des entiers de K (on fixe un tel prolongement). En particulier, la fibre X_0 a bonne reduction $\widetilde{X_0}$ en p. Alors l'espace des sections de $H^q_{dR}(X^{an}/S^{an})^\nabla$ au voisinage de 0 s'identifie à l'espace de cohomologie cristalline $H^q_{cris}(\widetilde{X_0}/W(\overline{\mathbb{F}}_p)) \otimes \mathbb{C}_p$ d'après [BO83]. (Si X_0 n'a pas bonne réduction, on conjecture généralement que X_0 a au moins réduction potentiellement semi-stable, et il convient de remplacer alors cohomologie cristalline par cohomologie cristalline logarithmique à la Hyodo-Kato).

Le choix d'une base de cet espace de cohomologie cristalline donne lieu à une matrice de "périodes p-adiques" $\Omega_p(x)$, liée aux évaluations p-adiques de $Y(x)$ par la formule

$$(*)_p \qquad Y(x) = \Omega_p(x)\Omega_p(0)^{-1}.$$

Toutefois pour donner un sens géométrique aux considérations de dépendance sur $\overline{\mathbb{Q}}$ entre périodes p-adiques, ou entre les évaluations des coefficients de $Y(x)$ en un point $\xi \in K$ p-adiquement proche de 0, il faudrait savoir donner une signification géométrique au $\overline{\mathbb{Q}}$-espace engendré par la base choisie de la cohomologie cristalline, et aux $\overline{\mathbb{Q}}$-espaces analogues attachés aux puissances de $\widetilde{X_0}$ (Künneth). On voudrait en particulier, en liaison avec la situation prédite par la conjecture de Grothendieck, que la classe cristalline des réductions des cycles algébriques sur un produit de puissances des fibres X_ξ et X_0 soient dans ces $\overline{\mathbb{Q}}$-espaces. On est donc conduit au problème suivant :

PROBLÈME. — *Construire une $\overline{\mathbb{Q}}$-structure dans la cohomologie cristalline (tensorisée avec $\mathbb{C}_p$) des puissances de $\widetilde{X_0}$, de telle sorte que les cycles algébriques soient rationnels relativement à cette $\overline{\mathbb{Q}}$-structure.*

Ce qui revient essentiellement à construire une cohomologie à coefficients dans $\overline{\mathbb{Q}}$ sur la catégorie des variétés sommes disjointes de puissances de $\widetilde{X_0}$, et un isomorphisme de comparaison entre cette cohomologie et la cohomologie cristalline (tensorisées avec $\mathbb{C}_p$). Les motifs pointent à l'horizon...

En attachant ensuite à toute fibre (ou tout produit fini de fibres) X_x, pour $x \in \mathbb{C}_p$ assez proche de 0, le $\overline{\mathbb{Q}}$-espace de cohomologie (problématique) de sa réduction, on obtiendrait un espace $H_B(X_x, \overline{\mathbb{Q}})$ analogue à la *cohomologie de Betti à coefficients* dans $\overline{\mathbb{Q}}$ dans la situation complexe; en désignant par U une petite boule autour de x, on aurait :

$$H^q_B(X_x, \overline{\mathbb{Q}}) \otimes_{\overline{\mathbb{Q}}} \mathbb{C}_p \cong H^q_{dR}(X^{an} \times_{S^{an}} U/U)^\nabla \hookrightarrow H^q_{dR}(X/S) \otimes_{\mathcal{O}_S} \mathcal{O}_U.$$

Remarque : *une fausse piste, le foncteur mystérieux.* Les travaux de J.-M. Fontaine, W. Messing et G. Faltings ont permis de construire ce que Grothendieck appelait le mystérieux foncteur reliant cohomologie étale et cohomologie cristalline. En fait, la construction vaut dans le cas relatif, du moins sous une hypothèse de bonne réduction, et peut se présenter ainsi [F89] : il existe un anneau différentiel filtré B_{dR}, et un isomorphisme de B_{dR}-modules à connexion filtrés :

$$\Gamma H^q_{dR}(X/S) \otimes_{\Gamma\mathcal{O}_S} B_{dR} \cong H^q_{et}(X_{\overline{\eta}}, \mathbb{Q}_p) \otimes_{\mathbb{Q}_p} B_{dR}\,.$$

D'autre part, tout plongement de $\overline{\mathbb{Q}_p}$ dans $\mathbb{C}$ induit un isomorphisme équivariant sous la monodromie :

$H^q_{et}(X_{\overline{\eta}}, \mathbb{Q}_p) \cong H^q_B(X^{an}_{\mathbb{C}} \times_{S^{an}_{\mathbb{C}}} \widehat{S}, \mathbb{Q}) \otimes_{\mathbb{Q}} \mathbb{Q}_p$ ($\widehat{S}$: revêtement universel de $S^{an}_{\mathbb{C}}$).

Une base de $H^q_B(X^{an}_{\mathbb{C}} \times_{S^{an}_{\mathbb{C}}} \widehat{S}, \mathbb{Q})$ étant choisie, l'isomorphisme composé donnerait naissance à une matrice de "périodes" $\Omega_{B_{dR}}$ liée à $Y(x)$ par la relation $Y(x) = \Omega_{B_{dR}} \cdot C$, où C est une matrice inversible à coefficients constants (au sens différentiel) dans B_{dR}. Mais le fait que B_{dR} contienne beaucoup trop de constantes différentielles (non réduites aux "scalaires") voue cette tentative d'interprétation géométrique des valeurs p-adiques de $Y(x)$, proposée dans l'introduction de [A89], à l'échec.

Signalons encore que, dans le cas constant, l'analogue de la conjecture de Grothendieck pour la matrice de "périodes" $\Omega_{B_{dR}}$ est faux [A90].

4. — Le cas des variétés abéliennes dégénérantes [A90]

C'est un cas qui déborde le cadre précédent – 0 est une singularité –, mais où l'analogue (partiel) du programme précédent peut être accompli plus aisément.

Une variété abélienne A sur une extension finie du corps complet $K((x))$ ou $\mathbb{Q}_p$ sera dite *dégénérante* si la composante neutre de la fibre spéciale de son modèle de Néron est un tore déployé $T \cong \mathbf{G}^g_m$. Ces variétés sont bien connues en géométrie rigide, puisque la variété rigide associée A^{rig} est un tore analytique.

a) Rappelons que si A est une variété abélienne complexe de dimension g, alors $A(\mathbb{C}) \cong \mathbb{C}^g/L$, où L est un réseau de rang $2g$, et on a un accouplement non-dégénéré $\oint_? ? : L \otimes H^1_{dR}(A) \to \mathbb{C}$. A l'aide de la fonction $\exp(2i\pi)$, on peut encore représenter $A(\mathbb{C})$ sous la forme $T(\mathbb{C})/M$, où M est un réseau de rang g, et T un tore de dimension g (paramétrisation de Jacobi). Soit M' le groupe $\mathrm{Hom}(\mathbf{G}_m, T)$, M'^v son dual. Alors L s'inscrit dans une suite exacte

(***) $$0 \longrightarrow 2i\pi M'^v \longrightarrow L \longrightarrow M \longrightarrow 0,$$

scindée par le choix d'une détermination du logarithme.

Le morphisme $M \hookrightarrow T$ se décrit par une application bilinéaire $M \times M' \to \mathbb{C}^*$; toute polarisation de A donne lieu à une "isogénie" $M \to M'$, d'où une application $q : M \otimes M \to \mathbb{C}^*$. Il s'avère que $-\log|q|$ est un produit scalaire sur $M_{\mathbb{R}}$.

b) Soit maintenant A une variété abélienne dégénérante de dimension g sur $\mathbb{C}_p$. Alors on peut représenter $A(\mathbb{C}_p)$ sous la forme $T(\mathbb{C}_p)/M$, avec M, T comme ci-dessus (paramétrisation de Tate). Toute polarisation de A donne lieu comme précédemment à une application $q : M \otimes M \to \mathbb{C}_p^*$, et il s'avère que $-\log|q|_p$ est un produit scalaire. On a $M = H_1(A^{rig}, \mathbb{Z})$, $M' = H^1((A^{dual})^{rig}, \mathbb{Z})$, et M^v se plonge canoniquement dans le dual du module de Tate de A.

On peut pousser plus loin l'analogie avec la situation complexe : dans [A90], on utilise le semi-endomorphisme de Frobenius sur la cohomologie cristalline d'un certain 1-motif fonctoriellement attaché à A pour construire un *réseau* L *de rang* $2g$, sur lequel agissent les endomorphismes de A, et qui s'inscrit dans une suite exacte

$$(***)_p \qquad 0 \longrightarrow (2i\pi)_p M'^v \longrightarrow L \longrightarrow M \longrightarrow 0,$$

scindée par le choix d'une détermination du logarithme p-adique; ici $(2i\pi)_p$ désigne un générateur du $\mathbb{Z}_p$-module des racines p-primaires de l'unité dans $\mathbb{C}_p^*$ (c'est l'analogue p-adique de $2i\pi$).

On construit en outre un accouplement canonique non-dégénéré :

$$\oint_? ? : L \otimes H^1_{dR}(A) \longrightarrow \mathbb{C}_p[(2i\pi)_p],$$

c) Soit $f : \underline{A} \to S\backslash\{s_0\}$ un schéma abélien de dimension relative g sur le complémentaire d'un K-point lisse s_0 dans une courbe affine[7] S sur K, et soit x une coordonnée locale en s_0. On suppose que la composante neutre de la fibre en $x = 0$ du modèle de Néron est un tore déployé T sur K. Il en est alors de même du schéma abélien dual, avec un tore T'. Posons $M = \mathrm{Hom}(\mathbf{G}_m, T')$, $M' = \mathrm{Hom}(\mathbf{G}_m, T)$.

Fixons $K \hookrightarrow \mathbb{C}$. Alors le sous-faisceau de $R_1 f^{an}_{\mathbb{C}} * \mathbb{Z}$ constant au voisinage de s_0 s'identifie à $2i\pi M'^v$; sa fibre en tout $s \in S(\mathbb{C})$, $s \neq s_0$, s'identifie au réseau $2i\pi M'^v_s$ de rang g associé en a) à la variété abélienne $\underline{A}_s$; de même M s'identifie au réseau M_s de rang g associé en a) à $\underline{A}_s$. D'autre part, le choix d'une détermination du logarithme, c'est-à-dire essentiellement d'un secteur U d'angle 2π d'une petite boule autour de $x = 0$, identifie $R_1 f^{an}_{\mathbb{C}} * \mathbb{Z}_{|U}$ à $L = 2i\pi M'^v \oplus M$; sa fibre en tout $s \in S(\mathbb{C})$, $s \neq s_0$, s'identifie au réseau $L_s = H_1(\underline{A}^{an}_{s\mathbb{C}}, \mathbb{Z})$ de rang $2g$.

---

[7] Pas nécessairement un ouvert de la droite, nous dévions des notations du § 2.

THÉORÈME 3. — *Pour tout* $\gamma \in 2i\pi M'^{v}$, *la série de Taylor* $y_\gamma(x)$ *de* $\frac{1}{2i\pi}\oint_{\gamma_s}\omega(s)$ *est une G-fonction. Pour tout* $\gamma \in L$, $\frac{1}{2i\pi}\oint_{\gamma_s}\omega(s)$ *s'écrit* $z_\gamma(x(s))\frac{\log a_\gamma x(s)^{n_\gamma}}{2i\pi} + y_\gamma(x(s))$, *où* $z_\gamma(x)$ *et* $y_\gamma(x)$ *sont des G-fonctions,* $a_\gamma \in K$, $n_\gamma \in \mathbb{Z}$.

Fixons d'autre part un premier p et $K \hookrightarrow \mathbb{C}_p$ tels que le schéma abélien f ait bonne réduction en p. Pour tout $s \in S(\mathbb{C}_p)$ assez proche mais distinct de s_0, M, resp. M', s'identifie au réseau M_s resp. M'_s de rang g associé en b) à la variété abélienne dégénérante $\underline{A}_s$. Le choix d'une détermination du logarithme p-adique (e.g. via $\log_p p = 0$) identifie $L = (2i\pi)_p M'^{v} \oplus M$ au réseau L_s de rang $2g$ associé en b) à $\underline{A}_s$. De plus, l'accouplement p-adique $\oint_?$? se prolonge en un accouplement *horizontal*

$$\oint_? ? : L \otimes H^1_{dR}(\underline{A}_U/U) \longrightarrow \mathcal{O}^*_U[(2i\pi)_p] \quad (U : \text{petite boule autour de } s).$$

On a alors, en parfaite analogie avec le théorème 3 (et avec les mêmes notations) :

THÉORÈME 4. — *Pour tout* $\gamma \in (2i\pi)_p M'^{v}$, $\frac{1}{(2i\pi)_p}\oint_{\gamma_s}\omega(s)$ *est l'évaluation p-adique de la G-fonction* $y_\gamma(x)$ *en* $x(s)$.

Pour tout $\gamma \in L$, *on a* $\frac{1}{(2i\pi)_p}\oint_{\gamma_s}\omega(s) = z_\gamma(x(s))\frac{\log_p a_\gamma x(s)^{n_\gamma}}{(2i\pi)_p} + y_\gamma(x(s))$.

Remarques :

i) Dans ce dernier théorème, on peut aussi remplacer l'accouplement $\oint_?$? par celui de Fontaine–Messing, à condition de remplacer le logarithme p-adique usuel par sa version B_{dR} introduite par Fontaine. Cela fournit donc un "pont" entre les théories p-adiques de Dwork et de Fontaine.

ii) Dans le cas où la variété abélienne dégénérante sur $\mathbb{C}_p$ est la jacobienne d'une courbe de Mumford, le réseau L mentionné au point b) est étroitement lié à un réseau de rang $2g$ construit antérieurement, et par de tout autres méthodes (fonctions thêta), par L. Gerritzen [Ge86]. [Le réseau considéré dans [A90], noté $\mathbb{Z}\mu''_1 \oplus \cdots \oplus \mathbb{Z}\mu''_g \oplus \mathbb{Z}m^v_1 \oplus \cdots \oplus \mathbb{Z}m^v_g$, est lié au réseau $\mathbb{Z}du_1/u_1 \oplus \cdots \oplus \mathbb{Z}du_g/u_g \oplus \mathbb{Z}\beta_1 \oplus \cdots \oplus \mathbb{Z}\beta_g$ de [Ge86] par les formules : $\beta_i = m^v_i$, $du_j/u_j = (2i\pi)_p\mu''_j + \Sigma(\log q_{ij})m^v_i$, où les $q_{ij} : M \times M \to \mathbb{C}^*_p$ $(j = 1, \ldots, g)$ sont les facteurs d'automorphie attachés à u_i].

iii) On peut compléter le théorème 3 en montrant l'existence d'un entier $N > 0$ tel que pour tout $\gamma \in 2i\pi M'^{v}$ (resp. M), $y_\gamma(Nx)$ (resp. $\exp(y_\gamma(Nx)/(Nx))$) soit à coefficients entiers algébriques.

5. — Le cas des courbes elliptiques avec bonne réduction [A]

Revenons à l'hypothèse de bonne réduction et examinons le problème du § 3 dans le cas non trivial le plus simple, celui d'une courbe elliptique X sur le corps de nombres K.

a) *Le cas où la réduction $\widetilde{X}$ est ordinaire*

Dans ce cas, on sait que $\widetilde{X}_{\overline{\mathbb{F}}_p}$ admet un relevé canonique $X^{can}/\overline{\mathbb{Q}}$ de type CM. On a : $H^1_{dR}(X)\otimes\mathbb{C}_p \cong H^1_{cris}(\widetilde{X_0}/W(\overline{\mathbb{F}}_p))\otimes\mathbb{C}_p \cong H^1_{dR}(X^{can})\otimes\mathbb{C}_p$, et on montre que $H^1_B(X,\overline{\mathbb{Q}}) := H^1_{dR}(X^{can})$ répond au problème du § 3.

b) *Réduction supersingulière*

Ici $D := \mathrm{End}\,\widetilde{X}_{\overline{\mathbb{F}}_p}$ est un ordre maximal d'une algèbre de quaternions sur $\mathbb{Q}$, donc $D_{\overline{\mathbb{Q}}} \cong M_2(\overline{\mathbb{Q}})$. Soient $E \subseteq D_{\mathbb{Q}}$ un corps quadratique (nécessairement imaginaire et déployant), et $\vec{v}$ un vecteur propre dans $H^1_{cris}(\widetilde{X_0}/W(\overline{\mathbb{F}}_p))\otimes\mathbb{C}_p$ pour l'action de E. Alors $D_{\overline{\mathbb{Q}}}\cdot\vec{v}$ est un $\overline{\mathbb{Q}}$–espace de dimension 2, contenant un E–vecteur propre $\vec{u}$ linéairement indépendant de $\vec{v}$ sur $\mathbb{C}_p$. Normalisant la base $(\vec{v},\vec{u})$ par une homothétie de sorte que $\vec{v}\wedge\vec{u}$ soit le générateur canonique de $H^2_{cris}(\widetilde{X_0}/W(\overline{\mathbb{F}}_p))$, on obtient une $\overline{\mathbb{Q}}$–*structure canonique* $H^1_B(X_{\mathbb{C}_p},\overline{\mathbb{Q}}) := \overline{\mathbb{Q}}\cdot\vec{v}\oplus\overline{\mathbb{Q}}\cdot\vec{u}$ dans $H^1_{cris}(\widetilde{X_0}/W(\overline{\mathbb{F}}_p))\otimes\mathbb{C}_p$ (elle ne dépend pas des choix auxilliaires de E et $\vec{v}$), qui répond au problème du § 3.

Remarque : il est probable que l'isomorphisme $H^1_{dR}(X)\otimes\mathbb{C}_p \cong H^1_B(X_{\mathbb{C}_p},\overline{\mathbb{Q}})\otimes\mathbb{C}_p$ est toujours transcendant. Du reste, pour X de type CM, on peut déduire de [O90] une expression des périodes p–adiques (i.e. des coefficients de la matrice de cet isomorphisme relativement à des bases de $H^1_{dR}(X)$ et $H^1_B(X_{\mathbb{C}_p},\overline{\mathbb{Q}})$ resp.) en termes de valeurs de la fonction Γ_p en des rationnels.

c) La *fonctorialité* de $H^1_B(X_x,\overline{\mathbb{Q}})$ permet de démontrer simplement des relations entre valeurs p–adiques de la matrice $Y(x)$ solution de l'équation de Picard–Fuchs relative à une famille de courbes elliptiques X/S.

Par exemple, supposons que la fibre en 0 ait multiplication complexe par $\mathbb{Q}(\sqrt{-d})$, et la fibre en $\xi\in K$ par $\mathbb{Q}(\sqrt{-d'}) \neq \mathbb{Q}(\sqrt{-d})$. Choisissons une base symplectique ω_1, ω_2 de sections de $H^1_{dR}(X/S)$ de telle sorte que ω_1 soit dans le cran 1 de la filtration de Hodge (i.e. la classe d'une forme différentielle relative régulière), et que $\omega_2(0)$ soit propre pour l'action de $e=\sqrt{-d}\in\mathrm{End}\,X_0$ sur $H^1_{dR}(X_0)$; alors $\omega_1(0)$, resp. $\omega_1(\xi)$, est propre sous l'action de e, resp. de $e'=\sqrt{-d'}\in\mathrm{End}\,X_\xi$, et il y a un unique élément σ de K tel que $\omega_2(\xi)+\sigma\,\omega_2(\xi)$ soit propre sous l'action de e'.

Soit p un nombre premier non ramifié dans $\mathbb{Q}(\sqrt{-d})$ et v une place de K divisant p (correspondant à un ensemble de plongements "équivalents" $K\hookrightarrow\mathbb{C}_p$) tel que ξ soit p–adiquement proche de 0 (de sorte que X_0 et X_ξ ont même réduction en p, nécessairement supersingulière). Examinons les images de e et e' dans $\mathrm{End}\,\widetilde{X_0}=\mathrm{End}\,\widetilde{X_\xi}$: $ee'+e'e$ commute à e et e', donc est un entier m_v. On a $(ee'-e'e)^2 = m_v^2-4dd'$. Comme $ee'-e'e$

anticommute à e, et n'est donc pas un entier, on a $|m_v| < 2\sqrt{dd'}$.

[Remarque : dans le cas d'une place archimédienne plutôt que p–adique, un argument analogue vaut, en considérant les images de e et e' dans $\mathrm{End}\, H^1_B(X_{0_{\mathbb{C}}}, \mathbb{Q})$ identifié à $H^1_B(X_{\xi\mathbb{C}}, \mathbb{Q})$ par le prolongement analytique dans une petite boule de centre 0].

De plus $ee' - e'e$ agit trivialement sur les formes différentielles invariantes, en caractérisant p; donc p divise $m_v^2 - 4dd'$. Soit de nouveau $(\vec{v}, \vec{u})$ la base symplectique de $H^1_B(X_{0_{\mathbb{C}_p}}, \overline{\mathbb{Q}}) \cong H^1_B(X_{\xi\mathbb{C}_p}, \overline{\mathbb{Q}})$ vérifiant $e \cdot \vec{v} = \sqrt{-d}\vec{v}$, $e \cdot \vec{u} = -\sqrt{-d}\vec{u}$. On a alors $e' \cdot \vec{v} + \frac{m_v}{2d}\sqrt{-d}\vec{v} \in \overline{\mathbb{Q}} \cdot \vec{u}$. Relativement aux bases $(\omega_1(0), \omega_2(0))$ (resp.$(\omega_1(\xi), \omega_2(\xi))$) et $(\vec{v}, \vec{u})$, la matrice de périodes $\Omega_p(0)$ s'écrit sous la forme $\begin{pmatrix} \varpi & 0 \\ 0 & \varpi^{-1} \end{pmatrix}$, resp. $\Omega_p(\xi)$ s'écrit $\begin{pmatrix} 1 & 0 \\ -\sigma & 1 \end{pmatrix} \begin{pmatrix} \varpi' & 0 \\ 0 & \varpi'^{-1} \end{pmatrix} M$ avec $M \in M_2(\mathbb{Q}(\sqrt{-d}, \sqrt{-d'}))$ [dans le cas d'une place v archimédienne, il conviendrait de multiplier les ϖ, ϖ' en première ligne par $2i\pi$]. L'évaluation v–adique $Y(\xi) = \Omega_p(\xi)\Omega_p(0)^{-1}$ a donc pour coefficients $Y_{11}(\xi) = M_{11}\varpi'/\varpi$, $Y_{12}(\xi) = M_{12}\varpi'\varpi$, $Y_{21}(\xi) = -M_{11}\sigma\omega'/\omega + M_{21}/\varpi'\varpi$, $Y_{22}(\xi) = M_{22}\varpi/\varpi' - M_{12}\sigma\varpi'\varpi$. Parce que M diagonalise l'action de e' dans $H^1_B(X_{\xi\mathbb{C}_p}, \overline{\mathbb{Q}})$, on obtient la relation $M_{11}M_{22} + M_{12}M_{21} = -m_v/2\sqrt{dd'}$; et comme d'autre part $e' \cdot \vec{v} + \frac{m_v}{2d}\sqrt{-d}\vec{v}$ est collinéaire à $\vec{u}$, on a $M_{11} = m_v/2\sqrt{-d}$.

Il n'est pas difficile de conclure alors à la relation suivante[8] entre évaluations v–adiques des coefficients de $Y(x)$ en ξ :

$$(m_v + 2\sqrt{dd'})Y_{11}(\xi)(Y_{22}(\xi) + \sigma Y_{12}(\xi)) = (m_v - 2\sqrt{dd'})Y_{12}(\xi)(Y_{21}(\xi) + \sigma Y_{11}(\xi))$$

où m_v est un entier borné en valeur absolue par $2\sqrt{dd'}$ et tel que $\sqrt{p(v)} \mid m_v + 2\sqrt{dd'}$ si v est ultramétrique de caractéristique résiduelle $p(v)$. Il y a en fait une telle relation pour chaque place v de K telle que $Y(x)$ converge v–adiquement en ξ.

Ces exemples en faveur de l'existence des $\overline{\mathbb{Q}}$–structures de Betti nous encouragent à explorer les dimensions supérieures.

6. — Motifs

Grosso modo, la théorie des motifs est à la géométrie algébrique ce que la théorie de Galois est à l'algèbre commutative.

Pour contourner l'excessive complexité de la catégorie des variétés projectives lisses sur un corps k fixé, il est souvent nécessaire d'enrichir

---

[8] Beukers a découvert ces relations simultanément et indépendamment [Be93] (sans toutefois en donner la forme explicite), par une méthode de relèvement d'isogénies. Il a en outre testé plusieurs exemples sur ordinateur.

ses morphismes. Dans les problèmes de classification, par exemple, il est commun de considérer non seulement les applications régulières, mais aussi les applications rationnelles. Il arrive d'avoir même à considérer toutes les correspondances algébriques ; A. Weil en a éloquemment montré l'intérêt en prouvant "l'hypothèse de Riemann" pour les courbes sur un corps fini.

Il y a une trentaine d'années, Grothendieck a eu l'idée flamboyante que la catégorie des variétés projectives lisses sur $k \subseteq \mathbb{C}$, avec pour morphismes les correspondances algébriques à coefficients rationnels modulo l'équivalence numérique, est "essentiellement"[9] équivalente à la catégorie des représentations semisimples de dimension finie sur $\mathbb{Q}$ d'un $\mathbb{Q}$–groupe pro–algébrique, – l'équivalence transformant produit cartésien en produit tensoriel –; et que pour k de caractéristique non nulle, la situation est similaire, quitte à remplacer "groupe" par "gerbe".

Définissons un *motif* comme un triplet (Z, n, e) (noté généralement $e\mathfrak{h}(Z)(n)$), où Z est un k–schéma projectif lisse, n un entier, et $e : Z \multimap Z$ une correspondance algébrique de degré 0 modulo l'équivalence numérique, vérifiant $e \circ e = e$ (idempotence).

Un morphisme de motifs $e\mathfrak{h}(Z)(n) \to e'\mathfrak{h}(Z')(n')$ est une correspondance algébrique de degré $n' - n$ modulo l'équivalence numérique de la forme $e' \circ f \circ e : Z \multimap Z'$.

[Rappelons qu'une correspondance algébrique $Z \multimap Z'$ de degré r (à coefficients rationnels) est une combinaison $\mathbb{Q}$–linéaire de sous–schémas intègres de $Z \times Z'$ de codimension $r + \dim Z$, ou une classe d'équivalence d'une telle combinaison ; et que, pour toute équivalence "adéquate", les correspondances algébriques se composent par la formule $g \circ f = p_{ZZ''*}(p^*_{ZZ'}f \cdot p^*_{Z'Z''}g)$, les degrés s'additionnant].

U. Jannsen a démontré que les motifs forment une catégorie $\mathbb{Q}$–linéaire *abélienne semisimple* [J92]. On la munit du produit tensoriel $e\mathfrak{h}(Z)(n) \otimes e'\mathfrak{h}(Z')(n') = (e \times e')\mathfrak{h}(Z \times Z')(n + n')$, et on la note $\mathcal{M}_k$.

Lorsque $e = id$, $n = 0$, on note le motif simplement $\mathfrak{h}(Z)$; le foncteur contravariant de *cohomologie motivique* associe $\mathfrak{h}(Z)$ à Z.

Faisons provisoirement l'hypothèse suivante (c'est l'une des "conjectures standard" de Grothendieck) :

(N) pour chacune des théories de cohomologies classique $H^{\cdot}$ (celles qui portent un nom), l'équivalence homologique coïncide avec l'équivalence numérique.

Il en découle que $H^{\cdot}$ (en particulier, chaque cohomologie étale ℓ–adique pour $\ell \neq \operatorname{car} k$) se factorise par la cohomologie motivique $\mathbb{Q}$–linéaire $\mathfrak{h}$, comme foncteur sur la catégorie des k–schémas projectifs lisses.

---

[9] *stricto sensu*, il faut ajouter formellent les noyaux des projecteurs.

Il en découle aussi que l'"isomorphisme de Lefschetz fort" L_Z^{d-i} : $H^i(Z) \to H^{2d-i}(Z)(d-i)$ (donné par le cup–produit itéré avec la classe d'une section hyperplane du k-schéma projectif Z de dimension d [D80] [KM74]) provient d'un isomorphisme de motifs; en particulier son inverse est algébrique, ce qui entraîne notoirement :

(C) les projecteurs de Künneth sont donnés par des correspondances algébriques.

Avec **(C)**, la $\otimes$–catégorie des motifs $\mathcal{M}_k$ se trouve graduée[10], et la théorie "tannakienne", initiée par Grothendieck à ce propos, montre alors que $\mathcal{M}_k$ est $\otimes$–équivalente à la catégorie des représentations de dimension finie d'une gerbe pro–algébrique sur $\mathbb{Q}$ (on dit que $\mathcal{M}_k$ est tannakienne sur $\mathbb{Q}$). Tout foncteur fibre[11] sur $\mathcal{M}_k$ s'interprète comme une cohomologie; et réciproquement si l'équivalence homologique coïncide avec l'équivalence numérique.

Sous **(N)**, et si de plus $k \subseteq \mathbb{C}$, la cohomologie de Betti réalise en fait une $\otimes$–équivalence entre $\mathcal{M}_k$ et la catégorie des représentations de dimension finie sur $\mathbb{Q}$ d'un $\mathbb{Q}$–groupe pro–réductif $\mathbf{G}_{mot}$, appelé *groupe de Galois motivique absolu.* L'image $G_{mot}(Z)$ de $\mathbf{G}_{mot}$ dans $GL(H^\cdot(Z))$ est le sous–groupe Zariski fermé qui fixe les classes algébriques parmi les tenseurs mixtes sur $H^\cdot(Z)$; $\mathbf{G}_{mot}$ s'exprime comme limite projective de ces groupes de Galois motiviques $G_{mot}(Z)$.

La conjecture **(P)** de Grothendieck sur les périodes exprime que l'isomorphisme canonique entre les foncteurs fibres $H^\cdot_{dR} \otimes \mathbb{C}$ et $H^\cdot_B \otimes \mathbb{C}$ est "générique" parmi tous les isomorphismes possibles.

7. — Réalisation de Betti; périodes p–adiques [A]

a) Reprenons les notations du § 3 : $f : X \to S$ est un morphisme projectif lisse ayant bonne réduction en la place v de K induite par un plongement fixé $K \hookrightarrow \mathbb{C}_p$, $\widetilde{X_0}$ est la réduction de la fibre $X_0 \cdots$ et considérons la catégorie tensorielle de motifs engendrée par $\mathfrak{h}(\widetilde{X}_{0_k})$ pour $k = \overline{\mathbb{F}}_p$ = clôture algébrique du corps résiduel de K en v; on la note $\widetilde{X}_{0_k}^{\otimes}$. Pour les k-variétés, l'énoncé **(C)** est vrai [KM74], donc $\widetilde{X}_{0_k}^{\otimes}$ est tannakienne sur $\mathbb{Q}$ [D90] [J92]. On en déduit :

THÉORÈME 5. — *Il existe un corps de nombres $F \subseteq \mathbb{C}_p$, et une cohomologie $H^\cdot(\cdot, F)$ à coefficients dans F pour les schémas sommes disjointes de puissances de $\widetilde{X}_{0_k}$ (d'où un F–groupe de Galois motivique $G_{mot}(\widetilde{X}_{0_k})$). En*

---

[10] on modifie alors la contrainte de commutativité évidente pour $\otimes$ par un signe obéissant à la règle de Koszul.

[11] =$\otimes$–foncteur exact à valeurs espaces vectoriels.

outre, si l'équivalence homologique cristalline coïncide avec l'équivalence numérique sur ces schémas, il existe un isomorphisme de cohomologies $\iota : H^{\cdot}_{cris}(\cdot/W(k)) \otimes \mathbb{C}_p \cong H^{\cdot}(\cdot, F) \otimes \mathbb{C}_p$, *bien défini à l'action près de* $G_{mot}(\widetilde{X}_{0_k})(\mathbb{C}_p)$.

Pour tout $x \in \mathbb{C}_p$ assez proche de 0 (de sorte que $\widetilde{X}_x \cong \widetilde{X}_{0_k}$), on obtient alors un foncteur fibre sur la catégorie tensorielle $X_x^{\otimes}$ engendrée par $\mathfrak{h}(X_x)$ en posant $H^{\cdot}_B(X_x^m, F) := H^{\cdot}(\widetilde{X}_{0_k}^m, F)$; nous l'appellerons *réalisation de Betti*. Il remplit le programme du § 3. La matrice de "périodes p-adiques" $\Omega_p(x)$, transposée de la matrice de l'isomorphisme composé $H^q_{dR}(X_x) \otimes \mathbb{C}_p \cong H^q_{cris}(\widetilde{X}_{0_k}/W(k)) \otimes \mathbb{C}_p \cong H^q_B(X_x, F) \otimes \mathbb{C}_p$ relativement à la base $\omega_1, \ldots, \omega_n$ de $\Gamma H^q_{dR}(X/S)$ et à une base de $H^{\cdot}_B(X_x, F)$ est solution de l'équation de Picard–Fuchs (*) lorsque x varie.

Remarques :

1) Il résulte des conjectures de Tate que $G_{mot}(\widetilde{X}_{0_k})$ devrait être un groupe diagonalisable définissable sur $\mathbb{Q}$, ce qui est connu pour les variétés abéliennes. On pourrait alors prendre pour F un corps déployant pour ses commutants dans diverses représentations.

2) On peut ôter un degré de liberté sur le choix de l'isomorphisme de comparaison ι en normalisant le déterminant. Par exemple, pour les courbes supersingulières, où $G_{mot}(\widetilde{X}_{0_k}) \cong \mathbf{G}_{mF}$, ι normalisé est unique : la $\overline{\mathbb{Q}}$-structure $H^{\cdot}(\widetilde{X}_{0_k}, F) \otimes \overline{\mathbb{Q}}$ sur $H^{\cdot}_{cris}(\widetilde{X}_{0_k}/W(k)) \otimes \mathbb{C}_p$ est celle définie au § 5b.

3) Par analogie avec la situation complexe, j'aimerais suggérer la conjecture

(P)$_p$ *pour* $x = \xi \in \overline{\mathbb{Q}}$ *(p–adiquement proche de 0), il existe un isomorphisme de comparaison normalisé* ι *(comme ci–dessus), tel que toute relation de dépendance algébrique sur* $\overline{\mathbb{Q}}$ *entre les coefficients de la matrice de périodes p–adiques* $\Omega_p(\xi)$ *provienne d'un cycle algébrique sur une puissance de* X_ξ.

Appliquée au produit $X_\xi \times X_0$, cette conjecture permettrait d'abandonner l'hypothèse que la place v figurant dans le "principe" de l'introduction est archimédienne (on pourrait échanger les rôles de v et w dans le raisonnement 7b).

Toutefois, dans le cas de connexions de Gauss–Manin non semisimples (éventuellement extensions de connexions de Gauss–Manin semisimples), des relations non triviales entre évaluations p-adiques peuvent apparaître sans contrepartie archimédienne : c'est un "artefact" dû au logarithme, illustré par l'exemple suivant (B. Dwork) : pour $y(x) = \log(1-x)$, $\xi = 1 - e^{2i\pi/7}$, on a $y(\xi)_7 = 0$ mais $y(\xi)_\infty \notin \overline{\mathbb{Q}}$ (quoique les périodes $\frac{1}{2i\pi}\Omega_{1,2} = \frac{\log(1-\xi)}{2i\pi}$ et $\frac{1}{(2i\pi)_7}(\Omega_7)_{1,2} = \frac{\log_7(1-\xi)}{(2i\pi)_7}$ du 1–motif $[\mathbb{Z} \to \mathbf{G}_m]$ soient rationnelles).

La conjecture ci-dessus rappelle celle de Leopoldt sur le *régulateur p-adique*. Soient d'ailleurs E un corps de nombres totalement réel, T le noyau de la norme $N_{E/\mathbb{Q}}$ dans le tore rationnel attaché à E, et M un réseau facteur direct et d'indice fini dans le groupe des unités de E. Considérons le 1-motif associé au plongement canonique $M \to T$. Alors pour un choix naturel des bases de cohomologie, la matrice Ω resp. Ω_p attachée à ce 1-motif est de la forme $\begin{pmatrix} 2i\pi I & R \\ 0 & I \end{pmatrix}$, resp. $\begin{pmatrix} (2i\pi)_p I & R_p \\ 0 & I \end{pmatrix}$, où R, resp. R_p, est la matrice de logarithmes dont le déterminant donne le régulateur (resp. $\cdots$ p-adique).

b) Dans cette section plus descriptive que démonstrative, nous voici enfin en mesure de justifier le *principe des relations globales* énoncé dans l'introduction.

Supposons d'abord que l'équation différentielle en question soit une équation de Picard-Fuchs (et non un quelconque facteur), attachée à un morphisme $X \to S$ comme au § 2. Il est alors équivalent de raisonner sur les $y_i^{(j)}(x)$ ou sur les coefficients de la matrice $Y(x)$. Considérons une relation exceptionnelle $Q_v(Y_{ij}(\xi)_v) = 0$ de dépendance algébrique sur K entre les évaluations v-adiques $Y_{ij}(\xi)_v$ ("exceptionnelle" voulant dire : ne provenant pas par spécialisation $x \to \xi$ d'une relation de dépendance algébrique à coefficients dans $K(x)$ entre les $Y_{ij}(x)$).

Nous allons montrer, en nous appuyant sur les conjectures **(P)** et **(N)** de Grothendieck, qu'il y a aussi une relation exceptionnelle $Q_w(Y_{ij}(\xi)_w) = 0$ entre les évaluations w-adiques $Y_{ij}(\xi)_w$, sauf peut-être dans une situation exceptionnelle décrite ci-dessous et peu plausible (qui ne se présente d'ailleurs pas dans le cas particulièrement intéressant où la composante neutre du groupe de Galois motivique de X_0 est un tore).

D'après la formule (**) et la conjecture de Grothendieck **(P)**, une telle relation (archimédienne) $Q_v(Y_{ij}(\xi)_v) = 0$ provient d'un cycle algébrique θ sur un produit $X_\xi^m \times X_0^{m'}$; plus précisément, elle s'obtient par élimination à partir des relations entre les périodes $\Omega_{v,ij}(\xi)$, $\Omega_{v,ij}(0)$ que donne la compatibilité de θ dans $H^q_{dR}(X_\xi)^{\otimes m} \otimes H^q_{dR}(X_0)^{\otimes m'}$ et $H^q_B(X_{\xi\mathbb{C}_v}, \mathbb{Q})^{\otimes m} \otimes H^q_B(X_{0\mathbb{C}_v}, \mathbb{Q})^{\otimes m'}$ respectivement.

Soit w une autre place de K ; si w est archimédienne, on supposera seulement que ξ est dans le disque de convergence w-adique de $Y(x)$; si w est p-adique, on va devoir supposer que $X \to S$ ait bonne réduction en w, et que ξ soit suffisamment proche de 0 pour que X_ξ et X_0 aient même réduction $\widetilde{X}_0$. En écrivant derechef la compatibilité des classes de θ dans $H^q_{dR}(X_\xi)^{\otimes m} \otimes H^q_{dR}(X_0)^{\otimes m'}$ et $H^q_B(X_{\xi\mathbb{C}_w}, F)^{\otimes m} \otimes H^q_B(X_{0\mathbb{C}_w}, F)^{\otimes m'}$ resp., on obtient des relations entre les périodes $\Omega_{w,ij}(\xi)$, $\Omega_{w,ij}(0)$ (quitte à remplacer K par une extension finie, on peut supposer que le corps des coefficients F de chaque cohomologie de Betti "w-adique" intervenant – il y en a un nombre fini – est contenu dans K).

Le problème est de montrer qu'on peut, grâce à ces relations, *éliminer* entre elles ces périodes suffisamment pour obtenir une relation entre les composantes de

$$Y(\xi)_w = \Omega_w(\xi)\Omega_w(0)^{-1}.$$

L'élimination effective s'avère très difficile au-delà du cas des courbes elliptiques (§ 5c), et un simple décompte de degrés de transcendance ne suffit pas. Il faut recourir à un argument géométrique d'espaces homogènes, comme suit.

Groupes en jeu : posons $V_\xi = H^q_{dR}(X_\xi)$, $V_0 = H^q_{dR}(X_0)$, $n = \dim V_\xi = \dim V_0$. Sur ces K–espaces vectoriels agissent linéairement le groupe de Galois motivique $G_{\xi,0} := G_{mot}(\mathfrak{h}^q(X_\xi) \oplus \mathfrak{h}^q(X_0))$ (version De Rham), et le groupe $G_{x\to\xi}$:= fixateur dans $GL(V_\xi \oplus V_0)$ des tenseurs mixtes qui proviennent par spécialisation $X \to \xi$ de cycles algébriques sur les puissances de $X_x \times X_0$ (en notant x le point générique de S).

On a $G_{\xi,0} \subseteq G_{x\to\xi}$, et $G_{x\to\xi}$ est réductif (c'est d'ailleurs une "forme tordue" de $G_{mot}(\mathfrak{h}^q(X_x) \oplus \mathfrak{h}^q(X_{0_{K(x)}}))$.

Considérons l'application birationnelle d'espaces affines

$$\phi : \mathrm{Hom}(V_\xi, K^n) \oplus \mathrm{Hom}(V_0, K^n) \dashrightarrow \mathrm{Hom}(V_\xi, V_0) \oplus \mathrm{Hom}(V_0, K^n)$$

donnée par $(h_\xi, h_0) \longmapsto (h_0^{-1} \circ h_\xi, h_0)$. Elle respecte l'action de $G_{x\to\xi}$ (action triviale sur K^n). Notons p_1, p_2, les projections ($G_{x\to\xi}$–équivariantes) sur chacun des facteurs $\mathrm{Hom}(V_\xi, V_0)$, $\mathrm{Hom}(V_0, K^n)$. Alors $p_2(G_{x\to\xi})$ s'identifie au groupe de Galois motivique $G_0 := G_{mot}(\mathfrak{h}^q(X_0))$, et $p_1(G_{x\to\xi})$ contient un sous–groupe fermé isomorphe au groupe de Galois différentiel "pointé en ξ", i.e. au fixateur $G_{dif,\xi} \subseteq GL(V_\xi)$ des tenseurs mixtes qui proviennent par spécialisation $x \to \xi$ de tenseurs horizontaux sous la connexion de Gauss–Manin (action triviale sur V_0). Ce qui fait apparaître $G_{\xi,0}$ et $G_{dif,\xi} \times 1$ comme sous–groupes Zariski–fermés de $p_1(G_{x\to\xi}) \times G_0$, et on a $p_2(G_{\xi,0}) = G_0$.

[En fait, on peut montrer que la suite $1 \to G_{dif,\xi} \times 1 \to G_{x\to\xi} \xrightarrow{p_2} G_0 \to 1$ est exacte – cela découle directement du th. 8 ci–dessous; en admettant ce résultat, on obtient une suite exacte de groupes réductifs

$$(****) \qquad 1 \longrightarrow G_{dif,\xi} \longrightarrow p_1(G_{x\to\xi}) \longrightarrow G'_0 \xrightarrow{\overline{p_2}} 1,$$

où G'_0 est le quotient de G_0 par $p_2(\ker p_1)$; en particulier $G_{dif,\xi} \cdot p_1(G_{\xi,0}) = p_1(G_{x\to\xi})$].

Espaces homogènes en jeu: soit $\mathfrak{Y} \subseteq \mathrm{Hom}(V_\xi, V_0)$ la sous–variété définie par les relations polynômiales entre coefficients de ${}^tY(x)$ spécialisées en $x = \xi$ (ceci prend un sens parce que ξ n'est pas une singularité), avec l'identification $V_0 = (\Gamma H^q_{dR}(X/S) \otimes K[[x]])^{\nabla}$. D'après la théorie de Galois différentielle, $\mathfrak{Y}$ est un espace homogène principal à droite sous $G_{dif,\xi}$. A fortiori, $\mathfrak{Y}$ est contenue dans un espace homogène sous $p_1(G_{x\to\xi})$.

Introduisons ensuite $\mathfrak{P}_v \subseteq \mathrm{Hom}(V_\xi, K^n) \oplus \mathrm{Hom}(V_0, K^n)$: K-adhérence de Zariski de $({}^t\Omega_v(\xi), {}^t\Omega_v(0))$, avec l'identification $K^n = H^q_B(X_{\xi\mathbb{C}_v}, \mathbb{Q}) \otimes K = H^q_B(X_{0\mathbb{C}_v}, \mathbb{Q}) \otimes K$ donnée par le prolongement horizontal; notation analogue : $\mathfrak{P}_w$ (en remplaçant $\mathbb{Q}$ par F). Alors ϕ induit un isomorphisme de $\mathfrak{P}_{v \text{ ou } w}$ sur une sous-variété de $\mathfrak{Y} \times p_2(\phi(\mathfrak{P}_{v,w}))$; il suit que $p_1(\phi(\mathfrak{P}_{v,w}))$ est la K-adhérence de Zariski de $Y(\xi)_{v,w}$. Quitte à remplacer K par une extension finie, on peut supposer que $p_1(\phi(\mathfrak{P}_{v,w}))$ possède un point rationnel $y_{v,w}$ sur K, dont on note $H_{v,w}$ le groupe d'isotropie dans $p_1(G_{x\to\xi})$.

La forme précise de la conjecture **(P)** stipule que $\mathfrak{P}_v$ (et donc $\phi(\mathfrak{P}_v)$) est l'adhérence[12] d'un espace homogène principal sous $G_{\xi,0}$. Ceci entraîne que $p_1(\phi(\mathfrak{P}_v))$ est l'adhérence de l'orbite de y_v sous $p_1(G_{\xi,0})$, avec pour groupe d'isotropie $H_v \cap p_1(G_{\xi,0})$. [On déduit de là que $\mathfrak{Y}$ est stable sous $G_{dif,\xi} \cdot p_1(G_{\xi,0}) = p_1(G_{x\to\xi})$; puis que p_2 induit un isomorphisme de H_v sur G'_0. En particulier H_v est réductif, de même que H_w qui lui est conjugué par un élément de $G_{dif,\xi}(K)$].

Quant à $\mathfrak{P}_w$, il est clair qu'elle est contenue dans l'adhérence de l'orbite de y_w sous $p_1(G_{\xi,0})$.

L'hypothèse qu'il existe une relation de dépendance algébrique sur K entre les évaluations v-adiques $Y_{ij}(\xi)_v$ ne provenant pas par spécialisation $x \to \xi$ d'une relation de dépendance algébrique à coefficients dans $K(x)$ entre les $Y_{ij}(x)$, c'est-à-dire que $p_1(\phi(\rho_v)) \neq \mathfrak{Y}$, entraîne donc que l'adhérence de l'orbite de y_v sous $p_1(G_{\xi,0})$ est distincte de $\mathfrak{Y}$ (c'est-à-dire que l'adhérence de $H_v \cdot p_1(G_{\xi,0})$ dans $p_1(G_{x\to\xi})$ ne contient pas $G_{dif,\xi}$). Alors de deux choses l'une :

α) $p_1(\phi(\rho_w)) \neq \mathfrak{Y}$, ou bien

β) l'orbite de y_w sous $p_1(G_{\xi,0})$ est dense dans $\mathfrak{Y}$, mais distincte de $\mathfrak{Y}$ (puisque l'orbite de y_v dans $\mathfrak{Y}$ est de dimension moindre). Ce cas ne me paraît pas plausible, mais je n'ai pas su l'écarter a priori [en utilisant (****), on voit tout de suite que le cas β) signifie que $H_w \cdot p_1(G_{\xi,0})$ contient un ouvert dense de $p_1(G_{x\to\xi})$, mais est distinct de $p_1(G_{x\to\xi})$; cela ne se produit pas si $H_w \cap p_1(G_{\xi,0})$ est réductif, par exemple si la composante neutre de G_0, et donc celle de H_w, est un tore].

Hors du cas β), cette analyse justifie *le principe* de l'introduction dans le cas des solutions de Gauss-Manin, ou plus généralement d'un facteur motivé (i.e. découpé par une correspondance algébrique relative), (sous

---

[12] ce serait un espace principal homogène, et non seulement l'adhérence d'un tel, si on avait pris la précaution de remplacer $\mathfrak{h}^q(X_x)$ par $\mathfrak{h}^q(X_x) \oplus \mathbb{Q}(1)$, mais peu importe.

réserve des conjectures de Grothendieck **(P)** et **(N)**, et d'une condition de bonne réduction) ; pour un quelconque facteur de Gauss–Manin, on utilise le corollaire du théorème 8 ci–dessous, d'où il résulte que la composante isotypique de ce facteur est motivée, du moins si $G_{x \to \xi}$ est connexe (le cas non connexe n'est guère plus difficile).

Enfin, en l'absence de la condition de bonne réduction (ou dans le cas d'extensions non–triviales de facteurs de Gauss–Manin), on peut espérer qu'une théorie convenable des motifs mixtes – à construire – permettrait d'étendre nos arguments. Les résultats du § 4 vont dans ce sens...

8. — Comment éviter les conjectures standard ? [A93]

a) Revenons au cadre général du § 6. Malheureusement, l'hypothèse **(N)** ne semble pas d'accès facile, surtout en caractéristique positive[13], (où elle ne paraît pas même connue pour les variétés abéliennes). On a vu que cette hypothèse entraîne l'algébricité de l'involution de Lefschetz, i.e. l'involution $*$ de $H^{\cdot}(Z)$ donnée[14] en chaque degré i par l'isomorphisme de Lefschetz L_Z^{d-i}, $0 \leq i \leq 2d$.

L'idée directrice de [A93], pour éviter **(N)**, est d'introduire formellement cette involution $*$ parmi les morphisme motivés.

Soit $\mathcal{V}$ une sous–catégorie pleine, stable par produits, sommes disjointes et composantes connexes, de la catégorie des k–schémas Z projectifs et lisses.

DÉFINITION. — *Un cycle motivé sur Z est un élément de $H^{\cdot}(Z)$ de la forme $p_{Z*}(\alpha \cdot *\beta)$, où α et β sont des $\mathbb{Q}$–cycles algébriques sur $Z \times Z'$, avec ζ' arbitraire dans $\mathcal{V}$, p_Z désignant la projection sur Z, et où $*$ est relative à une quelconque polarisation de $Z \times Z'$ de type "produit".*

Les cycles motivés forment un $\mathbb{Q}$–espace, gradué par la graduation moitié sur $H^{pair}(Z)$, contenant les cycles algébriques. Ces espaces ne dépendent pas de la cohomologie classique choisie $H^{\cdot}$ (à isomorphisme canonique près), du moins si k est de caractéristique nulle.

On montre que les opérations usuelles de la théorie des cycles algébriques s'étendent aux cycles motivés. En particulier, les correspondances motivées se composent. On peut alors définir sur les cycles motivés l'analogue $\equiv$ de l'équivalence numérique. L'analyse de cette équivalence conduit à introduire le sous–corps Q de F engendré par les valeurs $\int_Z \alpha \cdot *\alpha$, où α décrit les cycles algébriques sur Z, et Z parcourt $\mathcal{V}$, et à considérer les cycles et

---

[13] Voir toutefois [A93] § 2 pour une tentative dans ce sens ; signalons d'ailleurs que **(N)** entraîne la semi-simplicité de Frobenius.

[14] On néglige ici la torsion de Tate.

correspondances motivés à coefficients dans Q (si k est de caractéristique nulle, ou si $*$ est algébrique, on a bien entendu $Q = \mathbb{Q}$).

On définit alors la catégorie des *motifs modelés sur* $\mathcal{V}$, pour tout schéma Z dans $\mathcal{V}$, comme au §6, mais en remplaçant $\mathbb{Q}$-correspondances algébriques par correspondances motivées à coefficients dans Q (modulo $\equiv$). On la munit de $\otimes$ et la note $\mathcal{M}(\mathcal{V})$[15].

THÉORÈME 6. — *La $\otimes$-catégorie $\mathcal{M}(\mathcal{V})$ est tannakienne sur Q, graduée et semi-simple; en outre, si* car $k = 0$, *alors on a $Q = \mathbb{Q}$, $\equiv$ est l'identité, et toute cohomologie classique $H'^{\cdot}$ se factorise à travers la cohomologie motivique $\mathfrak{h}$, donnant naissance à un foncteur fibre gradué sur $\mathcal{M}(\mathcal{V})$.*

(On obtient alors, pour $k \subseteq \mathbb{C}$, une définition non conjecturale du groupe de Galois motivique.)

b) Ce théorème permet de spécialiser les motifs en inégale caractéristique. Soient K un corps de nombres, v une place p-adique de K et k la clôture algébrique du corps résiduel. Soient $\underline{\mathcal{V}}$ une catégorie stable par produits fibrés, sommes disjointes et composantes connexes, de schémas Z projectifs et lisses sur l'anneau des v-entiers de K, $\underline{\mathcal{V}}_K$ (resp. $\underline{\mathcal{V}}_k$) la catégorie des fibres génériques (resp. spéciales géométriques). Alors on a un foncteur de spécialisation $sp : \mathcal{M}(\underline{\mathcal{V}}_K) \to \mathcal{M}(\underline{\mathcal{V}}_k)$.

Pour associer à une cohomologie classique sur $\underline{\mathcal{V}}_k$, disons la cohomologie cristalline, un foncteur fibre sur $\mathcal{M}(\underline{\mathcal{V}}_k)$, il faut faire agir les correspondances motivées modulo $\equiv$ sur les espaces de cohomologie cristalline en respectant les fonctorialités idoines, c'est-à-dire relever de manière "cohérente" les classes mod. $\equiv$ en de vraies correspondances motivées. C'est possible, du moins lorsque $\underline{\mathcal{V}}$ est "engendrée" par un seul schéma $\mathcal{X}$ de fibre générique $\mathcal{X}_K$ connexe [A93][A] :

THÉORÈME 7. — *Il existe un foncteur fibre gradué $^{\spadesuit}H^*_{cris}$ sur $\mathcal{M}(\mathcal{X}_k^{\otimes})$ à valeurs dans les $W(k)_{\mathbb{Q}}$-espaces vectoriels, tel que pour tout m, $^{\spadesuit}H^*_{cris}(\mathcal{X}_k^m) = H^*_{cris}(\mathcal{X}_k^m/W(k))_{\mathbb{Q}}$, et tel que l'isomorphisme de Berthelot-Ogus induise un isomorphisme de foncteurs fibres gradués $H^*_{dR} \otimes \mathbb{C}_p \cong (^{\spadesuit}H^*_{cris} \otimes \mathbb{C}_p) \circ sp$ sur $\mathcal{M}(\mathcal{X}_K^{\otimes})$.*

Dans le cas d'une famille projective lisse $X \to S$ avec bonne réduction en v comme au §2, on peut appliquer ce résultat en prenant pour $\mathcal{X}$ un modèle de X_ξ, pour chaque $\xi \in S(K)$ p-adiquement proche de 0, séparément. La compatibilité des constructions pour plusieurs ξ est problématique : $\xi \neq 0$ étant fixé, j'ignore par exemple si la classe cristalline dans $H^*_{cris}(\widetilde{X}^m_{0_k}) = H^*_{cris}(\mathcal{X}_k^m) = {}^{\spadesuit}H^*_{cris}(\mathcal{X}_k^m)$ de la spécialisation de tout

[15] Voir note (8).

cycle motivé sur une puissance de X_0 coïncide avec la classe en théorie ♠H^*_{cris} ; c'est toutefois vrai pour les intersections de classes de diviseurs sans hypothèse supplémentaire sur $\mathcal{X}_k$, et pour tout cycle motivé si $\mathcal{X}_k$ est une variété abélienne. Cela permet d'appliquer la construction du §7 dans de nombreuses situations *en se passant de l'hypothèse* **(N)**, du moins lorsque $Q = \mathbb{Q}$.

c) Citons quelques applications du théorème 6 dans le cas d'un corps de base $k \subseteq \mathbb{C}$. Soient S un schéma réduit connexe de type fini sur $\mathbb{C}$, $f : X \to S$ un morphisme projectif et lisse, et θ une section globale du faisceau $R^{2p}f_*^{an}\mathbb{Q}(p)$.

Grothendieck a conjecturé que si la fibre θ_s est algébrique en un point $s \in S(\mathbb{C})$, il en est de même en tout point [G66]. En s'appuyant sur le "théorème de la partie fixe" de Deligne [D71] §4, on montre :

THÉORÈME 8. — *Si la fibre θ_s est motivée en un point $s \in S(\mathbb{C})$, il en est de même en tout point (pour un choix convenable de $\mathcal{V}$).*

Avec les notations du §7b, on en déduit :

COROLLAIRE. — *Lie $G_{dif,\xi}$ est un idéal de Lie $G_{x\to\xi}$; c'est aussi un sous-espace motivé de* End $H_{dR}(X_\xi)$.

Enfin, à l'aide d'un cas particulier du théorème de déformation 8 et de [A92], une argumentation suivant le fil de celle de Deligne pour prouver que tout cycle de Hodge l'est absolument sur les variétés abéliennes, permet de montrer :

THÉORÈME 9. — *Tout cycle de Hodge sur une variété abélienne complexe est motivé (pour $\mathcal{V}$ convenable).*

d) *Quelques mots sur la démonstration du théorème* 1 [A]

On se ramène à la situation du §2 définie sur un certain corps de nombres de base $K_\circ$, par le changement de variable $x \longmapsto 1/x^{\frac{1}{n}}$, et en remplaçant f par $1/x^{\frac{1}{n}} \circ f : X = \underline{A} \to S$ (à fibres non géométriquement connexes), n étant choisi de façon à tuer la ramification au-dessus de $x = 0$ (pour le nouveau choix de x). Les coefficients de la matrice $Y(x)$ sont alors des séries de puissances en x à coefficients dans K_0 (en fait des G-fonctions).

D'autre part, le nombre de places v de $K = K_\circ(\xi)$ divisant ξ exceptionnel est borné par hypothèse (par $n\delta^2$).

Le principe de la preuve consiste à construire des relations polynômiales $Q(Y_{ij}(\xi)_v) = 0$ de degré borné à coefficients dans une extension de degré borné de K entre les évaluations v-adiques en ξ des coefficients d'une matrice $Y(x)$ solution de l'équation de Picard–Fuchs, à multiplier entre elles ces relations (il y en a au plus $n\delta^2$), avant de conclure par le "principe de Hasse" de Bombieri cité dans l'introduction (compte–tenu de ce que les points ξ de hauteur et degré bornés sont en nombre fini).

L'hypothèse que les variétés abéliennes dans la fibre $x = 0$ sont de type CM entraîne que la composante neutre de $G_\circ = G_{mot}(X_\circ)$ est un tore.

Nos motifs seront modelés sur des variétés abéliennes ; nous utiliserons les conséquences suivantes du fait d'avoir affaire à des variétés abéliennes :

i) $Q = \mathbb{Q}$, car sur toute variété abélienne, l'involution de Lefschetz est algébrique (Lieberman–Grothendieck) ;

ii) $X_\circ$ a bonne réduction potentielle partout ; de là résulte l'existence d'un revêtement étale $S' \to S$, et pour toute place finie w, d'un ouvert U_w de Zariski de S'_{K_w} au–dessus duquel $f^{rig}_{U_w}$ a bonne réduction (Ogus).

La construction des relations "exceptionnelles" $Q(Y_{ij}(\xi)_v) = 0$ se fait en adaptant l'argument d'espaces homogènes de 7b.

Enfin, borner le degré de relations exceptionnelles revient à borner le degré de certains tenseurs "exceptionnels" invariants sous $G_{mot}(X_\xi)$; pour ce faire, on recourt au lemme suivant :

LEMME. — *Soit G un groupe algébrique semi–simple complexe. Il n'y a qu'un nombre fini de classes de conjugaison de sous–groupes fermés connexes $H \subseteq G$ contenant le centre de la composante neutre de leur normalisateur dans G.*

Manuscrit reçu le 22 janvier 1994

BIBLIOGRAPHIE

[A86] Y. ANDRÉ. — *Multiplication complexe dans un pinceau de variétés abéliennes*, Sém. de Théorie des Nombres de Paris, 1984–85, (C. Goldstein, ed.) Progress in Math. **63** Birkhäuser Boston (1986), 1–22.

[A89] Y. ANDRÉ. — *G-functions and Geometry*, Aspects of Math. vol. E13, Vieweg, Braunschweig/Wiesbaden (1989).

[A90] Y. ANDRÉ. — *p-adic Betti lattices*, in "p-adic analysis" Proceedings of the Trento conference, (F. Baldassarri, S. Bosch, B. Dwork, eds.) Springer L.N.M. **1454** (1990).

[A92] Y. ANDRÉ. — *Une remarque à propos des cycles de Hodge de type* CM, Sém. de Théorie des Nombres de Paris, 1989–90, (S. David, ed.) Progress in Math. **102** Birkhäuser Boston (1992), 1–7.

[A93] Y. ANDRÉ. — *Pour une théorie inconditionnelle des motifs*, soumis à publication, (première version prépubliée à l'Univ. Paris 6).

[A] Y. ANDRÉ. — *Réalisation de Betti des motifs p-adiques*, en préparation, (première partie prépubliée à l'I.H.E.S., Avril 1992).

[BO83] P. BERTHELOT, A. OGUS. — *F-isocrystals and the De Rham cohomology I*, Inv. Math. **72** (1983), 159–199.

[Be93] F. BEUKERS. — *Algebraic values of G-functions*, J. reine angew. Math. **434** (1993), 45–65.

[Bo81] E. BOMBIERI. — *On G-functions*, in Recent progress in analytic number theory, Durham **79**, Academic Press (1981) vol. 2, 1–67.

[D71] P. DELIGNE. — *Théorie de Hodge II*, Publ. Math. I.H.E.S. **40** (1971), 5–57.

[D80] P. DELIGNE. — *La conjecture de Weil II*, Publ. Math. I.H.E.S. **52** (1980), 137–252.

[D90] P. DELIGNE. — *Catégories tannakiennes*, in The Grothendieck Festschrift, Birkhäuser Boston (1990), vol II, 111–195.

[F89] G. FALTINGS. — *Crystalline cohomology and p-adic Galois representations*, in Algebraic analysis, Geometry and number theory, J.I. Igusa ed., proc. of the JAMI inaug. conf. John Hopkins Univ. (1989) 25–80.

[Ge86] L. GERRITZEN. — *Periods and Gauss–Manin connection for families of p–adic Schottky groups*, Math. Ann. **275** (1986) 425–453.

[G66] A. GROTHENDIECK. — *On the de Rham cohomology of algebraic varieties*, Publ. Math. I.H.E.S. **29** (1966), 93–103.

[GM78] H. GILLET, W. MESSING. — *Riemann–Roch and cycle classes in crystalline cohomology*, Duke Math. J. **45** (1978), 193–211.

[J92] U. JANNSEN. — *Motives, numerical equivalence, and semi–simplicity*, Inv. Math. **107** (1992), 447–452.

[KM74] N. KATZ, W. MESSING. — *Some consequences of the Riemann hypothesis for varieties over finite fields*, Invent. Math. I.H.E.S. **23** (1974), 73–77.

[O90] A. OGUS. — *A p–adic analogue of the Chowla–Selberg formula*, in "p–adic analysis" Proceedings of the Trento conference, (F. Badassarri, S. Bosch, B. Dwork, eds.) Springer L.N.M. **1454** (1990).

Yves André
UA 763 du C.N.R.S.
Université Paris 6
Collège de France
3, rue d'Ulm
75231 PARIS 05

Number Theory
Paris 1992–93

A refinement of the Faltings–Serre method

Nigel Boston(*)

1. — Introduction

In recent years the classification of elliptic curves over $\mathbb{Q}$ of various conductors has been attempted. Many results have shown that elliptic curves of a certain conductor do not exist. Later methods have concentrated on small conductors, striving to find them all and hence to verify the Shimura-Taniyama-Weil conjecture for those conductors. A typical case is the conductor 11. In [1], Agrawal, Coates, Hunt, and van der Poorten showed that every elliptic curve over $\mathbb{Q}$ of conductor 11 is $\mathbb{Q}$-isogenous to $y^2 + y = x^3 - x^2$. Their methods involved a lot of computation and the use of Baker's method. In [12], Serre subsequently applied Faltings' ideas to reprove this result in a much shorter way. He called this approach "the method of quartic fields".

In this paper I first seek to refine this method and to make it possible to classify elliptic curves over $\mathbb{Q}$ of conductor N for a large number of N. These N are all prime and so this work is indeed superceded by the result of Wiles that every semistable elliptic curve over $\mathbb{Q}$ is modular (if fixed). The advantage of my method is that it provides a much simpler approach (when it works). Like Wiles, I am using deformations of Galois representations but in a more elementary way. The second half of the paper indicates how the Faltings-Serre method can be used to describe spaces of Galois representations and gives the first applications of the method to mod p representations with $p \neq 2$.

The main result of the first half is Theorem 1 below. Note that there are extensive tables of class numbers and units of cubic fields due to Angell [2] and that information on quartic fields is not required

---

(*) Partially supported by NSF grant DMS 90-14522. I thank God for leading me to these results. I thank J.-P.Serre for generously sending me copies of his unpublished work.

THEOREM (1.1). — *Let N be a prime $\equiv 3 \pmod 8$, such that 3 divides neither $h(\mathbb{Q}(\sqrt{N}))$ nor $h(\mathbb{Q}(\sqrt{-N}))$. Let M be one of the cubic subfields of the unique cubic cyclic extension K of $\mathbb{Q}(\sqrt{-N})$ of conductor 2. Suppose that $h(M)$ is odd and that the minimum polynomial modulo N of a fundamental unit of M has a quadratic residue and a quadratic non-residue root.*

Then there is at most one $\mathbb{Q}$-isogeny class of elliptic curves over $\mathbb{Q}$ of conductor N with given trace of Frobenius at 2, a_2.

Remarks (1) There is a unique such field K because 2 is inert in $\mathbb{Q}(\sqrt{-N})$ and 3 does not divide $h(\mathbb{Q}(\sqrt{-N}))$.

(2) By Cohen–Lenstra heuristics, 47% of N should satisfy $3 \nmid h(\mathbb{Q}(\sqrt{N}))h(\mathbb{Q}(\sqrt{-N}))$. Apparently most (but not all, e.g. 571) of these N have $h(M)$ odd. Some of these satisfy the condition on the fundamental unit (e.g. $N = 11, 67, 179, ...$); some don't (e.g. $N = 19, 43, 163, ...$).

(3) The prime N may satisfy the hypotheses of the theorem but there be no elliptic curve over $\mathbb{Q}$ of conductor N, (e.g. $N = 227, 251, \ldots$ see [5]).

2. — The Basic Set-up and Elementary Properties

Let E be an elliptic curve over $\mathbb{Q}$ of conductor N. Let $\overline{\rho} : \mathrm{Gal}(\overline{\mathbb{Q}}/\mathbb{Q}) \to GL_2(\mathbb{F}_2)$ give the action of Galois on the 2-division points of E. The curve E has no rational points of order 2 since its conductor is neither 17 nor of the form $u^2 + 64$ (i.e. it is not a Setzer-Neumann curve) [13].

Using work of Brumer and Kramer [5] based on work of Serre [11], we can deduce various properties of E and $\overline{\rho}$. Firstly, plus or minus the discriminant of a semistable elliptic curve with no rational point of order 2 is never a perfect square. It follows that E has supersingular reduction at 2, since $\mathbb{Q}(\sqrt{\Delta})$ (Δ being the discriminant of E) is $\mathbb{Q}(\sqrt{N})$ or $\mathbb{Q}(\sqrt{-N})$ and so has no unramified cyclic cubic extensions by the hypotheses of the theorem.

Secondly, since E is supersingular modulo 2, its 2-division field is a cyclic cubic extension of $\mathbb{Q}(\sqrt{\Delta})$ unramified outside 2 and totally ramified at 2, and moreover 2 is inert in $\mathbb{Q}(\sqrt{\Delta})/\mathbb{Q}$. From this it follows that $\mathbb{Q}(\sqrt{\Delta}) = \mathbb{Q}(\sqrt{-N})$, that $\overline{\rho}$ is surjective, and that the 2-division field of E (i.e. the fixed field of $\ker \overline{\rho}$) is K.

3. — The Faltings-Serre Method [12]

Suppose that E' is another elliptic curve over $\mathbb{Q}$ with conductor N and the same trace of Frobenius at 2. Assume that E' is not isogenous to E. Let $\rho, \rho' : \mathrm{Gal}(\overline{\mathbb{Q}}/\mathbb{Q}) \to GL_2(\mathbb{Z}_2)$ give the action of Galois on the Tate modules $T_2(E), T_2(E')$ respectively. By Faltings [7], ρ and ρ' are not isomorphic. By section 2, their reductions modulo 2 are isomorphic.

Pick the largest α such that ρ and ρ' are isomorphic modulo 2^α. Replacing ρ' by a conjugate if necessary, we can assume that they are equal modulo 2^α.

Define $\sigma : \mathrm{Gal}(\overline{\mathbb{Q}}/\mathbb{Q}) \to M_2(\mathbb{F}_2)^0 o GL_2(\mathbb{F}_2)$ by

$$\sigma(x) = ((\rho'(x) - \rho(x))/2^\alpha (\mathrm{mod}\ 2),\ \overline{\rho}(x)),$$

where $M_2(\mathbb{F}_2)^0$ denotes the 2×2 matrices over $\mathbb{F}_2$ of trace zero (mapped to since det ρ equals det ρ'). Let $\tilde{K}$ be the fixed field of ker σ.

4. — The Proof of the Main Theorem

The idea of the Faltings-Serre method is to use it to produce a representation σ that can then be shown not to exist by methods of algebraic number theory (in particular tables of number fields). This then shows that there cannot be two non-isogenous curves with the properties stated in the main theorem.

PROPOSITION (4.1). — *The extension $\tilde{K}/K$ is unramified outside N.*

Proof : since E has supersingular reduction at 2, the theorem of Honda-Hill-Cartier [8] implies that the characteristic polynomial of the formal group associated to E at 2 is the same as the characteristic polynomial of the system of ℓ-adic representations at 2. This says that a_2 determines the formal group at 2 of E, which determines the 2-adic representation of a decomposition group D_2 at 2, i.e. $\rho|_{D_2} \cong \rho'|_{D_2}$. If $x \in D_2$ (so in particular if x is in an inertia group at 2), then $\sigma(x) = (0, \overline{\rho}(x))$.

It remains to show that such an extension $\tilde{K}/K$ cannot exist. The key idea is to use two results of Nicole Moser [10]. The first one is :

$$(1)\quad h(K) = (ah(M)^2 h(\mathbb{Q}(\sqrt{-N}))/3 \quad (a = 1 \text{ or } 3)$$

PROPOSITION (4.2). — *The class number $h(K)$ is odd.*

Proof : this follows from the above formula (1), from our hypothesis that $h(M)$ is odd, and from genus theory, which tells us that $h(\mathbb{Q}(\sqrt{-N}))$ is odd (since N is prime).

Secondly, Moser showed [10] that K has a Minkowski unit, i.e. a single generator of its unit group modulo torsion as a $\mathbb{Z}[\mathrm{Gal}(K/\mathbb{Q})]$-module. To apply this, consider by global class field theory the exact sequence of $\mathbb{F}_2[\mathrm{Gal}(K/\mathbb{Q})]$-modules :

$$0 \to B \to \overline{U} \to \oplus_{\wp|N} \overline{U}_\wp \to \overline{P} \to 0,$$

where $\overline{U}$ is the global units of K modulo squares, $\overline{U}_\wp$ is the local units of $K_\wp$ modulo squares, and $\overline{P}$ is the Galois group over K of a maximal elementary 2-abelian extension L unramified outside the primes of K above N.

Now $\dim_{\mathbb{F}_2} \overline{U} = 3, \dim_{\mathbb{F}_2} \overline{U}_\wp = 1$ implying that $\dim_{\mathbb{F}_2} \overline{P} = \dim_{\mathbb{F}_2} B$. Since $\tilde{K} \subseteq L$, it remains to show that $B = 0$. The existence of a Minkowski unit implies that $\overline{U} \equiv \{\pm 1\} \oplus V$, where V is an irreducible 2-dimensional $\mathbb{F}_2[\mathrm{Gal}(K/\mathbb{Q})]$-module. So we just need an element of V which is not in one of the kernels from $\overline{U} \to \overline{U}_\wp$. The image in V of the unit in the hypotheses of the theorem satisfies this.

5. — Examples

(1) $N = 11$. There is an elliptic curve over $\mathbb{Q}$ of conductor 11, namely (11_A) $y^2+y = x^3-x^2$. Let E be another such. Then [5], [13] $\overline{\rho}$ is determined, E has supersingular reduction at 2, and M has odd class number. In fact M is the cubic field of discriminant -44. By [6] a fundamental unit of M has minimum polynomial $x^3 + x^2 + x - 1$, which factors modulo 11 as $(x+3)^2(x+6)$. Since -3 is a quadratic non-residue modulo 11, theorem 1 shows that every elliptic curve over $\mathbb{Q}$ of conductor 11 with $a_2 = -2$ is isogenous to (11_A).

As in Serre's original letter [12], this classifies up to isogeny every elliptic curve over $\mathbb{Q}$ of conductor 11, because a similar argument to the above shows that an elliptic curve with good reduction outside 11 and $a_2 = 2$ (respectively $a_2 = 0$) is isogenous to (121_A) (respectively (121_D)).

(2) $N = 67$. There is an elliptic curve over $\mathbb{Q}$ of conductor 67, namely (67_A) $y^2 + y = x^3 + x^2 - 12x - 21$. Let E be another such. Then [5], [13] $\overline{\rho}$ is determined, E has supersingular reduction at 2, and M has odd class number. In fact M is the cubic field of discriminant -268. By [6] a fundamental unit of M has minimum polynomial $x^3 - 7x^2 + 13x - 1$, which factors modulo 67 as $(x+16)^2(x+28)$. Since -16 is a quadratic non-residue modulo 67, theorem 1.1 shows that every elliptic curve over $\mathbb{Q}$ of conductor 67 with $a_2 = 2$ is isogenous to (67_A).

The same argument as for $N = 11$ now applies, because the twist of (67_A) by the quadratic character associated to $\mathbb{Q}(\sqrt{-67})$ is an elliptic curve of conductor 67^2 with $a_2 = -2$ and with the same $\overline{\rho}$ and the curve of CM type relative to $\mathbb{Q}(\sqrt{-67})$ is an elliptic curve of conductor 67^2 with $a_2 = 0$ and the same $\overline{\rho}$.

6. — Deformation Spaces of Galois Representations

The homomorphisms ρ and ρ' are lifts of the same $\overline{\rho}$ to $\mathbb{Z}_2$. They therefore lie in the deformation space of lifts of $\overline{\rho}$ [3]. The Faltings-Serre method constructs from them a third lift σ to the dual numbers $\mathbb{F}_2[\epsilon]$ ($\epsilon^2 = 0$). We consider below some applications of this idea.

Let $\overline{\rho} : \mathrm{Gal}(\overline{\mathbb{Q}}/\mathbb{Q}) \to GL_2(\mathbb{F}_p)$ be an absolutely irreducible representation. Let $\mathcal{C}$ denote the category of complete, noetherian local rings with residue field $\mathbb{F}_p$. Objects of this category are rings of the form $\mathbb{Z}_p[[T_1, ..., T_r]]/I$. If R is such a ring, then two representations $\rho_1, \rho_2 : \mathrm{Gal}(\overline{\mathbb{Q}}/\mathbb{Q}) \to GL_2(R)$ will be called *strictly equivalent* if conjugate by an element of $\Gamma_2(R) := \ker(GL_2(R) \to GL_2(\mathbb{F}_p))$. A strict equivalence class of lifts of $\overline{\rho}$ is called a *deformation* of $\overline{\rho}$. Fix a finite set of rational primes S containing the primes ramified in $\overline{\rho}$.

Define a functor $\mathcal{F} : \mathcal{C} - - \to$ *Sets* by :

$$\mathcal{F}(R) = \{\text{deformations of } \overline{\rho} \text{ to } R \text{ unramified outside } S\}$$

Mazur [9] proved that $\mathcal{F}$ is representable, i.e. that there exists a representation $\xi : \mathrm{Gal}(\overline{\mathbb{Q}}/\mathbb{Q}) \to GL_2(\mathcal{R})$ (the universal deformation) lifting $\overline{\rho}$ and parametrizing lifts of $\overline{\rho}$ to R in $\mathcal{C}$ unramified outside S up to strict equivalence via $\mathrm{Hom}(\mathcal{R}, R)$. The set $\mathrm{Hom}(\mathcal{R}, \overline{\mathbb{Z}}_p)$ will be called the deformation space of lifts of $\overline{\rho}$.

At the joint AMS-LMS conference in Cambridge, England, in 1992, I suggested that deformation spaces of Galois representations should have some special properties. In particular, it appears that they are often coordinatized by their restrictions to various inertia subgroups I_ℓ ($\ell \in S$), namely (i) the restrictions to I_ℓ ($\ell \in S - \{p\}$) should indicate which component the lift is on and (ii) the restriction to I_p should indicate where the lift is on that component. This idea is now to use the Faltings-Serre method to prove some cases of (ii). The novelty of this approach lies in replacing the prime 2 by more general primes. See [3] for a further discussion of this.

Example (1) Let E be the elliptic curve $X_0(49)$. This is an elliptic curve over $\mathbb{Q}$ of conductor 49. In [4], it is calculated that the universal deformation ring of the Galois representation given by the 3-division points of E with $S = \{3, 7\}$ is $\mathbb{Z}_3[[T_1, T_2, T_3, T_4]]/((1 + T_4)^3 - 1)$. Thus its deformation space splits into three explicitly given components $\{T_4 = 0\}$, $\{T_4 = \omega - 1\}$, $\{T_4 = \omega^2 - 1\}$, where ω is a primitive 3rd root of unity, and as shown in [4] any lift to $\mathbb{Z}_3$ lies on the first component. Also, (i) above holds. In other words, the image of an inertia group at 7 determines on which component a representation over $\overline{\mathbb{Z}}_3$ lies.

Now let ρ and ρ' be two lifts of $\overline{\rho}$ to $\mathbb{Z}_3$ (so lying on the first component). Suppose that they agree on inertia at 3. We shall show that they are actually strictly equivalent (so give the same point in the deformation space).

Assume for now they are not strictly equivalent. Since $\overline{\rho}$ is absolutely irreducible, two lifts are strictly equivalent if and only if they are isomorphic. For suppose that $\rho' = A^{-1}\rho A$ with $A \in GL_2(\mathbb{Z}_p)$. Then A centralizes the image of $\overline{\rho}$ and so by Schur the image of A in $GL_2(\mathbb{F}_p)$ is a scalar matrix, i.e. $A = BC$ where B is scalar and $C \in \Gamma_2(\mathbb{Z}_p)$. But then $\rho' = C^{-1}\rho C$.

Since ρ and ρ' are not isomorphic, σ defines a homomorphism from $\mathrm{Gal}(\overline{\mathbb{Q}}/\mathbb{Q})$ into the semidirect product $M_2(\mathbb{F}_3) o GL_2(\mathbb{F}_3) (\cong GL_2(\mathbb{F}_3[\epsilon]), \epsilon^2 = 0)$ unramified outside $S = \{3, 7\}$. Letting K and $\tilde{K}$ denote, as before, the fixed fields of $\overline{\rho}$ and σ respectively, we get that $\tilde{K}/K$ is unramified outside 7. It is also unramified outside 3 because (i) holds, i.e. ρ and ρ' agree on inertia at 3. Such an everywhere unramified field extension of K does not exist, since its Galois group would be a quotient of the ideal class group killed by 3 with $\mathrm{Gal}(K/\mathbb{Q})$ acting via the adjoint action. This is excluded as explained in [4] by the work of Coates and Flach since 3 does not divide the numerator of a certain special value of the L-function of the symmetric square of E.

(2) Following [9], let p be a prime number of the form $27 + 4a^3$ and K be a splitting field over $\mathbb{Q}$ for $x^3 + ax + 1$. Embedding $\mathrm{Gal}(K/\mathbb{Q}) \cong S_3$ in $GL_2(\mathbb{F}_p)$, we obtain a representation $\overline{\rho} : \mathrm{Gal}(\overline{\mathbb{Q}}/\mathbb{Q}) \to GL_2(\mathbb{F}_p)$ unramified outside p. Letting $S = \{p\}$, Mazur showed that $\mathcal{R} \cong \mathbb{Z}_p[[T_1, T_2, T_3]]$.

Let ρ and ρ' be lifts of $\overline{\rho}$ to $\mathbb{Z}_p$ unramified outside S. Suppose that they agree on inertia at p, but are not strictly equivalent. Then they produce, as in (1), an unramified p-extension of the fixed field of ker $\overline{\rho}$, but as Mazur showed in [9], p-$h(K)$, a contradiction thereby proving (ii) in this case.

Manuscrit reçu le 17 août 1993

REFERENCES

[1] M. Agrawal, J. Coates, D. Hunt and A. van der Poorten. — *Elliptic curves of conductor* 11 Math. Comp. **35** (1980), 991–1002.

[2] I.O. Angell. — *A table of complex cubic fields.*

[3] N. Boston. — *Deformations of Galois representations,* (a monograph), in preparation .

[4] N. Boston and S.V. Ullom. — *Representations related to CM elliptic curves,* Math. Proc. Camb. Phil. Soc. **113** (1993), 71–85.

[5] A. Brumer and K. Kramer. — *The rank of elliptic curves,* Duke Math. J. **44**, n° 4 (1977), 715–742.

[6] B.N. Delone and D.K. Faddeev. — *The theory of irrationalities of the third degree,* AMS, Providence, RI, 1964.

[7] G. Faltings. — *Endlichkeitssätze für abelsche Varietäten über Zahlkörpern,* **73** (1983), 349–366.

[8] W. Hill. — *Formal groups and zeta-functions of elliptic curves,* Invent. Math. **12** (1971), 321–336.

[9] B. Mazur. — *Deforming Galois representations,* Proceedings of the March 1987 Workshop on “Galois groups over $\mathbb{Q}$” held at MSRI, Berkeley, California.

0] N. Moser. — *Unités et nombre de classes d'une extension galoisienne diédrale de* $\mathbb{Q}$, Abh. Math. Sem. Univ. Hamburg **48** (1979), 54–75.

1] J.-P. Serre. — *Propriétés galoisiennes des points d'ordre fini des courbes elliptiques,* Invent. Math. **15** (1972), 259–331.

2] J.-P. Serre. — Letter to Tate, Oct. 26, 1984.

[13] B. Setzer. — *Elliptic curves of prime conductor*, J. London Math. Soc. **10** (1975), 367–378.

Nigel BOSTON
Department of Mathematics
University of Illinois
273 Altgeld Hall, MC 382
1409 West Green street
Urbana, IL 61801
U.S.A.

Number Theory
Paris 1992–93

Sous-variétés algébriques de variétés semi-abéliennes sur un corps fini

John Boxall

RÉSUMÉ : Dans cet article nous étendons les résultats déjà obtenus dans [Bo1] concernant l'intersection dans une variété abélienne d'une sous-variété avec certains groupes de points de torsion aux variétés semi-abéliennes, le corps de base étant un corps fini.

SUMMARY : In this paper we extend results already proved in [Bo1] concerning intersections on abelian varieties of subvarities with certain groups of torsion points to semi-abelian varieties, the base field a finite field.

1. — Introduction

Soit k un corps fini, soit $\overline{k}$ une clôture algébrique de k, soit l un nombre premier différent de la caractéristique de k et soit G le groupe des racines de l'unité dans $\overline{k}$ dont l'ordre est une puissance de l. Nous nous proposons de montrer que l'ensemble des $\zeta \in G$ tels que $1 - \zeta \in G$ est fini. Plus généralement, soient $(a, b) \in k^2$ avec $ab \neq 0$: nous allons montrer que l'ensemble :

$$\{(\zeta, \eta) \in G^2 \mid a\zeta + b\eta = 1\}$$

est fini.

Dans ce but, désignons par Γ le groupe de Galois de $\overline{k}$ sur k et remarquons que si $(\zeta, \eta) \in G^2$ vérifie

$$a\zeta + b\eta = 1\,, \tag{1}$$

alors on a également $a\,\sigma\zeta + b\,\sigma\eta = 1$ pour tout $\sigma \in \Gamma$ et donc

$$a'\zeta + b'\eta = 1\,, \tag{2}$$

où l'on a posé :

$$a' = a\frac{\sigma\zeta}{\zeta} \quad \text{et} \quad b' = b\frac{\sigma\eta}{\eta}\,.$$

Or, les équations (1) et (2) ont au plus une solution (ζ, η) si $(\sigma\zeta, \sigma\eta) \neq (\zeta, \eta)$; il suffit donc pour conclure de montrer qu'il existe un sous-groupe fini G_0 de G ayant la propriété que pour tout $(\zeta, \eta) \in G^2$ avec $(\zeta, \eta) \notin G_0^2$, il existe $\sigma \in \Gamma$ tel que $\big((\sigma - 1)\zeta, (\sigma - 1)\eta\big)$ soit un élément de G_0^2 différent de $(1, 1)$. En effet, toute solution (ξ, η) n'appartenant pas à G_0^2 satisait à (1) et (2) avec $(a'a^{-1}, b'b^{-1}) \in G_0^2 \setminus (1, 1)$. Il y a donc au plus $|G_0|^2 + |G_0| - 1$ solutions.

Or, il est aisé de construire un tel sous-groupe G_0. Posons $l' = 4$ ou $l' = l$ selon que $l = 2$ ou l est impair. Soit k' l'extension de k dans $\overline{k}$ engendrée par les racines l'-ièmes de l'unité, soit l^e l'ordre de $G \cap k'^*$ et soit $\epsilon : \Gamma \longrightarrow \mathbb{Z}_l^*$ le caractère cyclotomique. On sait alors l'image de $\mathrm{Gal}(\overline{k}/k')$ par ϵ est $(1 + l^e\mathbb{Z}_l)^\times$. Soit donc $\zeta \in G$ avec $\zeta \notin k'$ et soit l^n, $(n > e)$, l'ordre de ζ. Si l'on choisit $\sigma \in \mathrm{Gal}(\overline{k}/k')$ de telle manière que $\epsilon(\sigma) \equiv 1 \pmod{l^n}$ mais $\epsilon(\sigma) \not\equiv 1 \pmod{l^{n-1}}$, alors $(\sigma - 1)\zeta$ appartient à k' et est différent de 1. On en tire que $G_0 = G \cap k'^*$ convient.

Voici une autre interprétation de ce résultat. Désignons comme d'habitude par $\mathbb{G}_m$ le groupe multiplicatif : le groupe G n'est autre chose que le groupe des points de torsion de $\mathbb{G}_m$ dont l'ordre est une puissance de l. Soit X la courbe dans $\mathbb{G}_m^2$ définie par l'équation $ax + by = 1$. Selon notre résultat $X(\overline{k}) \cap G$ est un ensemble fini qui peut être effectivement déterminé. Pour tout $Q = (\zeta, \eta) \in \mathbb{G}_m^2(\overline{k})$, on désigne par $T_Q X$ le translaté de X par le point $-Q = (\zeta^{-1}, \eta^{-1})$. Si donc $P \in X(\overline{k}) \cap G$ et si $\sigma \in \Gamma$, alors $\sigma P \in X(\overline{k})$ et donc $P = \sigma P - (\sigma P - P) \in T_{\sigma P - P} X(\overline{k})$, d'où $P \in X \cap T_{\sigma P - P} X(\overline{k})$. Autrement dit, si $P = (\zeta, \eta)$, alors (ζ, η) est une solution des équations (1) et (2) que, on le sait, n'ont qu'un nombre fini de solutions (éventuellement une au plus) lorsque $\sigma P \neq P$. Pour conclure il ne reste qu'à montrer que pour tout $P \in X(\overline{k}) \cap G$ en dehors d'un sous-groupe fini effectif G_0 on peut choisir σ de telle manière que $\sigma P \neq P$ et $\sigma P - P \in G_0$: d'après l'alinéa précédent, le choix $G_0 = G \cap k'^*$ convient.

De ce point de vue, ce résultat est capable d'importantes généralisations(*). Soit Σ une variété semi-abélienne, c'est-à-dire une extension d'une variété abélienne par un tore (pour plus de détails, le lecteur se reportera au début du §2). On suppose que Σ est définie sur k et on désigne par G un sous-groupe de $\Sigma(\overline{k})$ (puisque k est un corps fini, tout élément de $\Sigma(\overline{k})$ est d'ordre fini). Soit X une sous-variété fermée de Σ que l'on suppose définie sur k : comme précédemment, on désigne par $T_Q X$ le translaté de X par $-Q \in \Sigma(\overline{k})$. Pour simplifier on supposera que X est k-irréductible. Nous nous intéressons alors à déterminer $X(\overline{k}) \cap G$. Bien sûr, ceci ne sera possible que si G est d'une nature très particulière. Comme dans le résultat qui

(*) Après avoir terminé ce texte, je me suis aperçu que l'argument qui vient d'être présenté se trouve (pour une courbe plongée dans une variété abélienne) dans l'article de Raynaud ([R3], p3-4).

vient d'être démontré, on pourrait prendre comme G le groupe des points de torsion dont l'ordre est une puissance de l. Plus généralement, soit S un ensemble fini de nombres premiers et soit $\Omega(S)$ le monoïde multiplicatif engendré par S. Pour tout $n \in \Omega(S)$, on désigne par $\Sigma[n]$ le groupe de points de n-torsion de $\Sigma(\overline{k})$ et l'on pose $\Sigma_S = \cup_{n\in\Omega(S)}\Sigma[n]$. Nous démontrerons alors au §2 le théorème suivant :

THÉORÈME A. — *Avec les notations et les hypothèses qui viennent d'être introduites, il existe un ensemble fini de couples* $(P_i, B_i)_{i\in I}$, *où* $P_i \in \Sigma_S$, B_i *est une sous-variété semi-abélienne de* Σ *et* $T_{P_i}B_i \subseteq X$, *tel que :*

$$\overline{X(\overline{k}) \cap \Sigma_S} = \bigcup_{i\in I} T_{P_i}B_i.$$

On en tire immédiatement que, lorsque $G = \Sigma_S$, on a $X(\overline{k}) \cap G \subseteq \bigcup_{i\in I} T_{P_i}B_i(\overline{k})$. Ce dernier résultat a déjà été démontré dans [Bo1] lorsque Σ est une variété abélienne. Le résultat analogue lorsque k est un corps de nombres a été démontré par Bogomolov [Bg1], [Bg2]. Ensuite les travaux de Raynaud [R1], [R2], [R3], Hindry [H] et Faltings [F] ont établi le même résultat lorsque G est le groupe de tous les points de torsion sur $\overline{k}$, ou le groupe $\Sigma(k)$, ou même lorsque G est l'enveloppe divisible d'un sous-groupe de type fini de $\Sigma(\overline{k})$. Le cas où Σ est un tore et G un groupe de type fini (en caractéristique zéro toujours) a été traité par Laurent [La]. Lorsque G est l'enveloppe divisible d'un groupe de type fini et X est une courbe, Liardet [Li] a démontré que $X(\overline{k}) \cap G$ est soit un ensemble fini soit un ensemble formé de racines de l'unité et que cette dernière possibilité ne se produit que dans certains cas bien précis. Le travail récent de Ruppert [Ru] étudie les solutions en racines de l'unité de systèmes d'équations algébriques.

Lorsque k est un corps fini tout élément de $\Sigma(\overline{k})$ est de torsion et il est donc impossible d'étendre le théorème A au cas où $G = \Sigma(\overline{k})$. La situation, lorsque k est un corps de fonctions de caractéristique positive, a été étudiée par Voloch et Abramovich [Vo], [Ab-Vo].

Remarque 1 : si en plus on suppose que l'ensemble des $T_{P_i}B_i$ du théorème soit choisi de façon minimale, alors les $T_{P_i}B_i$ seront nécessairement les composantes irréductibles de $\overline{X(\overline{k}) \cap \Sigma_S}$. En particulier, ils sont uniquement déterminés par X et S.

Remarque 2 : notre démonstration montre que les sous-variétés semi-abéliennes B_i du théorème A peuvent être réalisées comme des stabilisateurs par translation de composantes irréductibles de fermés de la forme $T_{Q_1}X \cap T_{Q_2}X \cap \ldots \cap T_{Q_r}X$. On peut se demander si les sous-variétés abéliennes apparaissant dans les travaux de Raynaud, Hindry et Faltings puissent être construites de manière analogue.

Le §3 est consacré à une étude de l'effectivité dans la démonstration du théorème A.

Je remercie vivement L. Moret-Bailly pour la lecture approfondie d'une version préliminaire de ce travail.

2. — Variétés semi-abéliennes

Rappelons que par définition une *variété semi-abélienne* Σ est une extension d'une variété abélienne par un tore; elle est alors définie par une suite exacte

$$1 \to T \to \Sigma \to A \to 0,$$

où T est un tore et A une variété abélienne définis sur k.

A l'aide de la théorie de la structure des groupes algébriques, on voit aisément que tout sous-groupe algébrique connexe d'une variété semi-abélienne est une variété semi-abélienne. Il s'ensuit que si V est une sous-variété (fermée) de Σ, alors le stabilisateur $\tilde{B}_V$ de V par l'opération des éléments de $\Sigma(\overline{k})$ par translation est le produit d'une sous-variété semi-abélienne et d'un groupe fini. On désigne alors par B_V la composante neutre de $\tilde{B}_V$ et le groupe fini par H_V.

Pour toute sous-variété V de Σ, on désigne par $T_Q V$ le translaté de V par $-Q \in \Sigma(\overline{k})$. On peut alors écrire : $\tilde{B}_V = \cap_{Q \in V(\overline{k})} T_Q V$. On en tire que si V est irréductible, alors $\dim(B_V) \leq \dim V$ et si $\dim(B_V) = \dim V$ alors $\tilde{B}_V = B_V$ et V est le translaté de B_V par un élément de $\Sigma(\overline{k})$.

Soit à nouveau S un ensemble fini de nombre premiers, $\Omega(S)$ le monoïde multiplicatif engendré par S. Pour tout $n \in \Omega(S)$, on désigne par $\Sigma[n]$ le groupe des points d'ordre n de $\Sigma(\overline{k})$ et l'on pose $\Sigma_S = \cup_{n \in \Omega(S)} \Sigma[n]$. Soit Γ_S le groupe de Galois de $k(\Sigma_S)$ sur k. Pour tout $l \in S$ on désigne par $r(l)$ le rang du module de Tate $T_l(\Sigma) = \varprojlim_n \Sigma[l^n]$ (on a alors $r(l) = 2\dim A + \dim T$ si l est différent de la caractéristique de k et $0 \leq r(l) \leq \dim A$ si l est égal à la caractéristique de k). L'opération de Γ_S sur Σ_S induit une représentation de Γ_S dans le groupe des automorphismes continus $\mathrm{Aut}(\Sigma_S)$ de Σ_S. On fixe un choix des bases des $T_l(\Sigma)$, ce qui induit un isomorphisme continu entre $\mathrm{Aut}(\Sigma_S)$ et $\prod_{l \in S} \mathrm{GL}_{r(l)}(\mathbb{Z}_l)$; on obtient ainsi une représentation ρ de Γ_S dans ce dernier groupe. Posons $L = \prod_{l \in S} l'$, où l'on a écrit $l' = l$ si l est impair et $l' = 4$ si $l = 2$. Pour tout $n \in \Omega(S)$, on désigne par k_n le corps $k(\Sigma[n])$ et par Γ_n le groupe de Galois de $k(\Sigma_S)$ sur k_n. Soit N le plus grand élément de $\Omega(S)$ tel que $k_N = k_L$. On a alors, pour tout n divisible par L :

$$\rho(\Gamma_n) \subseteq \prod_{l \in S} \left(\mathrm{I} + l^{\mathrm{ord}_l(n)} \mathrm{M}_{r(l)}(\mathbb{Z}_l)\right)$$

(où I désigne la matrice identité et $\mathrm{M}_{r(l)}(\mathbb{Z}_l)$ désigne l'algèbre des matrices carrées d'ordre $r(l)$ à coefficients dans $\mathbb{Z}_l$). Soit θ un générateur topologique de Γ_N : k étant un corps fini tout élément de Γ_N s'écrit de manière unique dans la forme θ^b avec $b \in \hat{\mathbb{Z}} = \varprojlim_n \mathbb{Z}/n\mathbb{Z}$. Pour tout $l \in S$ désignons par Θ_l la l–composante de $\rho(\theta)$. D'après la définition de N, on a $\Theta_l \equiv \mathrm{I} \bmod l^{\mathrm{ord}_l(N)}$ mais $\Theta_l \not\equiv \mathrm{I} \pmod{l^{\mathrm{ord}_l(N)+1}}$ pour tout $l \in S$.

Pour tout $l \in S$, définissons la matrice Φ_l par $\Theta_l = \mathrm{I}+l^{\mathrm{ord}_l(N)}\Phi_l$. Comme θ ne laisse stable qu'un nombre fini d'éléments de Σ_S, Φ_l est inversible (i.e. un élément de $\mathrm{GL}_{r(l)}(\mathbb{Q}_l)$). Soit $e_l(P)$ l'exposant de l dans l'ordre de $P \in \Sigma_S$. Il s'ensuit alors que pour tout $l \in S$ on a :

$$e_l(P) - \mathrm{ord}_l(\text{dét}\,\Phi_l) - \mathrm{ord}_l(N) \le e_l(\theta P - P) \le e_l(P) - \mathrm{ord}_l(N).$$

LEMME 1. — *Soit b un entier strictement positif. Alors pour tout $P \in \Sigma_S$ on a :*

$$\begin{aligned} &e_l(P) - \mathrm{ord}_l(\text{dét}\,\Phi_l) - \mathrm{ord}_l(N) - \mathrm{ord}_l(b) \\ &\le e_l(\theta^b P - P) \le e_l(P) - \mathrm{ord}_l(N) - \mathrm{ord}_l(b)\,. \end{aligned}$$

Par conséquent, si l'on pose $\mu = \prod_{l\in S} l^{\mathrm{ord}_l\,(\det \Phi_l)}$, alors $\Sigma[n] \subseteq \Sigma_S(k_n) \subseteq \Sigma[\mu n]$ pour tout $n \in \Omega(S)$ divisble par N, $\Sigma_S(k_n)$ désignant les éléments de Σ_S rationnels sur k_n.

Démonstration : le cas $b = 1$ est déjà acquis. Si $b > 1$, on utilise l'équation

$$\begin{aligned} \Theta_l^b P - P = (\Theta_l^b - I)P &= b(\Theta_l P - P) + \sum_{r=2}^{b} C_b^r (\Theta_l - \mathrm{I})^r P \\ &= b l^{\mathrm{ord}_l(N)} \Phi_l P + \sum_{r=2}^{b} C_b^r l^{r\,\mathrm{ord}_l(N)} \Phi_l^r P\,. \end{aligned}$$

Puisque $l|N$ pour tout $l \in S$ et $4|N$ si $2 \in S$ on vérifie que la puissance exacte de l divisant $C_b^r l^{r\,\mathrm{ord}_l(N)}$ pour $2 \le r \le b$ est strictement supérieur à $\mathrm{ord}_l(b) + \mathrm{ord}_l(N)$. On en tire que $e_l\big((\Theta_l^b P - P)\big) = \mathrm{ord}_l(b) + \mathrm{ord}_l(N) + e_l\big((\theta P - P)\big)$ d'où l'encadrement de $e_l(\theta^b P - P)$.

Pour montrer la dernière assertion on pose $b = \frac{n}{N}$. Soit alors $P \in \Sigma_S(\overline{k})$ un point d'ordre $\mu n a$ avec $a \in \Omega(S)$ et $a > 1$ et soit l un élément de S divisant a. Alors $e_l(P) - \mathrm{ord}_l(\text{dét}\,\Phi_l) - \mathrm{ord}_l(N) - \mathrm{ord}_l(b) = \mathrm{ord}_l(a) \ge 1$ et donc θ^b ne laisse pas stable P, c'est-à-dire $P \notin \Sigma_S(k_n)$. On conclut que $\Sigma_S(k_n) \subseteq \Sigma[\mu n]$ et l'inclusion $\Sigma[n] \subseteq \Sigma_S(k_n)$ est triviale.

Remarque 1 : soit k'_m l'unique extension de k_N de degré m. Pour tout $n \in \Omega(S)$ divisible par N, $\theta^{\frac{n}{N}}$ est un générateur topologique de $\mathrm{Gal}(\overline{k}/k'_n)$.

On tire de la démonstration du lemme 1 que $\Theta_l^{\frac{n}{N}} \equiv \mathrm{I} \pmod{l^{\mathrm{ord}_l(n)}}$ mais $\Theta_l^{\frac{n}{N}} \not\equiv \mathrm{I} \pmod{l^{\mathrm{ord}_l(n)+1}}$. Il s'ensuit que $\theta^{\frac{n}{N}}$ est également un générateur topologique de k_n et donc que $k_n = k'_{\frac{n}{N}}$ pour tout $n \in \Omega(S)$ divisible par N.

Ce lemme nous permet d'aborder la démonstration du théorème A. Soit $n \in \Omega(S)$ divisible par N et soit V une sous-variété de Σ définie sur k, que l'on supposera k_n-irréductible. Supposons d'abord que $\tilde{B}_V = (0)$. Si $P \in V(\overline{k}) \cap \Sigma_S$, alors $\sigma P \in V(\overline{k}) \cap \Sigma_S$ pour tout $\sigma \in \Gamma$ et donc $P \in V \cap T_{\sigma P - P} V(\overline{k})$. Si $P \notin \Sigma[\mu n]$, le lemme 1 montre que P n'est pas définie sur k_n et donc n'est pas stable par $\theta^{\frac{n}{N}}$: on peut donc choisir une puissance σ de $\theta^{\frac{n}{N}}$ de telle façon que $0 \neq \sigma P - P \in \Sigma[\mu n]$. On trouve ainsi que

$$V(\overline{k}) \cap \Sigma_S \subseteq \Sigma[\mu n] \cup \bigcup_Q \left(V \cap T_Q V(\overline{k}) \cap \Sigma_S\right),$$

où Q parcourt $\Sigma[\mu n] \setminus (0)$. Puisque $\tilde{B}_V = (0)$ et V est k_n-irréductible, on conclut que toutes les composantes irréductibles des $V \cap T_Q V$ sont de dimension strictement inférieure à celle de V. En prenant l'adhérence de Zariski, on peut écrire :

$$\overline{V(\overline{k}) \cap \Sigma_S} = \{ \text{ partie finie de } \Sigma[\mu n] \ \} \bigcup_W \overline{W(\overline{k}) \cap \Sigma_S}, \tag{4}$$

où $\{W\}$ parcourt un ensemble de variétés définies sur $k_{\mu n}$ que l'on peut supposer $k_{\mu n}$-irréductibles.

On peut étendre cet argument au cas où $\tilde{B}_V \neq (0)$ en passant à la variété semi-abélienne quotient $\Sigma' = \Sigma / \tilde{B}_V$. Soient μ_V et N_V les entiers μ et N associés à Σ' et soit H_V un sous-groupe fini de Σ tel que $\tilde{B}_V = B_V \times H_V$. On a alors $\tilde{B}_{V,S} = B_{V,S} \oplus H_{V,S}$. Si $\epsilon(H_{V,S})$ désigne l'exposant de $H_{V,S}$ on obtient, pour tout $n \in \Omega(S)$ divisible par $N_V \epsilon(H_{V,S})$:

$$\overline{V(\overline{k}) \cap \Sigma_S} = \bigcup_P T_P B_V \cup \bigcup_W \overline{W(\overline{k}) \cap \Sigma_S}, \tag{5}$$

où P parcourt une partie de $\Sigma[\mu_V n]$ et W un ensemble fini de variétés définies et irréductibles sur $k_{\mu_V n}$.

On peut alors appliquer le même raisonnement à chacune des variétés W (en replaçant k par $k_{\mu_V n}$). Le théorème A peut alors être démontré par recurrence en la dimension de X.

En outre, comme les W sont des k_n-composantes des $V \cap T_Q V$, on conclut que $\overline{X(\overline{k}) \cap \Sigma_S}$ est une réunion finie de translatés de sous-variétés semi-abéliennes dont chacune est le stabilisateur d'une composante irréductible d'une intersection de translatés de X (remarque 2 de l'introduction).

Remarque 2 : lorsque Σ est une variété abélienne simple, on conclut que $V(\overline{k}) \cap \Sigma_S$ est une ensemble fini. Même lorsque Σ est une variété abélienne quelconque on peut parfois démontrer que $V(\overline{k}) \cap G$ est nécessairement un ensemble fini pour certain sous-groupes G de $\Sigma(\overline{k})$. Ceci est le cas par exemple lorsque G est le groupe des points tués par une puissance d'un idéal premier $\mathfrak{p}$ de degré un d'une sous-algébre commutative de rang 2 dim A de End $A \otimes \mathbb{Q}$ (voir [Bo2]). Rappelons que d'après un théorème bien connu de Tate [T], End $A \otimes \mathbb{Q}$ contient toujours une telle sous-algèbre lorsque k est un corps fini.

Remarque 3 : l'hypothèse que k soit un corps fini n'est pas essentielle dans notre démonstration du théorème 1. Si par exemple k est un corps de nombre (ou plus généralement un corps de type fini sur $\mathbb{Q}$ ou sur un corps fini) on peut vérifier une version modifiée du lemme 1 qui permet de conclure dans la même manière. On obtient ainsi une nouvelle démonstration des résultats originaux de Bogomolov ([Bg1], [Bg2]). Par contre, la méthode de Bogomolov ne s'applique apparemment pas sur un corps fini : en effet il utilise le fait que sur un corps de nombres $\rho(\Gamma)$ contient un sous-groupe d'indice fini du groupe des homothéties et cette propriété est fausse en général sur un corps fini.

3. — Un peu d'effectivité

D'après un résultat fondemental de Chow [Ch], toute variété semi-abélienne Σ est quasi-projective. Fixons une fois pour toutes un plongement de Σ dans un espace projectif $\mathbb{P}_k^M$. Le degré d'une sous-variété equidimensionnelle V dc Σ rélatif à ce plongement peut alors être défini comme le cardinal de l'intersection de l'adhérence de l'image de V avec un sous-espace projectif générique de codimension égale à la dimension de V. Si V est un fermé de Σ, on définit son degré comme étant la somme des degrés de ses composantes irréductibles. Le degré jouit alors des propriétés suivantes :

(a) Le degré d'une réunion finie de variétés est inférieure ou égale à la somme de leurs degrés ; en particulier, si W est une réunion de composantes irréductibles de V, alors $\deg W \leq \deg V$;

(b) Le degré d'une intersection finie de variétés est inférieure ou égale au produit de leurs degrés;

(c) Le degré est invariant par translation par un élément de $\Sigma(\overline{k})$ (car le morphisme défini par translation par un élément de $\Sigma(\overline{k})$ est plat).

(Pour plus de détails sur la notion de degré, le lecteur pourra consulter [Fu], notamment le §8).

Le but de ce paragraphe est de démontrer le théorème suivant :

THÉORÈME 2. — *Soit k un corps fini, soit $g \geq 1$ un entier, soit Σ une variété semi-abélienne de dimension g et soit S un ensemble fini de premiers. Alors*

$\tilde{B}$: on en tire que $\deg \tilde{B} \leq \deg (T_{Q_1} V \cap T_{Q_2} V) \leq (\deg V)^2$. Si au contraire $\dim B \leq \dim V - 2$, alors on montre à nouveau qu'il existe $Q_3 \in V(\overline{k})$ tel que $\tilde{B} \subseteq T_{Q_1} V \cap T_{Q_2} V \cap T_{Q_3} V$ et $\dim (T_{Q_1} V \cap T_{Q_2} V \cap T_{Q_3} V) \leq \dim V - 2$. En continuant ainsi, on montre que $\tilde{B}$ peut être écrit comme une réunion de composantes de l'intersection d'au plus $\dim V + 1$ translatés de V ce qui entraîne le lemme.

Reprenons la démonstration du théorème et étudions d'abord le terme $|\Sigma[\mu_V n]| \deg B_V$ dans l'inégalité (6). Puisque $\tilde{B}_V = B_V \times H_V$ et le degré est invariant par translation, on a $\deg \tilde{B}_V = (\deg B_V)|H_V(\overline{k})|$. D'après le lemme 3 on a : $\deg \tilde{B}_V \leq (\deg V)^{\dim V + 1}$. On conclut que $\deg B_V \leq (\deg V)^{\dim V+1}$ et également que $\epsilon(H_{V,S}) \leq |H_{V,S}| \leq (\deg V)^{\dim V+1}$. D'après ce qui précède, μ_V et N_V sont bornés en fonction de (k, g, S). Comme $|\Sigma'[a]| \leq a^{2g}$ pour tout $a \geq 1$ et pour tout $\Sigma' \in \mathcal{E}_g$ on en conclut qu'il y a une constante K_0 telle que pour tout $n \in \Omega(S)$ divisible par $N_V \epsilon(H_{V,S})$ on ait :

$$|\Sigma[\mu_V n]| \deg B_V \leq K_0 n^{2g} (\deg V)^{2(\dim V+1)}. \tag{7}$$

De la même manière que (6), on peut écrire

$$\sum_W \deg \overline{(W(\overline{k}) \cap \Sigma_S)} \leq \sum_W |\Sigma[\mu_W n]| \deg B_W + \sum_Y \deg \overline{(Y(\overline{k}) \cap \Sigma_S)},$$

où les Y sont de dimension $\leq \dim V - 2$ et $n \in \Omega(S)$ est divisible par le ppcm des $N_W \epsilon(H_{W,S})$ et $N_V \epsilon(H_{V,S})$ et est indépendant de W et de V. En appliquant (7) (avec V remplacé par W) on obtient :

$$\begin{aligned} \sum_W |\Sigma[\mu_W n]| \deg B_W &\leq K_0 n^{2g} \sum_W (\deg W)^{2(\dim W+1)} \\ &\leq K_0 n^{2g} \sum_W (\deg W)^{2(\dim V+1)} \\ &\leq K_0 n^{2g} \Big(\sum_W \deg W\Big)^{2(\dim V+1)} \end{aligned}$$

Comme

$$\sum_W \deg W \leq \sum_Q \deg(V \cap T_Q V) \leq |\Sigma[\mu_V n]|(\deg V)^2 \leq (\mu_V n)^{2g} (\deg V)^2,$$

on conclut que :

$$\begin{aligned} \deg \overline{V(\overline{k}) \cap \Sigma_S} &\leq K_0 n^{2g} (\deg V)^{2(\dim V+1)} \\ &\quad + (K_0' n^{2g})^{2\dim V+3} (\deg V)^{4(\dim V+1)} \\ &\quad + \sum_Y \deg \overline{Y(\overline{k}) \cap \Sigma_S}, \end{aligned}$$

K_0' étant une constante.

Après r étapes on trouve alors

$$\deg(\overline{V(\overline{k}) \cap \Sigma_S}) \leq K_r(n^{2g})^{1+2^{r-1}(\dim V+1)}(\deg V)^{2^r(\dim V+1)} + \\ + \sum_Z \deg(\overline{Z(\overline{k}) \cap \Sigma_S}),$$

où K_r est une constante et $\dim Z \leq \dim V - r$ pour tout Z. En particulier, lorsque $r = \dim V + 1$ la somme sur les Z devient vide, et l'on en tire :

$$\deg\left(\overline{V(\overline{k}) \cap \Sigma_S}\right) \leq K'(n^{2g})^{1+2^{\dim V}(\dim V+1)}(\deg V)^{2^{\dim V+1}(\dim V+1)},$$

où K' est une constante et n est divisible par le ppcm des $N_Z\epsilon(H_{Z,S})$ pour toutes les variétés Z qui interviennent à chaque étape. Or, on a $\epsilon(H_{Z,S}) \leq (\deg Z)^{\dim Z+1} \leq (\deg Z)^{\dim V+1}$. Enfin Z est une composante d'une intersection d'au plus $2^{\dim V - \dim Z}$ translatés de V : on a donc $\deg Z \leq (\deg V)^{2^{\dim V - \dim Z}} \leq (\deg V)^{2^{\dim V}}$ et enfin :

$$\epsilon(H_{Z,S}) \leq (\deg V)^{2^{\dim V}(\dim V+1)}.$$

Le ppcm des $N_Z\epsilon(H_{Z,S})$ sera alors majoré par une constante fois $(\deg V)^{2^{\dim V}(\dim V+1)|S|}$. On peut donc choisir comme n tout élément de $\Omega(S)$ plus grand qu'une constante fois $(\deg V)^{2^{\dim V}(\dim V+1)|S|}$. On obtient alors le théorème 2 en commençant avec $V = X$.

Il serait intérressant d'améliorer cette borne. Lorsque X est une courbe lisse et complète et Σ sa jacobienne on peut borner $X(\overline{k}) \cap \Sigma_S$ en fonction du genre de X (voir [Bo1]).

Remarque (et corrections à [Bo1], [Bo2]). Pour les variétés abéliennes, nous avons déjà enoncé dans [Bo1] (théorème 1) une borne explicite pour $\deg \overline{X(\overline{k}) \cap \Sigma_S}$. Or la démonstration est basée sur le lemme 1 qui est faux en général et doit être remplacé par le lemme 1 du présent article.

Autres corrections à [Bo1] : (i) lignes 6 et 7 de la démonstration du théorème 1 (p. 1065); il aurait fallu lire : ... on montre à l'aide des lemmes 3 et 4 que, si $P \in C \cap A_S$, il existe $\sigma \in \mathrm{Gal}(\overline{k}/k)$ avec ... (ii) ligne 3 de la démonstration du lemme 5 (p. 1066); il aurait fallu lire : Pour tout $\xi \in C(\overline{k})$ on peut trouver $\eta \in C(\overline{k})$ tel que les diviseurs ....

Corrections à [Bo2] : page 1 (en bas) : f ne s'étend pas en un morphisme $\mathcal{X} \to \mathbb{P}^1_{\mathfrak{O}}$ mais seulement en une application rationnelle dont l'ouvert de définition contient tous les points de codimension un. Page 6 (en bas) : la courbe affine $y^2 = x^5 - 11$ possède 50 points sur $\mathbb{F}_{41}$ (et non 64).

Manuscrit reçu le 3 novembre 1993
Version révisée reçue le 7 janvier 1994

Bibliographie

[Ab-Vo] D. ABRAMOVICH, J.-F. VOLOCH. — *Towards a proof of the Mordell-Lang conjecture in characteristic p.* Intern. Math. Research Notes **5** (1992) 103-115.

[Bg1] F. A. BOGOMOLOV. — *Sur l'algébricité des représentions l-adiques.* C. R. Acad. Sci. Paris. **290** série A (1980) 701-703.

[Bg2] F. A. BOGOMOLOV. — *Points of finite order on an abelian variety.* Math. USSR. Izv. **17** (1981) 55-72.

[Bo1] J. BOXALL. — *Autour d'un problème de Coleman.* C. R. Acad. Sci. Paris. **315** série A (1992) 1063-1066.

[Bo2] J. BOXALL. — *Valeurs spéciales de fonctions abéliennes.* Groupe de travail sur les problèmes diophantiens, Université de Paris VI, année 1990/1.

[Ch] W. L. CHOW. — *On the projective embedding of homogeneous varieties.* in : Symposium in honor of S Lefschetz, edité par R.H. Fox, D.C. Spencer et A.W. Tucker, Princeton University Press (1957).

[F] G. FALTINGS. — *Diophantine approximation on abelian varieties.* Annals of Math. **133** (1991) 549-576.

[Fu] W. FULTON. — Intersection theory. *Ergebnisse der Math. und ihrer Grunzgebiete,* 3. Folge, Band 2, Springer-Verlag (1984).

[H] M. HINDRY. — *Autour d'une conjecture de S Lang.* Invent. Math. **94** (1988) 575-603.

[La] M. LAURENT. — *Équations diophantiennes exponentielles.* Invent. Math. **78** (1984) 299-327.

[Li] P. LIARDET. — *Sur une conjecture de Serge Lang.* Astérisque **24-25** (1975) 187-210.

[R1] M. RAYNAUD. — *Courbes sur une variété abélienne et points de torsion.* Invent. Math. **71** (1983) 207-233.

[R2] M. RAYNAUD. — *Sous-variétés d'une variété abélienne et points de torsion,* dans Arithmetic and Geometry I, dédié à I Shafarevich, Birkhäuser, (1983) 327-352.

[R3] M. RAYNAUD. — *Around the Mordell conjecture for Function Fields and a Conjecture of Serge Lang,* dans Algebraic Geometry, Lecture notes in Maths. **1016**, Springer-Verlag (1982).

[Ru] W. M. RUPPERT. — *Solving algebraic equations in roots of unity.* J. Crelle. **435** (1993) 119-156.

[T] J. TATE. — *Endomorphisms of abelian varieties over finite fields.* Invent. Math. **2** (1966) 134-144.

[Vo] J.-F. VOLOCH. — *On the conjectures of Mordell and Lang in positive characteristic.* Invent. Math. **104** (1990) 643-646.

John BOXALL
Département de mathématiques
et de mécanique
Université de Caen
Esplanade de la Paix
14032 CAEN cedex
FRANCE
e-mail : boxall@math.unicaen.fr

Number Theory
Paris 1992–93

Propriétés transcendantes des fonctions automorphes

Paula Beazley Cohen

Le sujet de cet article est un travail en commun avec J. Wolfart et H. Shiga dont les détails sont donnés dans un manuscrit [CSW] intitulé "Criteria for complex multiplication and transcendence properties of automorphic functions", preprint du Johann Wolfgang Goethe-Universität, Frankfurt-am-Main. Le but du présent article est de servir d'introduction à ce manuscrit. Le point de départ est le résultat suivant, démontré par Th. Schneider en 1937 [Sch] :

THÉORÈME (Schneider). — *Soit $j = j(\tau)$ la fonction modulaire elliptique. Alors τ et $j(\tau)$ sont tous les deux des nombres algébriques si et seulement si τ est quadratique imaginaire.*

Rappelons que la fonction modulaire elliptique

$$j : \mathcal{H} \to \mathbb{C}$$

est holomorphe sur le demi-plan supérieure $\mathcal{H}$, méromorphe avec un développement de Fourier d'ordre -1 à l'infini et automorphe par rapport au groupe modulaire $\Gamma = PSL(2,\mathbb{Z}) = SL(2,\mathbb{Z})/\{\pm 1_2\}$, c'est-à-dire :

$$j\Big(\frac{m_1\tau + m_2}{n_1\tau + n_2}\Big) = j(\tau), \qquad m_1, m_2, n_1, n_2 \in \mathbb{Z}, m_1n_2 - m_2n_1 = 1.$$

La fonction j est normalisée de sorte que, dans son développement en puissances de $q = \exp(2i\pi\tau)$, le coefficient de q^{-1} soit égal à 1 et le terme constant soit égal à 744. De plus, on a une bijection,

$$\Gamma\backslash\mathcal{H} \simeq \{[\mathcal{E}_\tau]; \tau \in \mathcal{H}\}, \qquad \mathcal{E}_\tau = \mathbb{C}/(\mathbb{Z}\tau + \mathbb{Z}),$$

où $[\mathcal{E}_\tau]$ désigne la classe d'isomorphisme de $\mathcal{E}_\tau$ sur $\mathbb{C}$. L'invariant modulaire de $\mathcal{E} \in [\mathcal{E}_\tau]$ est donné par :

$$j(\mathcal{E}) = j(\tau).$$

Les valeurs de la fonction automorphe j paramétrisent les classes d'isomorphisme de courbes elliptiques : $j(\mathcal{E}) = j(\mathcal{E}')$ si et seulement si $\mathcal{E}$ et $\mathcal{E}'$ sont isomorphes sur $\mathbb{C}$. On a même : $j = j(\tau)$ *est algébrique si et seulement s'il existe une courbe elliptique* $\mathcal{E} \in [\mathcal{E}_\tau]$ *telle que* $\mathcal{E}$ *soit définie sur* $\bar{\mathbb{Q}}$. On écrira $\mathcal{E}/\bar{\mathbb{Q}}$ lorsque $\mathcal{E}$ est définie sur $\bar{\mathbb{Q}}$, c'est-à-dire lorsque les invariants $g_2 = g_2(\mathcal{L})$ et $g_3 = g_3(\mathcal{L})$ du réseau $\mathcal{L}$ où $\mathcal{E} = \mathbb{C}/\mathcal{L}$ sont des nombres algébriques. Il y a deux possibilités pour l'algèbre des endomorphismes $\mathrm{End}_o(\mathcal{E}_\tau) = \mathrm{End}(\mathcal{E}_\tau) \otimes_{\mathbb{Z}} \mathbb{Q}$. On a soit $\mathrm{End}_o(\mathcal{E}_\tau) = \mathbb{Q}$ soit $\mathrm{End}_o(\mathcal{E}_\tau) = \mathbb{Q}(\tau)$ avec τ quadratique imaginaire. Dans le deuxième cas $\mathcal{E}_\tau$ est donc à multiplication complexe (CM) par $\mathbb{Q}(\tau)$. On peut reformuler le théorème de Schneider de la façon suivante,

$$\text{soit } \mathcal{E}/\bar{\mathbb{Q}}, \text{ alors } \mathcal{E} \in [\mathcal{E}_\tau] \text{ avec } \tau \in \bar{\mathbb{Q}} \text{ si et seulement si } \mathcal{E} \text{ est à CM}.$$

Le travail de Wolfart, Shiga et moi-même donne une généralisation du résultat de Schneider aux dimensions supérieures : c'est-à-dire aux domaines $\mathcal{D}$, classifiés par Shimura et Siegel, qui paramétrisent certaines familles $\{A_z; z \in \mathcal{D}\}$ de variétés abéliennes polarisées d'un type d'endomorphisme donné. Qualitativement, avec les notations analogues à celles du cas elliptique, notre résultat dit que :

$$\text{soit } A/\bar{\mathbb{Q}} \text{ alors } A \in [A_z] \text{ avec } z \in \mathcal{D}(\bar{\mathbb{Q}}) \text{ si et seulement si } A \text{ est de type CM.}$$

Rappelons l'origine des domaines $\mathcal{D}$ [Shi]. Soit $(A, \mathcal{C})$ une variété abélienne simple avec polarisation $\mathcal{C}$. Alors, $\mathrm{End}_o(A)$ est une algèbre de division L sur $\mathbb{Q}$ munie d'une involution positive ρ induite par $\mathcal{C}$. De telles (L, ρ) ont été classifiées par Albert. Soit K le centre de L et $F = \{x \in K \mid x^\rho = x\}$ avec $g = [F : \mathbb{Q}]$, alors il y a quatre possibilités pour L :

TYPE (I) $L = F$

TYPE (II) L est une algèbre de quaternions totalement indéfinie sur F :
$$L \otimes_{\mathbb{Q}} \mathbb{R} \simeq M_2(\mathbb{R})^g$$

TYPE (III) L est une algèbre de quaternions totalement définie sur F :
$L \otimes_{\mathbb{Q}} \mathbb{R} \simeq H^g$ (où H est l'algèbre des quaternions hamiltoniens)

TYPE (IV) L est une algèbre centrale simple sur K où K est un corps CM donné par une extension quadratique totalement imaginaire de F.

Soient, n un entier positif et $\Phi : L \to M_n(\mathbb{C})$ une représentation complexe de L de dimension n (on aura toujours que $[L : \mathbb{Q}]$ divise $2n$) et soit $S = S(L, \Phi, \rho)$ l'ensemble des variétés abéliennes polarisées $(A, \mathcal{C})$, $\dim A = n$, (ici A n'est pas supposée simple) telles que :

$$L \subset \mathrm{End}_o(A)$$
$$\mathcal{C} \text{ soit compatible avec } \rho$$
$$\text{il existe un réseau } \Lambda \subset \mathbb{C}^n \text{ et un isomorphisme } A \simeq \mathbb{C}^n/\Lambda \text{ induisant } \Phi.$$

Une représentation Φ avec S non vide est appelée une représentation admissible. Dans le cas des types I, II, et III il n'y a qu'une seule Φ admissible à isomorphisme près. La construction de Shimura donne pour chaque L et chaque classe d'isomorphisme d'une Φ admissible un domaine $\mathcal{D}$ qui paramétrise des familles $\Sigma = \{(A_z, \mathcal{C}_z); z \in \mathcal{D}\}$ d'éléments de S. Pour chaque Σ il y a un groupe modulaire $\Gamma = \Gamma_\Sigma$ agissant sur $\mathcal{D}$ tel que l'on ait une bijection :

$$\Gamma \backslash \mathcal{D} \simeq \{[(A, \mathcal{C})], (A, \mathcal{C}) \in \Sigma\}.$$

On dit que $z \in \mathcal{D}$ est un point CM si $A_z \in \Sigma$ est de type CM. Rappelons qu'une variété abélienne est de type CM lorsqu'à isogénie près elle se casse en produits de puissances de variétés abéliennes simples B où les $\mathrm{End}_o(B)$ sont des corps CM de degré $2\dim(B)$ sur $\mathbb{Q}$. Le domaine $\mathcal{D}$ est supposé convenablement normalisé, c'est-à-dire il est un des domaines donnés dans [Shi, p. 162]. Les points CM de $\mathcal{D}$ sont alors algébriques : en les écrivant comme produits de matrices leurs coefficients sont tous des nombres algébriques. On désigne par $\mathcal{D}(\bar{\mathbb{Q}})$ les points algébriques de $\mathcal{D}$. Par les travaux de Borovoi, Deligne, Milne [Mi] et d'autres sur l'existence de modèles canoniques, on sait que la variété de Shimura $\Gamma \backslash \mathcal{D}$ a une structure de variété quasi-projective V définie sur $\bar{\mathbb{Q}}$. L'application canonique

$$J : \mathcal{D} \to V$$

est donc une généralisation de la fonction modulaire elliptique j. Wolfart, Shiga et moi-même avons démontré la généralisation suivante du résultat de Schneider :

Théorème. — *On a $z \in \mathcal{D}(\bar{\mathbb{Q}})$ et $J(z) \in V(\bar{\mathbb{Q}})$ si et seulement si z est un point* CM.

Pour simplifier les notations on va désigner par une seule lettre A une variété abélienne polarisée. Le Théorème est une conséquence du résultat suivant de Wolfart, Shiga et moi-même.

Résultat Principal. — *Soit A une variété abélienne polarisée définie sur $\bar{\mathbb{Q}}$. Alors il y a équivalence entre :*

(i) *A est isomorphe à A_z où $A_z \in \Sigma$ et $z \in \mathcal{D}(\bar{\mathbb{Q}})$,*

(ii) *A est de type* CM.

L'existence de V permet de définir le corps K des fonctions Γ-automorphes définies sur $\bar{\mathbb{Q}}$. Comme conséquence évidente du Théorème on a,

COROLLAIRE. — *Si $z \in \mathcal{D}(\bar{\mathbb{Q}})$ est un point non CM alors il existe un $f \in K$ telle que $f(z)$ soit transcendant.*

Avant de donner une idée de la démonstration de ce résultat passons à des exemples :

L'espace et les fonctions modulaires de Siegel : dans cet exemple on a $L = \mathbb{Q}$ avec, pour chaque $n \geq 1$, la représentation de $\mathbb{Q}$ à valeurs dans $M_n(\mathbb{C})$ donnée par $\Phi : \alpha \mapsto \alpha 1_n$, $\alpha \in \mathbb{Q}$. Le domaine $\mathcal{D}$ est donné par l'espace de Siegel

$$\mathcal{D} = \mathcal{H}_n = \{z \in M_n(\mathbb{C}) \mid {}^t z = z, \frac{1}{2i}(z - \bar{z}) > 0\}$$

sur lequel agit le groupe modulaire $\Gamma = Sp(2n, \mathbb{Z})$ où

$$Sp(2n,\mathbb{Z}) = \left\{ M = \begin{pmatrix} A & B \\ C & D \end{pmatrix} \in M_{2n}(\mathbb{Z}) \mid {}^t M \begin{pmatrix} 0 & 1_n \\ -1_n & 0 \end{pmatrix} M = \begin{pmatrix} 0 & 1_n \\ -1_n & 0 \end{pmatrix} \right\}.$$

Ici A, B, C, D sont dans $M_n(\mathbb{Z})$. On a $\Gamma = \Gamma_\Sigma$ pour une famille Σ de représentants des classes d'isomorphisme de variétés abéliennes de dimension n principalement polarisées. L'action de Γ sur $\mathcal{D}$ est donnée par $z \mapsto (Az + B)(Cz + D)^{-1}$. Soit A une variété abélienne principalement polarisée isomorphe à $\mathbb{C}^n/\Lambda$ où Λ est un réseau dans $\mathbb{C}^n$. On peut écrire

$$\Lambda = \mathbb{Z}(\int_{\gamma_1} \vec{\omega}) \oplus \ldots \oplus \mathbb{Z}(\int_{\gamma_{2n}} \vec{\omega})$$

où

$$\int_{\gamma_j} \vec{\omega} = {}^t\Big(\int_{\gamma_j} \omega_1, \ldots, \int_{\gamma_j} \omega_n\Big) \in \mathbb{C}^n$$

pour une base $\omega_1, \ldots, \omega_n$ de $H^o(A, \Omega)$ sur $\mathbb{C}$ et une base $\gamma_1, \ldots, \gamma_{2n}$ de $H_1(A, \mathbb{Z})$ sur $\mathbb{Z}$. On peut choisir ces bases de $H^o(A, \Omega)$ et de $H_1(A, \mathbb{Z})$ de telle façon à ce que pour les matrices des périodes

$$\Omega_1 = \Big(\int_{\gamma_1} \vec{\omega}, \ldots, \int_{\gamma_n} \vec{\omega}\Big), \qquad \Omega_2 = \Big(\int_{\gamma_{n+1}} \vec{\omega}, \ldots, \int_{\gamma_{2n}} \vec{\omega}\Big)$$

on ait $z = \Omega_1^{-1}\Omega_2$ dans $\mathcal{H}_n$. On appelle z un module de A. Clairement, A est isomorphe à $A_z = \mathbb{C}^n/(z.\mathbb{Z}^n + \mathbb{Z}^n)$ et A_z est munie d'une polarisation principale $\mathcal{C}_z$. On a une bijection

$$\Gamma \backslash \mathcal{H}_n \simeq \{[(A_z, \mathcal{C}_z)], z \in \mathcal{H}_n\}$$

L'hypothèse que la polarisation soit principale n'est pas cruciale : pour l'enlever, il faudrait modifier le groupe Γ et à chaque variété abélienne polarisée de dimension n on peut associer un module (modulo Γ) dans $\mathcal{H}_n$. Le Résultat principal donne donc :

Proposition 1. — *Soient A une variéte abélienne polarisée définie sur $\bar{\mathbb{Q}}$ et $z \in \mathcal{H}_n$ un module de A. Alors $z \in \mathcal{H}_n(\bar{\mathbb{Q}})$ si et seulement si A est de type* CM.

Le Théorème permet de déduire des résultats de transcendance sur les fonctions automorphes de Siegel définie sur $\bar{\mathbb{Q}}$. Soit K le corps de ces fonctions dont l'existence est une conséquence de celle de V, ou si l'on veut K est le corps des fonctions automorphes de Siegel qui sont des quotients de formes automorphes de Siegel à coefficients de Fourier algébriques. Alors le Corollaire donne :

Proposition 2. — *Si $z \in \mathcal{H}_n(\bar{\mathbb{Q}})$ alors toute fonction automorphe de Siegel dans K et définie à z prend une valeur algébrique à z si et seulement si z est un point CM.*

Un exemple de Freitag montre qu'en général les résultats donnés par le Corollaire, comme celui de la Proposition 2, sur la transcendance des valeurs des fonctions automorphes aux points algébriques non CM ne peuvent pas être améliorés sans faire des hypothèses supplémentaires sur le point non CM.

Exemple de Freitag [Fr] : prenons $n = 2$ dans l'exemple des espaces de Siegel. Alors Freitag a montré que le corps K des fonctions sur $\mathcal{H}_2$ automorphe par rapport à $\Gamma = Sp(4, \mathbb{Z})$ et définies sur $\bar{\mathbb{Q}}$ est engendré par trois fonctions f_1, f_2, f_3 telles que sur

$$\mathcal{D}' = \left\{ z = \begin{pmatrix} \tau_1 & 0 \\ 0 & \tau_2 \end{pmatrix} \in \mathcal{H}_2 \mid \tau_1, \tau_2 \in \mathcal{H} \right\},$$

on ait :

$$f_1(z) \equiv 0, f_2(z) \equiv g(\tau_1) + g(\tau_2), f_3(z) \equiv g(\tau_1)^{-1} + g(\tau_2)^{-1}$$

où $g(\tau) = \frac{G_6^2(\tau)}{G_4^3(\tau)}$ avec $G_4 = G_4(\tau)$ et $G_6 = G_6(\tau)$, $\tau \in \mathcal{H}$ les séries d'Eisenstein (normalisées) de poids 4 et de poids 6. On peut choisir $z \in \mathcal{D}'$ avec τ_1 un point CM et τ_2 un point algébrique mais non CM (et $g(\tau_1)$, $g(\tau_2) \neq 0$). Le point z sera alors un point algébrique non CM auquel deux fonctions dans K algébriquement indépendantes prennent une valeur algébrique. Seulement une troisième fonction dans K algébriquement indépendante de ces deux fonctions et définie en z prendra une valeur transcendante en z.

Les fonctions modulaires de Hilbert : la démonstration du Résultat principal peut donner un corollaire plus fort. En effet, on n'a pas toujours besoin de savoir que tous les coefficients du point non CM z sont des nombres algébriques, pour avoir un $f \in K$ tel que $f(z)$ soit transcendant. Dans le cas des fonctions modulaires de Hilbert on a par exemple :

PROPOSITION 3. — *Soit F un corps totalement réel de degré $g = [F : \mathbb{Q}]$ et Θ l'anneau des entiers de F. Le group $\Gamma = PSL(2, \Theta)$ agit sur $\mathcal{H}^g$ et on désigne par K le corps des fonctions Γ-automorphes définies sur $\bar{\mathbb{Q}}$. Alors, si $z = (z_1, \ldots, z_g) \in \mathcal{H}^g$ n'est pas un point CM et si $z_i \in \mathcal{H}(\bar{\mathbb{Q}})$ pour un seul $i = 1, \ldots, g$, il existe un $f \in K$ avec $f(z)$ transcendant.*

On va donner une démonstration (différente de celle qui figure dans [CSW]) de la Proposition 3 à la fin de cet article. La condition qu'il s'agit d'un seul $i = 1, \ldots, g$ dans la Proposition 3 peut donner lieu à des points dans l'espace de Siegel $\mathcal{H}_g$ dans l'image d'un plongement de $\mathcal{H}^g$ dans $\mathcal{H}_g$ dont tous les coefficients sont transcendants mais auxquels les fonctions modulaires de Siegel définies sur $\bar{\mathbb{Q}}$ ne prennent pas toutes des valeurs algébriques.

Idée de la démonstration du Résultat principal : les détails de la démonstration du Résultat principal sont donnés dans notre manuscrit [CSW]. Le fait que (ii) implique (i) est une conséquence de la normalisation choisie pour $\mathcal{D}$ (d'ailleurs si B est une variété abélienne de type CM, alors par la théorie de la multiplication complexe, B est isomorphe sur $\mathbb{C}$ à une variété abélienne définie sur $\bar{\mathbb{Q}}$). Comme dans le cas elliptique, c'est la démonstration que (i) implique (ii), c'est-à-dire que si A_z n'est pas de type CM alors $z \notin \mathcal{D}(\bar{\mathbb{Q}})$ qui utilise la théorie des nombres transcendants. La démonstration de Schneider utilisait une fonction auxiliaire, polynôme en certaines fonctions elliptiques de Weierstrass. Notre démonstration en dimension supérieure utilise le Théorème du sous-groupe analytique de G. Wüstholz [Wü] (donc une fonction auxiliaire donnée par des fonctions abéliennes est implicite) et des renseignements très explicites tirés des travaux de Shimura, notamment de [Shi].

Prenons le cas où A est simple, $A \in [A_z]$, $A_z \in \Sigma$, $z \in \mathcal{D} = \mathcal{D}(L, \Phi)$ avec $L = \mathrm{End}_o(A)$. On suppose que A_z n'est pas de type CM et donc que $\dim(\mathcal{D}) > 0$. On démontre ensuite que l'hypothèse que $z \in \mathcal{D}(\bar{\mathbb{Q}})$ entraîne une contradiction. Soit $\omega_1, \ldots, \omega_n$ une base du $\bar{\mathbb{Q}}$-espace vectoriel $H^0(A, \Omega_{\bar{\mathbb{Q}}})$. Alors il existe $\gamma_1, \ldots, \gamma_m \in H_1(A, \mathbb{Z})$, $m = \frac{2n}{[L:\mathbb{Q}]}$ avec, pour $\int_{\gamma_j} \vec{\omega} = {}^t\left(\int_{\gamma_j} \omega_1, \ldots, \int_{\gamma_j} \omega_n\right) \in \mathbb{C}^n$,

$$\Lambda_{\mathbb{Q}} = \Lambda \otimes_{\mathbb{Z}} \mathbb{Q} = \sum_{j=1}^{m} \Phi(L) \int_{\gamma_j} \vec{\omega}.$$

Comme A est définie sur $\bar{\mathbb{Q}}$ on a $\Phi(L) \subset M_n(\bar{\mathbb{Q}})$. Donc les éléments de L induisent des rélations de dépendance linéaire sur $\bar{\mathbb{Q}}$ entre les périodes $\int_\gamma \omega$, où $\omega \in H^0(A, \Omega_{\bar{\mathbb{Q}}})$, $\gamma \in H_1(A, \mathbb{Z})$.

D'autre part, dans la construction de Shimura le domaine $\mathcal{D}$ se décompose en un produit de $g = [F : \mathbb{Q}]$ (avec les notations déjà données) domaines irréductibles

$$\mathcal{D} = \mathcal{D}_1 \times \ldots \times \mathcal{D}_g ,$$

et donc le point z s'ecrit :

$$z = (z_\nu)_{\nu=1}^{g}.$$

Les z_ν sont des matrices à coefficients algébriques (par hypothèse) qui sont des quotients de matrices

$$z_\nu = (\Omega_1^\nu)^{-1}(\Omega_2^\nu).$$

Les matrices $\Omega_k^\nu, k = 1,2$ ont leurs coefficients des combinaisons $\bar{\mathbb{Q}}$-linéaires des $\int_j \omega_i, i = 1,\ldots,n, j = 1,\ldots,m$: c'est implicite dans la construction de Shimura. Si z_ν est une matrice à coefficients algébriques il y a donc une relation de dépendance $\bar{\mathbb{Q}}$-linéaire entre les périodes $\int_j \omega_i, i = 1,\ldots,n, j = 1,\ldots,m$ et on peut montrer que cette relation est non triviale.

C'est ici qu'intervient l'argument de transcendance qui dit que les relations non triviales de dépendance linéaire sur $\bar{\mathbb{Q}}$ entre les $\int_j \omega_i, i = 1,\ldots,n, j = 1,\ldots,m$ proviennent toutes de rélations non triviales de dépendance linéaire sur $\Phi(L)$ entre les $\int_j \vec{\omega}, j = 1,\ldots,m$. En effet, dans [CSW] nous démontrons un résultat que G. Wüstholz a annoncé sans démonstration dans [Wü,ICM] :

LEMME. — *Soit A une variété abélienne définie sur $\bar{\mathbb{Q}}$ et isogène au produit direct $A_1^{k_1} \times \ldots \times A_N^{k_N}$ de variétés abéliennes simples A_μ, dim$(A_\mu) = n_\mu$, définies sur $\bar{\mathbb{Q}}$ et deux-à-deux non isogènes. Alors le $\bar{\mathbb{Q}}$-espace vectoriel V_A engendré par toutes les périodes des différentielles dans $H^0(A, \Omega_{\bar{\mathbb{Q}}})$ est de dimension :*

$$\dim_{\bar{\mathbb{Q}}}(V_A) = \sum_{\mu=1}^{N} \frac{2n_\mu^2}{\dim_{\mathbb{Q}} \mathrm{End}_o(A_\mu)}.$$

Comme les $\int_{\gamma_j} \vec{\omega}, j = 1,\ldots,m$ sont indépendants sur $\Phi(L)$ l'hypothèse $z \in \mathcal{D}(\bar{\mathbb{Q}})$ entraine la contradiction voulue (en fait même l'hypothése $z_\nu \in \mathcal{D}_\nu(\bar{\mathbb{Q}})$ pour un seul ν aurait été suffisante : voir la Proposition 3 pour une application de cette remarque). Pourtant, lorsque L est strictement contenu dans $\mathrm{End}_o(A)$ on utilise l'hypothèse plus forte sur z.

Si L est strictement contenu dans $\mathrm{End}_o(A)$ et A est simple alors A est isomorphe à $A_{z'}$ où $A_{z'}$ appartient à une famille $\Sigma' \subset \Sigma$ d'éléments d'un $S(L', \Phi', \rho')$ où $L' = \mathrm{End}_o(A_{z'})$. Soit $D' = D(L', \Phi')$. Si $\dim(\mathcal{D}') = 0$ alors A est de type CM. Si $\dim(\mathcal{D}') > 0$ alors la discussion précédante montre que le point z' ne peut pas être algébrique. Pour déduire la transcendance de z de celle de z', il faut utiliser des propriétés de rationnalité de certains plongements modulaires, définis sur $\bar{\mathbb{Q}}$, de $\mathcal{D}'$ dans $\mathcal{D}$. Un argument facile de plongement modulaire permet aussi de traiter le cas où A n'est pas simple.

Afin d'illustrer l'idée de la démonstration du Résultat principal démontrons la Proposition 3.

Démonstration de la Proposition 3 : le quotient $PSL(2,\Theta)\backslash\mathcal{H}^g$ est l'espace des modules d'une famille Σ de variétés abéliennes polarisées de dimension g dans $S(F,\Phi,\rho)$ avec $\Phi : F \hookrightarrow M_g(\mathbb{C})$ donnée par :

$$\Phi : \theta \mapsto \mathrm{diag}(\sigma_1(\theta),\ldots,\sigma_g(\theta)),$$

où $\sigma_1,\ldots,\sigma_g$ sont les plongements galoisiens de F dans $\mathbb{R}$. On est donc dans le cas du TYPE (I) avec $m=2$. Soit A une variété abélienne définie sur $\bar{\mathbb{Q}}$ et isomorphe à $A_z \in \Sigma$ où $z \in \mathcal{D}$. L'espace $H^0(A,\Omega_{\bar{\mathbb{Q}}})$ se décompose en g sous-espaces propres pour l'action induite de F. Tous ces sous-espaces sont de dimension 1 sur F. Soient $\omega_1,\ldots,\omega_g$ des générateurs correspondants et soient γ_1,γ_2 des éléments de $H_1(A,\mathbb{Z})$ qui engendrent $H_1(A,\mathbb{Q})$ sur F. On peut choisir cette base de telle sorte que :

$$z_i = \frac{\int_{\gamma_1}\omega_i}{\int_{\gamma_2}\omega_i}$$

soit dans $\mathcal{H}$ et que $z=(z_i)_{i=1,\ldots,g}$ soit le point qui correspond à A_z. Si $z_i \in \bar{\mathbb{Q}}$ alors on sait que z_i n'est pas dans $\sigma_i(F)$ et donc par le Lemme il doit y avoir un élément M de $\mathrm{End}_o(A)$ qui n'est pas dans $\Phi(F)$. Mais les éléments de $\Phi(F)$ commutent aux éléments de $\mathrm{End}_o(A)$. Donc $L_\Phi = \Phi(F)(M)$ est un sous-corps de $\mathrm{End}_o(A)$ totalement imaginaire de degré $2\dim(A)$ et une extension CM de F, d'où il vient que A est à multiplication complexe.

Manuscrit reçu le 7 février 1994

REFERENCES

[CSW] P. B. COHEN, H. SHIGA, J. WOLFART. — *Criteria for complex multiplication and transcendence properties of automorphic functions*, preprint J. W. Goethe-Universität, Frankfurt am Main (1993).

[Mi] J. S. MILNE. — *Canonical models of (mixed) Shimura varieties and automorphic vector bundles*, "Automorphic forms, Shimura varieties and L-functions", Vol. 1, ed. by L. Clozel, J. S. Milne, Ann Arbor 1988, Academic Press (1990), 283–414.

[Sch] Th. SCHNEIDER. — *Arithmetische Untersuchungen elliptischer Integrale*, Math. Ann. **113** (1937), 1–13.

[Shi] G. SHIMURA. — *On analytic families of polarized abelian varieties and automorphic functions*, Ann. Math. **78** (1963), 149–192.

[Wü] G. WÜSTHOLZ. — *Algebraische Punkte auf analytischen Untergruppen algebraische Gruppen*, Ann. of Math. **129** (1989), 501–517.

[Wü,ICM] G. WÜSTHOLZ. — *Algebraic Groups, Hodge Theory and Transcendence*, Proc. ICM Berkeley 1986, Vol. 1, AMS, (1987), 476–483.

Paula Beazley COHEN
UA 747 CNRS
Collège de France
3 rue d'Ulm
F–75005 Paris
France

Number Theory
Paris 1992–93

Supersingular primes common to two elliptic curves

E. Fouvry and M. Ram Murty

1. — Introduction

Let E be an elliptic curve over $\mathbb{Q}$. Denote by $\pi_0(x, E)$ the cardinality of the set of the supersingular primes of E less than x. It has been conjectured by Lang and Trotter [L–T] that, when E has no complex multiplication, the following holds, when $x \to \infty$

$$\pi_0(x, E) \sim C_E \frac{\sqrt{x}}{\log x} \tag{1.1}$$

where C_E is a positive constant depending only on E, precisely defined in terms of $\mathrm{Gal}(\mathbb{Q}(E_{\mathrm{tors}}), \mathbb{Q})$.

The first significant step towards (1.1) is due to Elkies ([El1]), who proved that each elliptic curve over $\mathbb{Q}$, has infinitely many supersingular primes. This result was improved by the authors, who proved

THEOREM A ([F–M] Théorème 1). — *Let E be an elliptic curve over $\mathbb{Q}$. Then, for every positive δ, there exists $x_0(\delta, E)$ such that, for $x > x_0(\delta, E)$, the following holds :*

$$\pi_0(x, E) > \frac{\log_3 x}{(\log_4 x)^{1+\delta}}.$$

Here $\log_k$ is the k-fold iterated logarithm function. Note that the best upperbound for $\pi_0(x, E)$ is due to Elkies and Murty ([El2], [El3]) and has the shape $\pi_0(x, E) = O_E(x^{\frac{3}{4}})$, for any non CM–curve E, with the convention that CM means that the curve has complex multiplication.

In the direction of (1.1), we must also quote another result which asserts, very vaguely speaking, that the Lang–Trotter Conjecture is true *on average*. More precisely, let a, b be two integers with $4a^3 + 27b^2 \neq 0$ and let $E_{a,b}$ be the elliptic curve defined by the equation

$$y^2 = x^3 + ax + b,$$

then we have :

THEOREM B ([F–M] Théorème 6). — *For every positive* ε, *we have, for* $x \to \infty$, *the asymptotic relation*

$$\sum_{|a|\le A}\sum_{|b|\le B} \pi_0(x, E_{a,b}) \sim \left(\frac{\pi}{3}\cdot\frac{\sqrt{x}}{\log x}\right).(4AB)$$

uniformly for $A \ge x^{\frac{1}{2}+\varepsilon}$, $B \ge x^{\frac{1}{2}+\varepsilon}$, $AB \ge x^{\frac{3}{2}+\varepsilon}$.

A familiar way to write (1.1) is to say that the probability for a prime p to be supersingular for E is

$$\frac{C_E}{2}\cdot\frac{1}{\sqrt{p}}, \tag{1.2}$$

and we are led to the problem of the primes supersingular for two given elliptic curves E and E'. We say that two elliptic curves over $\mathbb{Q}$ are *in general position* when none of them has complex multiplication and when they are not isogenous over $\bar{\mathbb{Q}}$. Let us recall that, when E has complex multiplication, we have $\pi_0(x,E) \sim \frac{x}{2\log x}$, and if E and E' are isogenous over $\bar{\mathbb{Q}}$, they have the same supersingular primes apart from the prime divisors of the conductors of these curves; in other words, we have

$$\pi_0(x,E,E') = \pi_0(x,E) - O_E(1) = \pi_0(x,E') - O_{E'}(1), \tag{1.3}$$

where $\pi_0(x,E,E')$ is the cardinality of the set of primes, less than x, supersingular for E and E'.

Then, following (1.2), it is natural to think that if E and E' are in general position, the probability for a prime p to be supersingular for E and E' should be

$$\frac{C_{E,E'}}{p},$$

where $C_{E,E'}$ is a positive constant depending only on E and E'.

Such a probabilistic assumption of independence appears in [L–T] page 37, and leads to the conjecture

$$\pi_0(x,E,E') \sim C_{E,E'}\log_2 x \quad (x\to\infty). \tag{1.4}$$

when E and E' are supposed to be in general position. This conjecture seems extremely hard to prove – for the moment, nobody knows how to prove that $\pi_0(x,E,E') \to \infty$ when $x \to \infty$ – since the set of primes in question is very, very sparse (heuristically as sparse as the following set connected with the Mersenne conjecture : $\{p \le x; 2^p - 1 \text{ is a prime}\}$).

The bulk of this paper is to prove that (1.4) is true *on average*, in the same philosophy as Theorem B proves (1.1) on average. We will prove

THEOREM 1. — *For every positive ε, we have for $x \to \infty$, the asymptotic relation*

$$\sum_{|a|\le A}\sum_{|a'|\le A'}\sum_{|b|\le B}\sum_{|b'|\le B'} \pi_0(x, E_{a,b}, E_{a',b'}) \sim \frac{35}{96}\cdot \log_2 x \cdot (16AA'BB') \tag{1.5}$$

holds uniformly for $A,\ A' \ge x^{\frac{1}{2}+\varepsilon}$, $B,\ B' \ge x^{\frac{1}{2}+\varepsilon}$, $AB,\ A'B' \ge x^{\frac{3}{2}+\varepsilon}$.

Let $S(A, A', B, B')$ be the sum studied in (1.5). To be allowed to say that Theorem 1 proves (1.4) on average, we must check that the contribution to $S(A, A', B, B')$ of the pairs of curves $(E_{a,b}, E_{a',b'})$ which are not in general position is negligible. So we denote respectively by $S^{\mathrm{CM}}(A, A', B, B')$ and $S^{\mathrm{iso}}(A, A', B, B')$ the contribution of those pairs with $E_{a,b}$ having complex multiplication and with $E_{a,b}$ and $E_{a',b'}$ isogenous over $\bar{\mathbb{Q}}$.

The first contribution satisfies :

$$S^{\mathrm{CM}}(A, A', B, B') \le |\{(a,b); |a| \le A, |b| \le B, E_{a,b} \text{ is } CM\}|\cdot \\ \cdot \sum_{|a'|\le A'}\sum_{|b'|\le B'} \pi_0(x, E_{a',b'}).$$

There are thirteen families of elliptic curves with complex multiplication, they can be written as

$$E_{0,t};\ E_{t,0};\ E_{\alpha_i t^2, \beta_i t^3}\ \ (t \in \mathbb{Z}^*,\ 1 \le i \le 11)$$

where the (α_i, β_i) are eleven pairs of integers. Theorem B implies under the conditions of Theorem 1

$$S^{\mathrm{CM}}(A, A', B, B') = O\Big(\max(A, B).A'B'\frac{\sqrt{x}}{\log x}\Big) = O(AA'BB'),$$

which is clearly negligible, compared to the expected main term.

The term $S^{\mathrm{iso}}(A, A', B, B')$ requires a more delicate treatment. We shall prove later the following :

LEMMA 1. — *Let E be an elliptic curve over $\mathbb{Q}$. Then, for A and B tending to infinity, we have*

$$|\{(a,b); 0 < |a| \le A,\ 0 < |b| \le B, E_{a,b} \text{ isogenous over } \bar{\mathbb{Q}} \text{ to } E\}| \\ = O(\min(A^{\frac{1}{2}}, B^{\frac{1}{3}}) \log^9(2AB)),$$

where the "O" is independant of E.

This lemma and Theorem B directly imply

$$\begin{aligned}
&S^{\text{iso}}(A, A', B, B') \\
&\quad \leq \max_{E/\mathbb{Q}} |\{(a,b); |a| \leq A, |b| \leq B, E_{a,b} \text{ isogenous over } \bar{\mathbb{Q}} \text{ to } E\}| \cdot \\
&\quad\quad \cdot \sum_{|a'| \leq A'} \sum_{|b'| \leq B'} \pi_0(x, E_{a',b'}) \\
&\quad = O(\min(A^{\frac{1}{2}}, B^{\frac{1}{3}}).A'B'\sqrt{x}\log^{10}(2AB)) = O(AA'BB')
\end{aligned}$$

which is also negligible.

Proof of Lemma 1 : since $E_{a,b}$ and E are isogenous over $\bar{\mathbb{Q}}$, the isogeny is defined on a number field K of relative degree at most 12 over $\mathbb{Q}$ ([M–W] Lemma 6.1 for instance). Now, we use a strong theorem of Masser and Wustholz ([M–W] Theorem), asserting, that if there is an isogeny between two elliptic curves E and E' over a number field k of degree at most d over $\mathbb{Q}$, then there exists between them a "simple" isogeny, i.e. with a degree less than $c(w(E'))^4$, where c depends only on d and $w(E')$ is the maximum of 1 and of the logarithmic Weil height of the curve E'. In our situation, we deduce that, if E and $E_{a,b}$ are isogenous, there is an isogeny of degree $O(\log^4(2AB))$. A simple trick of algebra ([M–W] Lemma 6.2) allows us to suppose that this isogeny is cyclic.

Let j_E be the invariant of E. The modular polynomial of order n, $\Phi_n(X, j_E)$ ([La] pages 55–59) detects the existence of a cyclic isogeny of degree n between $E_{a,b}$ and E. More precisely, such an isogeny exists if and only if we have

$$\Phi_n\left(\frac{6912a^3}{4a^3 + 27b^2}, j_E\right) = 0.$$

The degree in X of $\Phi_n(X, j_E)$ is $n \prod_{p|n}(1 + \frac{1}{p}) = O(n \log 2n)$, which implies that the equation $\Phi_n(X, j_E) = 0$ has at most $O(n \log 2n)$ solutions in $\mathbb{Q}$. Let $\frac{u}{v}$ be such a root, which we can suppose different from 0 and 1728 since $ab \neq 0$. The equation in a and b

$$\frac{6912a^3}{4a^3 + 27b^2} = \frac{u}{v}$$

has $O(\min(A^{\frac{1}{2}}, B^{\frac{1}{3}}))$ roots in the rectangle $[-A, A] \times [-B, B]$. Then gathering these solutions when $\frac{u}{v}$ varies and when n varies, we complete the proof of Lemma 1.

The above discussion gives another formulation of Theorem 1 :

THEOREM 2. — *Let* † *indicates that* $E_{a,b}$ *and* $E_{a',b'}$ *are in general position. Then under the conditions of Theorem 1, we have*

$$\sum_{|a|\leq A}\sum_{|a'|\leq A'}\sum_{|b|\leq B}\sum_{|b'|\leq B'}{}^{\dagger}\pi_0(x,E_{a,b},E_{a',b'}) \sim \left(\tfrac{35}{96}.\log_2 x\right)\sum_{|a|\leq A}\sum_{|a'|\leq A'}\sum_{|b|\leq B}\sum_{|b'|\leq B'}{}^{\dagger}1. \tag{1.6}$$

A more difficult question seems to give an asymptotic formula similar to (1.6), but where $E_{a,b}$ and $E_{a',b'}$ takes only one value in each isogeny class.

2. — From supersingular primes to class numbers

The starting point of our proof is

LEMMA 2. — *Let* $p \geq 5$ *be a prime. The number of isomorphism classes of elliptic curves over* $\mathbb{F}_p$ *with* $p+1$ *points is equal to* $H(-4p)$ *(number of of isomorphism classes of positive quadratic forms, not necessary primitive, with discriminant* $-4p$*).*

Such a statement appears at several places in the literature, for instance in [Bi], page 58. Since the quadratic forms counted by $H(-4p)$ are not necessarily primitive, we have the equality

$$H(-4p) = h(-p) + h(-4p) \tag{2.1}$$

with the convention that the h symbol is equal to zero when it is not defined. We denote by $y^2 = x^3 + a_i x + b_i$ $(1 \leq i \leq H(-4p))$ the equations which define representatives of the above isomorphism classes of elliptic curves over $\mathbb{F}_p$. We can suppose that these equations are minimal relative to p, for instance by imposing the conditions $0 \leq a_i < p$, $0 \leq b_i < p$. If the equation defining $E_{a,b}$ is minimal for p we deduce that p is supersingular for $E_{a,b}$ if and only if there exists $t \in \mathbb{F}_p^*$ and $1 \leq i \leq H(-4p)$, such that

$$a \equiv a_i t^4,\ b \equiv b_i t^6 (\text{mod } p). \tag{2.2}$$

For $a_i b_i \not\equiv 0(\text{mod } p)$, the image of the application $t \in \mathbb{F}_p^* \to (a_i t^4, b_i t^6)$ has cardinality $\frac{p-1}{2}$ for $p \geq 5$. We gather the above observations into

LEMMA 3. — *Let* $p \geq 5$*, then there exists a set* $\mathcal{E}_p$ *of residue classes mod* $p \times$ *mod* p *with the following properties*

i) $|\mathcal{E}_p| = \frac{H(-4p)p}{2} + O(p)$

ii) *if* $(a,b) \in \mathbb{Z}^2 - \{(0,0)\}$ *and if* k *is the largest integer such that* $p^{4k}|a$ *and* $p^{6k}|b$, *then* p *is a supersingular prime for* $E_{a,b}$, *if and only if* $(ap^{-4k}, bp^{-6k}) (\mathrm{mod}\ p)$ *belongs to* $\mathcal{E}_p$.

It is now easy to transform $S(A,B,A',B')$ into

$$S(A,B,A',B') = \sum_{p\le x}\Big(\sum_{(u,v)\in\mathcal{E}_p} \sum_{\substack{a\equiv u (\mathrm{mod}\ p)\\ |a|\le A}} \sum_{\substack{b\equiv v(\mathrm{mod}\ p)\\ |b|\le B}} 1 + O\Big(\Big(\frac{A}{p^4}+1\Big)\Big(\frac{B}{p^6}+1\Big)\Big)\Big)\times$$

$$\Big(\sum_{(u,v)\in\mathcal{E}_p} \sum_{\substack{a'\equiv u (\mathrm{mod}\ p)\\ |a'|\le A'}} \sum_{\substack{b'\equiv v(\mathrm{mod}\ p)\\ |b'|\le B'}} 1 + O\Big(\Big(\frac{A'}{p^4}+1\Big)\Big(\frac{B'}{p^6}+1\Big)\Big)\Big) + O(ABA'B'),$$

where the errors terms come from the non-minimal equations and from the primes 2 and 3. Using the trivial equality

$$\sum_{\substack{a\equiv u(\mathrm{mod} p)\\ |a|\le A}} 1 = \frac{2A}{p} + O(1),$$

we get the equality

$$\begin{aligned} S(A,B,A',B') &= \sum_{p\le x} |\mathcal{E}_p|^2 \frac{2A}{p}.\frac{2B}{p}.\frac{2A'}{p}.\frac{2B'}{p} + \\ &+O\Big(x^4\log^2 x + (A+B+A'+B')x^3\log^2 x + \\ &\qquad +(AB+\cdots+A'B')x^2\log^2 x \\ &\qquad +(ABA'+\cdots+BA'B')x\log^2 x + ABA'B'\Big), \end{aligned} \tag{2.3}$$

by using Lemma 3 ii) and by using the classical upperbound $h(-d) = O(\sqrt{d}\log d)$ (note that some log–factors could be spared by using average bounds of that quantity by methods of the next Proposition). We postpone to paragraph IV an improvement of (2.3) by appealing to the theory of exponential sums. The equation i) of Lemma 3 splits (2.3) into main term and error term :

$$S = MT + ET \tag{2.4}$$

with

$$MT(A,B,A',B') = 4AA'BB' \sum_{p\le x} \frac{H^2(-4p)}{p^2}$$

and

$$ET(A,B,A',B') = O(ABA'B') \tag{2.5}$$

under the conditions

$$A, A', B, B' \ge x\log x. \tag{2.6}$$

3. — Class numbers on average

To evaluate the main term in (2.4), we will prove the

PROPOSITION. — *For $x \longrightarrow \infty$ we have*

$$\sum_{p\le x} \frac{H^2(-4p)}{p^2} \sim \frac{35}{24} \log_2 x.$$

Such a formula, one more time illustrates the well known fact that on average $h(-d)$ behaves like $\sqrt{d}$ and we will treat it by the values of L–functions at the point 1. Formula (2.1) transforms our sum into

$$\sum_{p\le x} \frac{h^2(-p)}{p^2} + 2\sum_{p\le x} \frac{h(-p)h(-4p)}{p^2} + \sum_{p\le x} \frac{h^2(-4p)}{p^2} = T_{1,1}(x) + 2T_{1,4}(x) + T_{4,4}(x) \tag{3.1}$$

say. We will concentrate on the typical sum $T_{1,1}(x)$, for which we will prove

$$T_{1,1}(x) \sim \frac{5}{24} \log_2 x\,; \tag{3.2}$$

but the proof of this formula can be obtained, after integrating by parts, from the following

$$T^*_{1,1}(x) := \sum_{p\le x} \frac{h^2(-p)}{p} \sim \frac{5}{24} \frac{x}{\log x}. \tag{3.3}$$

The classical Dirichlet class number formula, coupled with the Polya–Vinogradov formula gives an expression of $h(-p)$ as a finite sum

$$h(-p) = \frac{\sqrt{p}}{\pi} \sum_{n\le U} \frac{\chi_{-p}(n)}{n} + O\Big(\frac{p \log p}{U}\Big),$$

for any $U > 1$ and χ_{-p}, the Kronecker symbol associated to $-p$ for $p \equiv 3 (\mathrm{mod}\ 4)$. Squaring this equality, we obtain for $\sqrt{x} \le U \le x$

$$T^*_{1,1}(x) = \frac{1}{\pi^2} \sum_{p\le x} \sum_{n_1\le U} \sum_{n_2\le U} \frac{\chi_{-p}(n_1 n_2)}{n_1 n_2} + O\Big(\frac{x^{\frac{3}{2}} \log x}{U}\Big). \tag{3.4}$$

To treat the triple sum in (3.4), we put $n_1 n_2 = ml^2$ with m squarefree and $d_U(n)$ the modified divisor function

$$d_U(n) := |\{(n_1, n_2); n_1 \le U,\ n_2 \le U,\ n_1 n_2 = n\}|,$$

the main term in (3.4) is now

$$\frac{1}{\pi^2}\sum_m \frac{\mu^2(m)}{m}\sum_{p\le x}\sum_l \frac{\chi_{-p}(m)\chi_{-p}^2(l)}{l^2}d_U(ml^2).$$

The most important contribution comes from $m = 1$, it has the shape

$$\frac{1}{\pi^2}\sum_{l\le U}\frac{d_U(l^2)}{l^2}\Big(\sum_{\substack{p\le x,\, p\nmid l\\ p\equiv 3 (\mathrm{mod}\ 4)}} 1\Big) = \Big(\frac{1}{2\pi^2}+o(1)\Big).\frac{x}{\log x}.\Big(\sum_{l=1}^{\infty}\frac{d(l^2)}{l^2}+O(U^{-\frac14})\Big).$$

But we have the equality

$$\sum_{l=1}^{\infty}\frac{d(l^2)}{l^2} = \prod_p\sum_{k=0}^{\infty}\frac{2k+1}{p^{2k}} = \prod_p\frac{1+1/p^2}{(1-1/p^2)^2} = \frac{\zeta^3(2)}{\zeta(4)} = \frac{5\pi^2}{12}.$$

Thus (3.4) may be written as

$$\begin{aligned}&\tfrac{5}{24}\tfrac{x}{\log x}(1+o(1))+\\ (3.5)\qquad & O\Big(\sum_l\frac{1}{l^2}\Big|\sum_{p\le x}\sum_{m\ne 1}\frac{\mu^2(m)\chi_{-p}(m)\chi_{-p}^2(l)}{m}d_U(ml^2)\Big|\Big).\end{aligned}$$

Since $m \ne 1$, we hope cancellation from summation over p and m of the terms $\chi_{-p}(m)$. We call $W(x,l)$ the double sum over these variables in the error term of (3.5). After cutting the ranges of summation, we have :

$$W(x,l) = O(\log^2 x\{\sup_{\substack{M>1\\ 1\le P\le x}} |W(l,M,P)| + x^{\varepsilon}\})$$

with

$$W(l,M,P) = \sum_{P<p\le 2P}\sum_{M<m\le 2M}\tfrac{\mu^2(m)\chi_{-p}(m)}{m}d_U(ml^2)$$

and the error term x^{ε} coming from the p dividing l. In $W(l,M,P)$, m is never divisible by 4, so we put $m' = m$ or $m' = m/2$ if m is odd or even; if we fix the congruence of p and m' modulo 8, we see that the Kronecker symbol is expressed in terms of the Jacobi symbol

$$\chi_{-p}(m) = \Big(\frac{m'}{p}\Big)\cdot\varepsilon_1 = \Big(\frac{p}{m'}\Big).\varepsilon_2$$

where ε_1 and ε_2 are of absolute value 1, are independent of m and p. We now appeal to a general upperbound for a double sum over Jacobi symbols :

LEMMA 4. — *Let (a_m) and (b_n) be complex numbers. Then we have*

$$\sum_{M<m\le 2M}\ \sum_{N<n\le 2N} a_m b_n\Big(\frac{m}{n}\Big) \ll ||a||_2\, ||b||_2(M^{\frac12} + M^{\frac14}N^{\frac{23}{32}}),$$

where the sum is over odd square free integers m and n.

Remark that the reciprocity law for Jacobi symbols implies that, in the above sum the variables m and n play a similar rôle and the following upperbound will be sufficient for our purpose :

$$\sum_{M<m\leq 2M}\ \sum_{N<n\leq 2N} a_m b_n \left(\frac{m}{n}\right) \ll ||a||_2\ ||b||_2 (MN)^{\frac{1}{2}} (\min(M,N))^{-\delta}, \tag{3.6}$$

for some absolute positive δ.

Our proof is quite standard and mixes Cauchy–Schwarz inequality and Burgess bound for character sum ([Bu] Theorem 2 with $\varepsilon = 1/32$) . This proof is a slight generalisation of [H–B], Lemma 4. We have

$$\begin{aligned}\Big|\sum_m a_m \sum_n b_n \left(\frac{m}{n}\right)\Big| &\leq ||a||_2 \Big\{\sum_m \Big|\sum_n b_n \left(\frac{m}{n}\right)\Big|^2\Big\}^{\frac{1}{2}} = \\ &= ||a||_2 \Big\{\sum_{n_1}\sum_{n_2} b_{n_1}\bar{b}_{n_2} \sum_m \left(\frac{m}{n_1 n_2}\right)\Big\}^{\frac{1}{2}} \\ &\ll ||a||_2 (||b||_2^2 M + \sum_{n_1}\sum_{n_2} |b_{n_1} b_{n_2}| M^{\frac{1}{2}} N^{\frac{7}{16}})^{\frac{1}{2}} \\ &\ll ||a||_2 ||b||_2 M^{\frac{1}{2}} + ||a||_2 ||b||_2 M^{\frac{1}{4}} N^{\frac{23}{32}}.\end{aligned}$$

Note that the classical Polya–Vinogradov would be sufficient to prove (3.6), if the situation $M^{1-\delta_1} < N < M^{1+\delta_1}$ (with δ_1 a very little positive constant) was excluded, Burgess bound is used to cover that case. This bound, which depends on a result of Weil, could be replaced by the much more accessible (Corollary of [F–I]), if that last result was stated for general moduli, not necessarily prime moduli.

We now use (3.6) to bound $W(l, M, P)$:

$$\begin{aligned} W(l,M,P) &\ll M^{-1}\Big(\sum_m d_U^2(ml^2)\Big)^{\frac{1}{2}} M^{\frac{1}{2}} P(\min(M,P))^{-\delta} \\ &\ll d(l^2) P \log^{\frac{3}{2}} x\, (\min(M,P))^{-\delta} \ll l^{\frac{1}{2}} x \log^{-10} x \end{aligned} \tag{3.7}$$

as soon as we have $M \geq (\log x)^{\frac{100}{\delta}}$. In the case where $M \leq (\log x)^{\frac{100}{\delta}}$, (3.7) is a direct application of the famous Siegel–Walfisz Theorem on the distribution of primes in arithmetic progressions. Inserting this bound in (3.5), summing over l, we obtain (3.3) and by the way (3.2).

The other terms $T_{1,4}(x)$ and $T_{4,4}(x)$ are evaluated along the same techniques. We write

$$T_{1,4}^*(x) = \frac{2}{\pi^2} \sum_{p\leq x} \sum_{n_1 \leq U} \sum_{n_2 \leq U} \frac{\chi_{-p}(n_1)\chi_{-4p}(n_2)}{n_1 n_2} + O\Big(\frac{x^{\frac{3}{2}} \log x}{U}\Big).$$

In the above sum, we may suppose that n_2 is odd, this implies that $\chi_{-4p}(n_2) = \chi_{-p}(n_2)$. If we write

$$d_U^*(n) = |\{(n_1, n_2); n_1 \leq U,\ n_2 \leq U,\ n_1 n_2 = n, n_2 \text{ odd}\}|,$$

we obtain the equality

$$\begin{aligned} T_{1,4}^*(x) &\sim \frac{2}{\pi^2} \sum_{l \leq U} \frac{d_U^*(l^2)}{l^2} \Big(\sum_{\substack{p \leq x, p \nmid l \\ p \equiv 3 (\text{mod } 4)}} 1 \Big) \\ &\sim \frac{2}{\pi^2} \cdot \frac{4}{3} \cdot \prod_{p \geq 3} \frac{1 + 1/p^2}{(1 - 1/p^2)^2} \cdot \frac{x}{2 \log x} \sim \frac{1}{4} \cdot \frac{x}{\log x} \end{aligned} \tag{3.8}$$

Similarly, we have the equality

$$T_{4,4}^*(x) = \frac{4}{\pi^2} \sum_{p \leq x} \sum_{n_1 \leq U} \sum_{n_2 \leq U} \frac{\chi_{-4p}(n_1) \chi_{-4p}(n_2)}{n_1 n_2} + O\Big(\frac{x^{\frac{3}{2}} \log^2 x}{U} \Big)$$

and we may suppose that both n_1 and n_2 are odd. We define

$$d_U^{**}(n) = |\{(n_1, n_2); n_1 \leq U,\ n_2 \leq U,\ n_1 n_2 = n, n_1 \text{ and } n_2 \text{ are odd}\}|,$$

and we obtain

$$\begin{aligned} T_{4,4}^*(x) &\sim \frac{4}{\pi^2} \sum_{l \leq U} \frac{d_U^{**}(l^2)}{l^2} \Big(\sum_{p \leq x, p \nmid l} 1 \Big) \\ &\sim \frac{4}{\pi^2} \prod_{p \geq 3} \frac{1 + 1/p^2}{(1 - 1/p^2)^2} \cdot \frac{x}{\log x} \sim \frac{3}{4} \cdot \frac{x}{\log x}. \end{aligned} \tag{3.9}$$

Gathering (3.1), (3.2), (3.3), (3.8) and (3.9), we get

$$T(x) \sim \Big(\frac{5}{24} + \frac{1}{2} + \frac{3}{4} \Big) \frac{x}{\log x} = \frac{35}{24} \frac{x}{\log x},$$

which ends the proof of Theorem 1.

4. — Use of exponential sums

The aim of this paragraph is to weaken the conditions (2.6) over A, A', B, B' down to the conditions appearing in Theorem 1. We improve the use of the formula

$$\sum_{(u,v) \in \mathcal{E}_p} \sum_{\substack{a \equiv u (\text{mod } p) \\ |a| \leq A}} \sum_{\substack{b \equiv v (\text{mod } p) \\ |b| \leq B}} 1 + O\Big(\Big(\frac{A}{p^4} + 1 \Big) \Big(\frac{B}{p^6} + 1 \Big) \Big)$$

giving the number of $E_{a,b}$ with $|a| \leq A$, $|b| \leq B$ for which a given prime $p \geq 5$ is supersingular. The idea, already appearing in [F–M], is to say that $\mathcal{E}_p$ is not too chaotic, (i. e. if $E_{a,b}$ belongs to $\mathcal{E}_p$, then E_{au^4, bu^6} is also in $\mathcal{E}_p$); it is now possible to appeal to the theory of exponential sums to detect the conditions $|au^4| \leq A$, $|bu^6| \leq B$. We extract from [F–M], paragraph 7 the following lemma :

LEMMA 5. — *Let $0 < \alpha, \beta < p$, such that the given prime $p \geq 5$ is supersingular for $E_{\alpha,\beta}$. then the number of $E_{a,b}$, ($|a| \leq A$, $|b| \leq B$) isomorphic to $E_{\alpha,\beta}$ with $p^4 \nmid a$ and $p^6 \nmid b$ is*

$$\frac{4AB}{p^2}.\frac{p-1}{2} + O\Big((\sqrt{p}\log^2 p)\Big(1 + \frac{A}{p} + \frac{B}{p}\Big)\Big).$$

Let us denote by $\mathcal{F}_p$ a subset of representative classes of isomorphism of the form $E_{\alpha,\beta}$ with $p \nmid \alpha\beta$. Note that the cardinality of that subset is $H(-4p) - O(1)$. By inserting the result of Lemma 5 in the formula (2.3), we now obtain

$$S(A,B,A',B') =$$
$$\sum_{p\leq x}\Big(|\mathcal{F}_p|\Big(\frac{4AB}{p^2}.\frac{p-1}{2} + O\Big(\sqrt{p}\log^2 p\Big(1+\frac{A}{p}+\frac{B}{p}\Big)\Big)\Big) + O\Big(\frac{AB}{p}+A+B\Big)\Big)\times$$
$$\Big(|\mathcal{F}_p|\Big(\frac{4A'B'}{p^2}.\frac{p-1}{2} + O\Big(\sqrt{p}\log^2 p\Big(1+\frac{A'}{p}+\frac{B'}{p}\Big)\Big)\Big) + O\Big(\frac{A'B'}{p}+A'+B'\Big)\Big),$$

where the error terms come from the curves $E_{a,b}$ and $E_{a',b'}$ with $p|ab$ and $p|a'b'$. Using now the relation $|\mathcal{F}_p| = O(\sqrt{p}\log p)$, we get

$$\begin{aligned} S(A,B,A',B') &= \sum_{p\leq x}\Big(|\mathcal{F}_p|\frac{4AB}{p^2}\cdot\frac{p-1}{2} + O\Big(p\log^3 p\Big(\frac{A}{p}+1\Big)\Big(\frac{B}{p}+1\Big)\Big)\Big)\times \\ &\quad\Big(|\mathcal{F}_p|\frac{4A'B'}{p^2}\cdot\frac{p-1}{2} + O\Big(p\log^3 p\Big(\frac{A'}{p}+1\Big)\Big(\frac{B'}{p}+1\Big)\Big)\Big) \\ &= 16ABA'B'\Big(\sum_{p\leq x}\frac{|\mathcal{F}_p|^2}{p^2}\Big) + O(ABA'B'), \end{aligned}$$

under the conditions $A, A' \geq x^{\frac{1}{2}+\varepsilon}$, $B, B' \geq x^{\frac{1}{2}+\varepsilon}$, $AB, A'B' \geq x^{\frac{3}{2}+\varepsilon}$. Since we have $|\mathcal{F}_p|^2 = H^2(-4p) + O(\sqrt{p}\log p)$, the treatment of the main term is straightforward by the Proposition.

Manuscrit reçu le 29 septembre 1993

REFERENCES

[Bi] B. BIRCH. — How the number of points of an elliptic curve over a fixed prime field varies, *J. London Math. Soc.* **43** (1968), 57–60.

[Bu] D.A. BURGESS. — On characters sums and L-series. II, *Proc. London Math. Soc.(3)* **13**, (1963), 524–536.

[El1] N. ELKIES. — The existence of infinitely many supersingular primes for every elliptic curve over $\mathbb{Q}$, *Inv. Math.* **89** (1987), 561–567.

[El2] N. ELKIES. — Supersingular primes of a given elliptic curve over a number field, *Ph. D. Thesis*, Harvard University, (1987).

[El3] N. ELKIES. — Distribution of Supersingular Primes, *Astérisque-Journées Arithmétiques de Luminy 1989* **198–199–200** (1991), 127–132.

[F–M] E. FOUVRY and R. MURTY. — On the distribution of supersingular primes, (preprint).

[F-I] J. FRIEDLANDER and H. IWANIEC. — A mean-value theorem for character sums, *Michigan Math. J.* **39** (1992), 153–159.

[H–B] D.R. HEATH-BROWN. — The size of Selmer groups for the congruent number problem, *Inv. Math.* **111** (1993), 171–195.

[La] S. LANG. — *Elliptic functions*, Addison-Wesley, (1973).

[L–T] S. LANG and H. TROTTER. — *Frobenius in GL_2 extensions*, Lecture Notes in Mathematics **504**, Springer Verlag, (1976).

[M–W] D.W. MASSER and G. WUSTHOLZ. — Estimating isogenies on elliptic curves, *Inv. Math.* **100** (1990), 1–24.

[Mu] R. MURTY. — Recent developments in the theory of elliptic curves, *Proceedings of the Ramanujan Centennial International Conference*, (1987), 45–54.

Etienne FOUVRY
Mathématique- Bâtiment 425
Université de Paris-Sud
F-91405 ORSAY Cedex

Ram MURTY
Department of Mathematics
Mc GILL University
MONTREAL, PQ
CANADA H3A 2K6

Number Theory
Paris 1992–93

Arithmetical lifting and its applications

Valeri Gritsenko[*]

1. — Introduction and formulation of the main results

Let $F(Z)$ be a Siegel modular form of weight k with respect to $Sp_4(\mathbb{Z})$. By definition F is a holomorphic function on the Siegel upper half-plane

$$\mathbb{H}_2 = \left\{ Z = \begin{pmatrix} \tau & z \\ z & \omega \end{pmatrix} \in M_2(\mathbb{C}),\ \mathrm{Im}\,(Z) > 0 \right\},$$

that satisfies the functional equation

$$(1)\quad F|_k\, g(Z) := J(g, Z)^{-k} F(g < Z >) = F(Z), \quad J(g, Z) = \det(CZ + D),$$

for any $g = \begin{pmatrix} A & B \\ C & D \end{pmatrix} \in Sp_4(\mathbb{Z})$. The Fourier-Jacobi expansion of F is its Fourier development with respect to the variable ω

$$(2)\qquad F(\tau, z, \omega) = f_0(\tau) + \sum_{m \geq 1} f_m(\tau, z) \exp(2\pi i\, m\omega),$$

where $\tau = u + iv$ $(v > 0)$ belongs to the usual upper half-plane $\mathbb{H}_1$ and $z \in \mathbb{C}$. The Satake compactification of the quotient space $Sp_4(\mathbb{Z}) \setminus \mathbb{H}_2$ has two boundary components : the curve $SL_2(\mathbb{Z})\backslash\mathbb{H}_1$ and the point ∞. The function $f_0(\tau)$ is equal to the restriction of the modular form to the boundary curve and the expansion (2) corresponds to the Fourier expansion with respect to the maximal parabolic subgroup defining the boundary curve (see [P-S]). The functions $f_m(\tau, z)$ are examples of Jacobi modular forms of index m (see [EZ]). In this paper we construct a lifting from the space of Jacobi modular forms of index t in the space of modular forms on the Siegel upper-half space $\mathbb{H}_2$ with respect to the so-called paramodular group $\Gamma[t]$:

$$(3)\qquad \begin{aligned} &\textit{Lifting} : \{\text{Jacobi forms } f_t : \mathbb{H}_1 \times \mathbb{C} \to \mathbb{C}\} \to \\ &\qquad \to \{\text{modular forms } F : \Gamma[t] \setminus \mathbb{H}_2 \to \mathbb{C}\}, \end{aligned}$$

[*] Partly supported by Forschungsschwerpunkt "Arithmetik Mannheim-Heidelberg".

where by $\Gamma[t]$ we denote the following subgroup of the rational symplectic group

$$\Gamma[t] = \left\{ \begin{pmatrix} * & t* & * & * \\ * & * & * & t^{-1}* \\ * & t* & * & * \\ t* & t* & t* & * \end{pmatrix} \in Sp_2(\mathbb{Q}) \right\}, \tag{4}$$

where t is a natural number and all $*$ denote integral numbers.

This group appears in the following algebro-geometric context. Let S be an abelian variety of dimension two with polarization of type $(1, t)$ ($t \in \mathbb{N}$). We may write S as a two dimensional complex torus

$$S \approx \mathbb{C}^2/(Z, T)\mathbb{Z}^4$$

where (Z, T) is the period matrix, $Z \in \mathbb{H}_2$ and $T = \begin{pmatrix} 1 & 0 \\ 0 & t \end{pmatrix}$. The polarization with respect to this basis is given by the bilinear form $J_t = \begin{pmatrix} 0 & T \\ -T & 0 \end{pmatrix}$. The integral symplectic group of this skew-symmetric form

$$Sp(J_t, \mathbb{Z}) = \{g \in M_4(\mathbb{Z}) : \ gJ_t{}^t g = J_t\}$$

is called the *parasymplectic* (or *paramodular*) group. It is easy to see, that this group is conjugated to the group $\Gamma_t^{ab} \subset Sp_4(\mathbb{Q})$ and $\Gamma_t^{ab} = {}^t\Gamma[t]$, more exactly,

$$I_t^{-1} Sp(J_t, \mathbb{Z}) I_t = \Gamma_t^{ab} = \left\{ \begin{pmatrix} * & * & * & t* \\ t* & * & t* & t* \\ * & * & * & t* \\ * & t^{-1}* & * & * \end{pmatrix} \in Sp_2(\mathbb{Q}) \right\} = {}^t\Gamma[t],$$

where $I_t = \operatorname{diag}(1, 1, 1, t)$ and all $*$ denote integral numbers. We shall also keep the name "paramodular group" for $\Gamma[t]$.

The quotient space

$$\mathcal{A}_t = \Gamma_t^{ab} \setminus \mathbb{H}_2$$

is the coarse moduli space of abelian surfaces with polarization of type $(1, t)$ (see [I], [HKW]). $\mathcal{A}_t$ has a structure of a quasi-projective algebraic variety. For p=1 the variety $\mathcal{A}_1$ is the moduli space of abelian surfaces with principal polarization and it is rational (Igusa). For p=5 this variety is connected with the famous Horrocks-Mumford vector bundle (see [HKW]) and is also rational. It is known, that $\mathcal{A}_2$, $\mathcal{A}_3$, $\mathcal{A}_7$ are rational.

The first application of the lifting (3) is the following theorem about geometrical type of the variety $\mathcal{A}_t$.

THEOREM 1. — *Let $\widetilde{\mathcal{A}}_p$ be a non-singular model of a compactification of the moduli space $\mathcal{A}_p$ of abelian surfaces with polarization of type $(1, p)$. The variety $\widetilde{\mathcal{A}}_p$ is not unirational for any prime p, greater than 11.*

The lifting (3) has a purely arithmetical description. We shall construct the lifted form F using a representation of a Hecke ring of a parabolic subgroup of $Sp_4(\mathbb{Z})$ on the graded space of Jacobi modular forms. This Hecke ring is defined in §2. We shall see, that the lifted form F is in a sense a generalization of the classical theta-function $\theta(\tau) = \sum_{n\in\mathbb{Z}} \exp{(2\pi i n^2\tau)}$ (see §3). This analogy will give us the second important application : *a new integral representation of the Spin L-function of Siegel modular forms.*

Let us recall the definition of the Spin L-function. The Hecke ring $\mathcal{H}(\Gamma)$ of the integral symplectic group $\Gamma = Sp_4(\mathbb{Z})$ is generated by the following elements

$$T(p) = \Gamma \mathrm{diag}(1,1,p,p)\Gamma,\ T_{1,p} = \Gamma \mathrm{diag}(1,p,p^2,p)\Gamma,\ (\Delta_p)^{\pm 1} = (\Gamma p E_4 \Gamma)^{\pm 1},$$

where p is a prime number. The local factors of the Spin (or Andrianov) L-function are connected with the following polynomials $Q_p(X)$ of degree four over the Hecke ring $\mathcal{H}(\Gamma)$

$$Q_p(X) = 1 - T(p)X + p(T_{1,p} + (p^2+1)\Delta_p)X^2 - p^3\Delta_p T(p)X^3 + p^6\Delta_p^2 X^4.$$

Let $F(Z)$ be a Siegel modular form, which is an eigenfunction of all Hecke operators. Then one defines L-function $Z_F(s)$ of the modular form F

$$Z_F(s) = \prod_{p - prime} Q_{p,F}(p^{-s})^{-1},$$

where the polynomial $Q_{p,F}(X)$ is obtained from the polynomial $Q_p(X)$ by exchanging the elements of Hecke ring in the coefficients of this polynomial with their corresponding eigenvalues.

If we denote by $\alpha_0, \alpha_1, \alpha_2$ the Satake parameters of the one dimensional representation of the local Hecke ring $\mathcal{H}_p(\Gamma)$ defined by eigenvalues of the function $F(Z)$, then the local factor $Q_{p,F}(X)$ has the following form

$$Q_{p,F}(X) = (1-\alpha_0 X)(1-\alpha_0\alpha_1 X)(1-\alpha_0\alpha_2 X)(1-\alpha_0\alpha_1\alpha_2 X).$$

$Z_F(s)$ is the Spin L-function in the Langlands classification (see [L]).

The analytical continuation of this L-function was constructed in [A], but the proof contains cumbersome calculations and takes 50 pages in the Russian Mathematical Survey. In §4 and §5 we construct new integral representations of $Z_F(s)$ as a Rankin-Selberg convolution of a given cusp form with the lifting of some of its Fourier-Jacobi coefficients. It shall give a new short proof of Andrianov's result together with additional information about poles of this function obtained in the papers of Evdokimov and Oda.

THEOREM 2 (See [A], [Ev], [O2]). — *Let F be a cusp form of weight k with respect to $Sp_4(\mathbb{Z})$ and F be an eigenfunction of all Hecke operators. Then the function*

$$Z_F^*(s) = (2\pi)^{-2s}\Gamma(s)\Gamma(s-k+2)Z_F(s)$$

can be continued meromorphically to the whole s-plane with only two possible poles at $s = k-2,\ k$ and satisfies the functional equation

$$Z_F^*(2k-2-s) = (-1)^k Z_F^*(s).$$

Moreover $Z_F^*(s)$ is entire function except the case of F being a Maass modular form of even weight k. For such a Maass form the L-function $Z_F^*(s)$ has two simple poles at $s = k-2,\ k$ with the residue $\pi^{2-k} < F, F > / < f_1, f_1 >$ at $s = k$, where $< F, F >$ is the scalar square of the Siegel modular form F and $< f_1, f_1 >$ the scalar square of its first Fourier-Jacobi coefficient.

2. — The Spin L-function and Dirichlet series

The Fourier-Jacobi coefficients $f_m(\tau, z)$ of a modular form F (see (2)) are modular forms of weight k with respect to congruence subgroups of $SL_2(\mathbb{Z})$ for any fixed z. For fixed τ they are Jacobi functions, that we usually use to construct embeddings of the elliptic curve $\mathbb{C}/\tau\mathbb{Z}+\mathbb{Z}$ in the projective spaces. Taking these two properties together one may say, that f_m is a modular form with respect to the *Jacobi group* $\Gamma^J = SL_2(\mathbb{Z}) \ltimes H(\mathbb{Z})$, where $H(\mathbb{Z})$ is the integral Heisenberg group, i.e., the following central extension

$$0 \to \mathbb{Z} \to H(\mathbb{Z}) \to \mathbb{Z}\times\mathbb{Z} \to 0.$$

The Jacobi group is isomorphic to the following maximal parabolic subgroup of $Sp_4(\mathbb{Z})$

$$\text{(5)}\quad \Gamma_\infty = \left\{\begin{pmatrix} * & 0 & * & * \\ * & * & * & * \\ * & 0 & * & * \\ 0 & 0 & 0 & * \end{pmatrix}\right\} = \left\{\begin{pmatrix} a & 0 & b & 0 \\ 0 & 1 & 0 & 0 \\ c & 0 & d & 0 \\ 0 & 0 & 0 & 1 \end{pmatrix}\right\} \ltimes \left\{\begin{pmatrix} 1 & 0 & 0 & l \\ -q & 1 & l & r \\ 0 & 0 & 1 & q \\ 0 & 0 & 0 & 1 \end{pmatrix}\right\},$$

where $q, l, r \in \mathbb{Z}$ and $\begin{pmatrix} a & b \\ c & d \end{pmatrix} \in SL_2(\mathbb{Z})$. Using this realization of the Jacobi group as the parabolic subgroup Γ_∞ we may give the following definition of Jacobi modular forms.

DEFINITION. — *A holomorphic function*

$$\phi(\tau, z) : \mathbb{H}_1 \times \mathbb{C} \to \mathbb{C}$$

is called a Jacobi form of index m and weight k if the function $\widetilde{\phi}(Z) = \phi(\tau, z) exp(2\pi i m\omega)$ on the Siegel upper half-plane $\mathbb{H}_2$ is a modular form of weight k with respect to the parabolic group Γ_∞, i.e.,

1. $\widetilde{\phi}|_k M = \widetilde{\phi}$ *for any* $M \in \Gamma_\infty$;
2. *The function has the usual Fourier expansion*

$$\phi(\tau, z) = \sum_{\substack{n,l \in \mathbb{Z},\ n \geq 0 \\ 4nm \geq l^2}} f(n, l) \exp(2\pi i (n\tau + lz)).$$

This definition is equivalent to the definition given in [EZ]. We call a function ϕ a Jacobi cusp form if we have the strict inequality $4nm > l^2$ in the last summation. We shall denote the space of all Jacobi forms or all Jacobi cusp forms of index m and weight k by $\mathfrak{M}^J_{k,m}$ or $\mathfrak{S}^J_{k,m}$. The construction of the lifting will be described in terms of the Hecke ring of the parabolic subgroup Γ_∞. Note here, that Γ_∞ is not reductive! We shall consider this ring as a non-commutative extension of the Hecke ring of $Sp_4(\mathbb{Z})$. First of all let us recall the definition of an abstract Hecke ring.

Definition A pair (Γ, G), where Γ is a subgroup of a semigroup G, is called a Hecke pair if any double coset $\Gamma g \Gamma$ $(g \in G)$ is the union of a finite number of left and right cosets relative to Γ. The *Hecke ring* $\mathcal{H}(\Gamma, G)$ of the pair (Γ, G) is the Γ-invariant subspace of the $\mathbb{Q}$-vector space consisting of all formal finite linear combinations $X = \sum_i a_i \Gamma g_i$ $(a_i \in \mathbb{Q},\ g_i \in G)$, where a representation of the group Γ on this space is defined by the right multiplication $X \to X \cdot \gamma = \sum_i a_i \Gamma(g_i \gamma)$. For any two elements of this space $X = \sum_i a_i \Gamma h_i$ and $Y = \sum_j b_j \Gamma g_j$ their product is defined by $X \cdot Y = \sum_{i,j} a_i b_j \Gamma(h_i g_j)$. The product is independent of the choice of representatives g_i, h_j and $\mathcal{H}(\Gamma, G)$ is an associative ring.

The elements $\Gamma g \Gamma = \sum_i \Gamma g_i$ $(g \in G)$ form a basis of the vector space $\mathcal{H}(\Gamma, G)$ and our definition is equivalent to the standard definition of the Hecke ring.

Let us define two Hecke rings

$$\mathcal{H}(\Gamma) = \mathcal{H}_{\mathbb{Q}}(Sp_4(\mathbb{Z}), GSp_4(\mathbb{Q})) \qquad \text{and} \qquad \mathcal{H}(\Gamma_\infty) = \mathcal{H}_{\mathbb{Q}}(\Gamma_\infty, G\Gamma_\infty(\mathbb{Q})),$$

where

$$GSp_4(\mathbb{Q}) = \{g \in M_4(\mathbb{Q}) :\ {}^t g J_1 g = \mu(g) J_1, \quad \mu(g) \in \mathbb{Q}^+\}$$

is the group of symplectic similitudes and $G\Gamma_\infty(\mathbb{Q})$ its parabolic subgroup of type Γ_∞. If $X \in \mathcal{H}(\Gamma)$, then according to the elementary divisor theorem one can represent X in the form $X = \sum_i a_i \Gamma g_i$, where $g_i \in G\Gamma_\infty(\mathbb{Q})$ and $a_i \in \mathbb{Q}$. It easy to see that the map

$$Im : X = \sum_i a_i \Gamma g_i \to \sum_i a_i \Gamma_\infty g_i \tag{6}$$

is a homomorphic embedding of the Hecke ring $\mathcal{H}(\Gamma)$ into $\mathcal{H}(\Gamma_\infty)$ (see [G1]) and we shall identify the ring $\mathcal{H}(\Gamma)$ with its image in $\mathcal{H}(\Gamma_\infty)$.

We have the following representation of the ring $\mathcal{H}(\Gamma_\infty)$ on the space of functions, which are invariant with respect to $|_k$-action (see (1)) of the parabolic subgroup Γ_∞,

$$F \to F|_k X = \sum_i a_i\, \mu(g_i)^{2k-3} J(g_i, Z)^{-k} F(g_i < Z >), \tag{7}$$
$$(X = \sum_i a_i \Gamma_\infty g_i \in \mathcal{H}(\Gamma_\infty)).$$

If F is a Siegel modular form of weight k with respect to Γ and $X \in \mathcal{H}(\Gamma) \subset \mathcal{H}(\Gamma_\infty)$ we obtain the representation of the ring $\mathcal{H}(\Gamma)$ on the finite dimensional space of Siegel modular forms (*Hecke operators*). It is known that eigenfunctions of all Hecke operators form a basis of the space of all Siegel modular forms.

We have identified the Hecke ring of the symplectic group with its image (see (6)) in the ring $\mathcal{H}(\Gamma_\infty)$, which also contains two subrings isomorphic to the Hecke ring $\mathcal{H}(SL_2(\mathbb{Z}))$:

$$\mathcal{H}(\Gamma) \xrightarrow{Im} \mathcal{H}(\Gamma_\infty) \xrightarrow{j_\pm} \mathcal{H}(SL_2)\,. \tag{8}$$

It is enough to define the embeddings $j_\pm$ for the generators

$$T(p) = SL_2(\mathbb{Z})\mathrm{diag}(1,p)SL_2(\mathbb{Z}) \quad and \quad T(p,p) = SL_2(\mathbb{Z})\mathrm{diag}(p,p)SL_2(\mathbb{Z})$$

of the ring $\mathcal{H}(SL_2(\mathbb{Z}))$. By definition we have

$$j_-(T(p)) = T_-(p) = \Gamma_\infty \mathrm{diag}(1,p,p,1)\Gamma_\infty,$$
$$j_+(T(p)) = T_+(p) = \Gamma_\infty \mathrm{diag}(1,1,p,p)\Gamma_\infty,$$
$$j_-(T(p,p)) = \Lambda_-(p) = \Gamma_\infty \mathrm{diag}(p,p^2,p,1)\Gamma_\infty,$$
$$j_+(T(p,p)) = \Lambda_+(p) = \Gamma_\infty \mathrm{diag}(p,1,p,p^2)\Gamma_\infty.$$

The statement that the mapping j_- is a homomorphic embedding is clear, because there is a one-to-one correspondence between the left cosets in the decomposition of the double cosets $T(p)$, $T(p,p)$ and $T_-(p)$, $\Lambda_-(p)$ (see [G1] and [G5] where more general embeddings have been constructed). The mapping j_+ is dual to the embedding j_- with respect to the involution $*$ of the Hecke ring $\mathcal{H}(\Gamma_\infty)$

$$* : \ \Gamma_\infty g \Gamma_\infty \to \Gamma_\infty \mu(g) g^{-1} \Gamma_\infty.$$

The next lemma is a special case of a general result proved in [G1].

LEMMA 1. — *The polynomial $Q_p(X)$ splits over the ring $\mathcal{H}(\Gamma_\infty)$:*

$$Q_p(X) = j_-(Q_p^{SL}(X))\big(1 + p\Delta_p(\nabla_p - p)X^2\big)\, j_+(Q_p^{SL}(X)),$$

where the first and the third factors are the $j_\pm$-images of the Hecke polynomial $Q_p^{SL}(X) = 1 - T(p)X + pT(p,p)X^2$ for the group $SL_2(\mathbb{Z})$ and

$$\nabla_p = \sum_{r \in p^{-1}\mathbb{Z}/\mathbb{Z}} \Gamma_\infty \nabla(r) = \sum_{r \in p^{-1}\mathbb{Z}/\mathbb{Z}} \Gamma_\infty \begin{pmatrix} 1 & 0 & 0 & 0 \\ 0 & 1 & 0 & r \\ 0 & 0 & 1 & 0 \\ 0 & 0 & 0 & 1 \end{pmatrix}.$$

Proof : using the elementary divisor theorem for the symplectic group we can calculate the images (6) of the generators of $\mathcal{H}(\Gamma)$ in the Hecke ring of the parabolic subgroup $\mathcal{H}(\Gamma_\infty)$:

$$T(p) = T_-(p) + T_+(p),\; T_{1,p} = \Lambda_-(p) + \Lambda_+(p) + \Gamma_\infty \mathrm{diag}(1, p, p^2, p)\Gamma_\infty + \Delta_p(\nabla_p - 1).$$

The coefficients $T_-(p)$ and $\Lambda_-(p)$ of the polynomial $j_-(Q_p^{SL}(t)) = 1 - T_-(p)t + p\Lambda_-(p)t^2$ have the same decompositions as sums of left cosets as the elements $T(p)$ and $T(p,p)$ in the Hecke ring of $SL_2(\mathbb{Z})$ and it is easy to verify that the following identities hold :

$$\Lambda_-(p)T_+(p) = p^2\Delta_p T_-(p), \quad T_-(p)(\nabla_p - p) = 0, \quad \Lambda_-(p)(\nabla_p - p) = 0.$$

Using the antiautomorphism $*$, we have that

$$T_-(p)\Lambda_+(p) = p^2\Delta_p T_+(p), \quad (\nabla_p - p)T_+(p) = 0, \quad (\nabla_p - p)\Lambda_+(p) = 0.$$

Taking into consideration the identity

$$T_-(p)T_+(p) = p\Gamma_\infty \mathrm{diag}(1, p, p^2, p)\Gamma_\infty + (p^3 + p^2)\Delta_p,$$

which one can easily check, we obtain the factorization of the lemma.

There are two representations of the Hecke ring $\mathcal{H}(\Gamma_\infty)$ on the space of the Fourier-Jacobi coefficients of Siegel modular forms. The first one is the representation "$|_k$" on the space of all Jacobi forms of weight k (homogeneous modular forms with respect to Γ_∞), defined in (7), and the second is the representation on the space of Fourier coefficients of Γ_∞-invariant functions $F(Z)$

$$f_m\|_k X := \text{ the } m^{th} \text{ Fourier–Jacobi coefficient of the function } F|_k X\,.$$

The following formulae are clear from the definitions (see [G1] for more general statements)

$$
(9)\qquad
\begin{aligned}
&f_m(\tau,z)||_k\,T_+(n)=\widetilde{f}_{mn}|_k\,T_+(n)(Z)\exp(-2\pi i\,m\omega),\\
&f_m(\tau,z)||_k\,T_-(n)=\begin{cases}\widetilde{f}_{m/n}|_k\,T_-(n)(Z)\exp(-2\pi i\,m\omega), & \text{if } m\equiv 0 \bmod n,\\ 0, & \text{otherwise},\end{cases}
\end{aligned}
$$

where by $T_\pm(n)$ we denote the $j_\pm$-images of the standard Hecke element

$$T^{SL}(n)=\sum_{\substack{ab=n\\ a|b}} SL_2(\mathbb{Z})\mathrm{diag}(a,b)SL_2(\mathbb{Z}).$$

To make our notation shorter we set

$$
(10)\qquad
\begin{aligned}
f_{mn}|_k\,T_+(n) &:= \big(\widetilde{f}_{mn}|_k\,T_+(n)\big)(Z)\exp(-2\pi i\,m\omega),\\
f_m|_k\,T_-(n) &:= \big(\widetilde{f}_m|_k\,T_-(n)\big)(Z)\exp(-2\pi i\,mn\omega).
\end{aligned}
$$

These are Jacobi forms of index m and mn respectively.

COROLLARY 1. — *Let*

$$F(\tau,z,\omega)=\sum_{m\ge 1} f_m(\tau,z)\,\exp(2\pi i\,m\omega)$$

be a Siegel cusp form of weight k which is an eigenfunction of all Hecke operators. Then for any natural n and prime p the following identity holds in the ring of formal power series

$$Q_{p,F}(X)\sum_{\delta\ge 0} f_{np^\delta}|_k\,T_+(p^\delta)\,X^\delta =$$

$$\big(f_n+f_{\frac{n}{p}}|_k\,T_-(p)X+pf_{\frac{n}{p^2}}|_k\,\Lambda_-(p)X^2\big)|_k\,(1+p(\nabla_p-p)\Delta_pX^2),$$

where

$$f_m|_k\,(1+p(\nabla_p-p)\Delta_pX^2)=\begin{cases} f_m, & \text{if } m\equiv 0 \bmod p,\\ (1-p^{2k-4}X^2)\,f_m, & \text{otherwise.}\end{cases}$$

Proof : taking the j_+-image of the formal power Hecke series for $SL_2(\mathbb{Z})$ we have $\sum_{\delta\ge 0} T_+(p^\delta)X^\delta = j_+(Q_p^{SL}(X))^{-1}$. The function $F(Z)$ is an eigenfunction, thus $Q_{p,F}(X)f_n = f_n||_k\,Q_p(X)$ and we obtain with help of Lemma (11) 1 the following identities in the ring of the formal power series

$$Q_{p,F}(X)\sum_{\delta\ge 0} f_{np^\delta}|_k\,T_+(p^\delta)\,X^\delta = Q_{p,F}(X)\sum_{\delta\ge 0} f_n||_k\,T_+(p^\delta)\,X^\delta =$$

$$f_n||_k\,Q_p(X)(1-T_+(p)X+p\Lambda_+(p)X^2)^{-1} =$$

$$f_n||_k\,(1-T_-(p)X+p\Lambda_-(p)X^2)(1+p(\nabla_p-p)\Delta_pX^2).$$

To finish the proof we can use the formulae (9).

COROLLARY 2. — *Let $F(Z)$ be the same as in the previous corollary. Let t be a natural number such that $f_t \not\equiv 0$ and all Fourier-Jacobi coefficients $f_{t/d}$ of the modular form F, where $d > 1$ is a divisor of t, are identically equal to 0. Then the following identity holds for sufficiently large $Re(s)$*

$$L(2s - 2k + 4, \chi_t) \sum_{n \geq 1} f_{tn}(\tau, z)|_k T_+(n)\, n^{-s} = f_t(\tau, z) Z_F(s),$$

where $L(2s-2k+4, \chi_t)$ is the Dirichlet L-function with the principal Dirichlet character modulo t.

Proof : one has to apply the identities of previous corollary successively for all primes p with $X = p^{-s}$ and to take into account the standard estimation $|f_m(\tau, z)| = O((v/m)^{-\frac{k}{2}} e^{2\pi m y^2/v})$ of Fourier-Jacobi coefficients (see [KS]).

A generalization of these results to the case of Sp_n can be found in [G1].

3. — Jacobi lifting.

In full analogues with (1) one can define the space $\mathfrak{M}_k(\Gamma[t])$ of all modular forms of weight k with respect to the paramodular group $\Gamma[t]$ (see (4)).

In this section we construct an injective map from the space of Jacobi forms of index $t \geq 1$ and weight k (i.e., from the space of modular forms on the parabolic subgroup Γ_∞) into the space of modular forms with respect to the paramodular group of level t.

THEOREM 3. — *Let $\phi(\tau, z)$ be a Jacobi form of weight k and index $t \geq 1$ with the following Fourier expansion*

$$\phi(\tau, z) = \sum_{\substack{n,l \in \mathbb{Z} \\ 4nt \geq l^2}} f(n, l) \exp(2\pi i\,(n\tau + lz)).$$

If the zeroth Fourier coefficient $f(0,0)$ of the Jacobi form ϕ is not 0, we also suppose that the weight $k \geq 4$. Then the following function (see (10))

$$G_\phi(\tau, z, \omega) = f(0,0)E_k(\tau) + \sum_{m=1}^{\infty} m^{2-k} \left(\phi\,|_k\, T_-(m)\right)(\tau, z) \exp(2\pi i\, tm\omega)$$

is a modular form of weight k with respect to the paramodular group $\Gamma[t]$, where $E_k(\tau) = -\frac{2k}{B_k} + \sum_{n \geq 1} \sigma_{k-1}(n) \exp(2\pi i\, n\tau)$ is the Eisenstein series of weight k on $SL_2(\mathbb{Z})$.

Let us make some remarks about this theorem. If index $t=1$, the map $\phi \to G_\phi$ coincides with well-known the Maass or the Saito-Kurokava lifting (see [EZ]). The theorem shows that the Maass lifting is only the first member in the infinite series of liftings connected with Jacobi forms. Thus for any Siegel modular form

$$F(\tau,z,\omega)=\sum_{m\geq 0} f_m(\tau,z)\exp(2\pi i m\omega)$$

we can construct a infinite series of lifted functions F_{f_m}, that defined a "section" of the following infinite product

$$F\to \coprod_{m\in\mathbb{N}} \mathfrak{M}_k(\Gamma[m]).$$

We may rewrite at least formally the definition of the form G_ϕ using multiplicative notations. Let $f(0,0)=0$ and $\widetilde{\phi}(Z)=\phi(\tau,z)\exp(2\pi i\, t\omega)$. Then

$$(11)\quad G_\phi(Z)=\widetilde{\phi}\,|_k\sum_{m=1}^{\infty} m^{2-k}T_-(m)=\widetilde{\phi}\,|_k\prod_p(1-T_-(p)p^{2-k}+T_-(p,p)p^{3-2k})^{-1},$$

where the p-factor in the infinite product is the j_--image $j_-(Q_p^{SL}(p^{2-k}))$ of the Hecke polynomial for $SL_2(\mathbb{Z})$ (see Lemma 1).

It is interesting, that we can rewrite the classical theta-function in the same terms. To this end let us define the Hecke rings $\mathcal{H}(SL_2)=\mathcal{H}(SL_2(\mathbb{Z}),SL_2(\mathbb{Q}))$ and $\mathcal{H}(\Gamma_0)=\mathcal{H}(\Gamma_0,\Gamma_0(\mathbb{Q}))$ of the special linear group and its parabolic subgroup $\Gamma_0=\left\{\begin{pmatrix}\pm1 & b\\ 0 & \pm1\end{pmatrix},\ b\in\mathbb{Z}\right\}$. Like in the case of $Sp_4(\mathbb{Z})$ (see (6)) we can define an embedding $\mathcal{H}(SL_2)\to\mathcal{H}(\Gamma_0)$. We may continue the comparison with (8) and define an embedding of the multiplicative semigroup $\mathbb{N}^{-1}$ or, more generaly, the polynomial ring $\mathbb{Q}[x^{-1}]$ into $\mathcal{H}(\Gamma_0)$. ($\mathbb{Q}[x^{-1}]$ is isomorphic to the Hecke ring $\mathcal{H}(\{1\},\mathbb{N}^{-1})$ of the trivial group, consisting only of the unity!) By definition

$$n^{-1}\xrightarrow{j_-}[n^{-1}]=\Gamma_0\begin{pmatrix}n & 0\\ 0 & n^{-1}\end{pmatrix}\Gamma_0=\Gamma_0\begin{pmatrix}n & 0\\ 0 & n^{-1}\end{pmatrix}\in\mathcal{H}(\Gamma_0).$$

We can interpret $\mathbb{Z}$-periodic functions of the complex variable τ as automorphic functions with respect to the parabolic subgroup $\Gamma_0\subset SL_2(\mathbb{Z})$ (compare with the definition of the Jacobi forms). If we take the representation of the Hecke ring $\mathcal{H}(\Gamma_0)$ on the space of $\mathbb{Z}$-periodic functions (automorphic with respect to Γ_0) we obtain, for instance, that $\exp(2\pi i\,\tau)|\,[n^{-1}]$

$= \exp(2\pi i\, n^2\tau)$. As a consequence, we can represent the classical theta-function as a sum over a semigroup of the Hecke operators $\{[n^{-1}],\ n\in\mathbb{N}\}$ instead of as a sum over the lattice $\mathbb{Z}$. Namely,

$$\theta(\tau)=\sum_{n\in\mathbb{Z}}\exp(2\pi i\, n^2\tau)=1+2\sum_{[n^{-1}]\in\mathcal{H}(\{1\},\mathbb{N}^{-1})}\exp(2\pi i\,\tau)|\,[n^{-1}],$$

or using some formal notation

$$\theta(\tau)=1+2\exp(2\pi i\,\tau)|\prod_p(1-[p^{-1}])^{-1}=1+2\exp(2\pi i\,\tau)|\,j_-(\zeta(1)).$$

From this point of view the lifting (11) is a generalization of the last formal identity.

Proof of Theorem 3 : the function $G_\phi(Z)$ is the sum of Jacobi forms of indices mt for $m\geq 0$ (the Eisenstein series is a Jacobi form of index 0). Thus G_ϕ is invariant with respect to the action of the subgroup Γ_∞ and, moreover, with respect to $\Gamma_\infty[t]=\Gamma_\infty(\mathbb{Q})\cap\Gamma[t]$. Let us calculate the Fourier expansion of G_ϕ :

$$G_\phi(Z)=f(0,0)E_k(\tau)+\sum_{m\geq 1}\sum_{ad=m}\frac{m^{k-1}}{d^k}\sum_{\substack{b \bmod d\\ 4tn\geq l^2}} f(n,l)\exp(2\pi i\,(n\frac{a\tau+b}{d}+laz+tm\omega))$$

$$=f(0,0)E_k(\tau)+\sum_{m\geq 1}\Big(f(0,0)\sigma_{k-1}(m)\exp(2\pi i\,mt\omega)$$

$$+\sum_{ad=m}a^{k-1}\sum_{\substack{4tdn_1\geq l^2\\ n_1\neq 0}}f(dn_1,l)\exp(2\pi i\,(n_1a\tau+alz+adt\omega))\Big)$$

$$=f(0,0)\Big(-\frac{2k}{B_k}+\sum_{m\geq 1}\sigma_{k-1}(m)\big(\exp(2\pi i\,m\tau)+\exp(2\pi i\,mt\omega)\big)\Big)$$

$$+\sum_{\substack{4tmn\geq l^2\\ m\geq 1,n\geq 1}}\sum_{a|(n,l,m)}a^{k-1}f\Big(\frac{nm}{a^2},\frac{l}{a}\Big)\exp(2\pi i\,(n\tau+lz+mt\omega)).$$

This expansion shows us that $G_\phi(\tau,z,\omega)$ is invariant with respect to exchanging of the variables $(\tau\to t\omega,\ \omega\to t^{-1}\tau)$. The element

$$W_t=\begin{pmatrix}{}^tU_t & 0\\ 0 & U_t\end{pmatrix},\quad \text{where } U_t=\begin{pmatrix}0 & \sqrt{t}^{-1}\\ \sqrt{t} & 0\end{pmatrix},$$

realizes this transformation. Hence

$$G_\phi|_k W_t = (-1)^k G_\phi. \tag{12}$$

Moreover we have $G_\phi|_k J_t = G_\phi$, where J_t is the element from the definition of the parasymplectic group (see §1), since

$$W_t I W_t I = J_t, \qquad \text{where } I = \begin{pmatrix} 0 & 0 & 1 & 0 \\ 0 & 1 & 0 & 0 \\ -1 & 0 & 0 & 0 \\ 0 & 0 & 0 & 1 \end{pmatrix} \in \Gamma_\infty.$$

It is easy to see that the element J_t and the group $\Gamma_\infty[t]$ generate the paramodular group $\Gamma[t]$. The theorem is proved.

From the definition of the function G_ϕ follows the following

COROLLARY. — *The lifting*

$$J:\ \mathfrak{M}^J_{k,t} \to \mathfrak{M}_k(\Gamma[t]), \quad J(\phi) := G_\phi$$

is injective and satisfies the following commutative relation

$$J(\phi)|_k T_-(m) = J(\phi|_k T_-(m)).$$

We would like to compare the lifting of Theorem 3 with the analytical theta-lifting connected with dual reductive pairs (see, for example, [Ku]). The space of Jacobi forms $\mathfrak{M}^J_{k,m}$ is isomorphic to a subspace of modular forms of a half-integral weight with respect to an appropriate congruence subgroup of $SL_2(\mathbb{Z})$ (see [EZ]). There is an isogeny between the paramodular group $\Gamma[t]$ and a special orthogonal group of type $SO(2,3)$ (see [G4]). Let us consider the theta-lifting for the pair $(\widetilde{SL_2}, SO(2,3))$, i.e., the integral operator with a theta-function of an even quadratic form of signature $(2,3)$ as a kernel (see [O1], [RS], [Ko]). It will give us a map from the modular forms of a half-integral weight into the space of modular forms with respect to *a congruence subgroup* of $\Gamma[t]$. We would get the *full* paramodular group $\Gamma[t]$ only for an unimodular even quadratic form of signature $(2,3)$, which does not exist!

The next defect of the theta-lifting is non-existence of the theta-integral for modular forms of small weights. We shall see below, that in order to prove Theorem 1 about the moduli spaces we need modular forms of weight 3. Moreover, it is not easy to construct the theta-lifting of Eisenstein series, but in context of the arithmetical lifting we can take not only the Eisenstein series, but we can also lift a constant function that gives us so-called singular modular forms (see [G4]).

To finish this short discussion we would like to add, that Theorem 3 is a particular example of a general lifting from the space of Jacobi forms defined on $\mathbb{H}_1 \times \mathbb{C}^n$ (see [G3] and [G4]).

Now we shall prove Theorem 1.

Basis to the geometric theory of automorphic forms is the fact that automorphic forms of special weights correspond to sections of canonical line bundles on algebraic varieties. Let $F \in \mathfrak{M}_3(\Gamma[t])$ be a modular form of weight 3. The holomorphic differential form on the Siegel upper-half plane $\omega_F = F(Z) \wedge dZ = F(\tau, z, \omega) d\tau \wedge dz \wedge d\omega$ is $\Gamma[t]$–invariant and defines an element of the zeroth cohomology group $H^0(\mathcal{A}_t, \Omega_3(\mathcal{A}_t))$, where $\Omega_3(\mathcal{A}_t)$ is the sheaf of canonical differential forms on $\mathcal{A}_t$. The complex variety $\mathcal{A}_t$ is not compact and has a lot of singularities. Due to Freitag we have the following simple criterion about continuation of canonical differential froms on a singular variety to its non-singular model.

LEMMA (Freitag). — *The element $\omega \in H^0(\mathcal{A}_t, \Omega_3(\mathcal{A}_t))$ could be extended to a canonical differential form on a non-singular model $\widetilde{\mathcal{A}}_t$ of a compactification of the variety $\mathcal{A}_t$ if and only if the differential form ω is square integrable.*

See [F], Hilfsatz 3.2.1.

It is known, that ω_F is square-integrable for the cusp modular form F. Thus we have the following identity for the geometrical genus of the variety $\widetilde{\mathcal{A}}_t$

$$p_g(\widetilde{\mathcal{A}}_t) = h^{3,0}(\widetilde{\mathcal{A}}_t) = \dim_{\mathbb{C}} \mathfrak{S}_3(\Gamma^{ab}[t])$$

(see §1). If $F \in \mathfrak{M}_k(\Gamma[t])$, then $F|_k J_1 \in \mathfrak{M}_k(\Gamma^{ab}[t])$, since $J_1 g J_1 = {}^t g^{-1}$ for any $g \in \Gamma[t]$. Theorem 3 gives us examples of modular forms with respect to $\Gamma[t]$. It is easy to show that the Satake compactification of $\Gamma[p] \setminus \mathbb{H}_2$ has two one-dimensional components, which are isomorphic to $SL_2(\mathbb{Z}) \setminus \mathbb{H}_1$ (see [HKW]). Thus the restrictions of the lifting G_ϕ ($\phi \in \mathfrak{M}^J_{k,p}$) to the boundary of $\Gamma[p] \setminus \mathbb{H}_2$ are modular forms of weight k with respect to $SL_2(\mathbb{Z})$. They are identically equal to 0 for odd k and we automatically get cusp forms. Consequently, using the lifting of Theorem 3 we have the following estimation for the geometrical genus of the moduli variety

$$p_g(\widetilde{\mathcal{A}}_t) \geq \dim_{\mathbb{C}} \mathfrak{M}^J_{3,p} = \sum_{j=1}^{m-1} \{2j+2\}_{12} - \left\lfloor \frac{j^2}{4m} \right\rfloor = O(m)\,,$$

with

$$\{2j+10\}_{12} = \begin{cases} \lfloor \frac{k}{12} \rfloor & \text{if } k \not\equiv 2 \bmod 12, \\ \lfloor \frac{k}{12} \rfloor - 1 & \text{if } k \equiv 2 \bmod 12. \end{cases}$$

The formula for the dimension of the space of Jacobi forms has been obtained in [EZ] (see also [SZ]). For a prime number $p > 11$ we have $p_g(\widetilde{\mathcal{A}}_p) > 0$, that proves Theorem 3.

Corollary from Theorem 3 gives us even more.

THEOREM 3bis. — *The variety $\widetilde{\mathcal{A}}_t$ is not unirational if t has a prime divisor greater than* 11.

Proof : let us take a Jacobi form ϕ of weight 3 and index p. In accordance with (9) $\phi\,|_k\, T_-(m) \in \mathfrak{M}^J_{3,pm}$ and the lifting $J(\phi|_k T_-(m)) = J(\phi)|_k T_-(m)$ is a cusp form of weight 3 with respect to $\Gamma[pm]$.

It is possible to prove, that the variety $\mathcal{A}_t$ could be unirational only for finite numbers of t. The maximal such t is 36. This subject will be developed in more detail in the separate publication [G6] (see also [G4]).

4. — Analytical continuation of $Z_F(s)$

In this section we shall construct an analytical continuation of L-function $Z_F(s)$ using a variant of Rankin-Selberg convolution of two Siegel modular forms, proposed in [KS]. The first function in this convolution will be the eigenfunction $F(Z)$ and the second one will be a lifting of some Fourier-Jacobi coefficient f_t. The lifting $J(f_t)$ has been defined (see (11)) as action of the "infinite product" on the Jacobi form f_t, hence it is not a surprise that the Rankin-Selberg convolution of this form with an eigenfunction has an Euler product. In order to get the functional equation of $Z_F(s)$ we consider in §5 more "advanced" variant of this integral, in which we take a convolution with respect to the parabolic subgroup of the maximal normal extension of the paramodular group $\Gamma[t]$.

LEMMA 2. — *Let $F(Z)$ be a cusp form which is an eigenfunction of all Hecke operators and t be the same as in Corollary 1 of Lemma 1. Let $G_t = J(f_t)$ be the lifting of the Fourier-Jacobi coefficients f_t of F. For Re $s > 3$ the following identity holds*

$$\pi^{k-2} t^{-(s+k-2)} < f_t, f_t > Z_F^*(s+k-2)$$
$$= L^*(2s, \chi_t) \int_{\Gamma_{00}(t)\backslash \mathbb{H}_2} F(Z)\overline{G_t(Z)} E_{0t}(Z,s) |Y|^{k-3} dX dY,$$

where

$$< f_t, f_t >= \int_{(\tau,z)\in\Gamma_\infty\backslash\mathbb{H}_1\times\mathbb{C}} f_t(\tau,z)\overline{f_t(\tau,z)} v^{k-3} \exp(-4\pi t y^2 v^{-1}) du dv dx dy$$

is the scalar product of Jacobi forms, $E_{0t}(Z,s)$ is the Eisenstein series of the congruence subgroup $\Gamma_{00}(t) = Sp_2(\mathbb{Z}) \cap \Gamma[t]$

$$E_{0t}(Z,s)= \sum_{\gamma\in\Gamma_\infty\backslash\Gamma_{00}(t)} |Y(\gamma < Z >)|^s v(\gamma < Z >)^{-s}, \quad (Z=X+iY=\begin{pmatrix} \tau & z \\ z & \omega \end{pmatrix}),$$

$\tau = u+iv$, $z = x+iy$, $|Y| = \det Y$ and $L^*(2s,\chi_t) = \pi^{-s}\Gamma(s)L(2s,\chi_t)$ (see Corollary 2).

Proof : the integral on the right hand side is the Rankin-Selberg convolution of two modular forms on $\Gamma_{00}(t)$. As usual in this method, we may pass to the integral over a fundamental domain of the parabolic subgroup Γ_∞ of $\Gamma_{00}(t)$

$\Gamma_\infty\backslash\mathbb{H}_2=\{0\le u_1 \le 1\}\times\{v_1\ge y^2v^{-1}\}$, where $\tau=u+iv$, $z=x+iy$, $\omega=u_1+iv_1$.

After taking the integral over this domain one obtains

$$\int_{\Gamma_{00}(t)\backslash\mathbb{H}_2} \ldots = (4\pi t)^{-(s+k-2)}\Gamma(s+k-2)\sum_{m\ge1} < f_{tm}, m^{2-k} f_t|_k T_-(m) > m^{-(s+k-2)}.$$

The operators $T_-(m)$ and $T_+(m)$ are connected by the duality $*$ (see §1), thus using the standard Hermitian consideration it is not difficult to prove that $< f_{tm}, m^{2-k} f_t|_k T_-(m) > = < f_{tm}|_k T_+(m), f_t >$ (see [F], Chapter IV, and [KS]). We can finish the proof using Corollary 2.

Proof of analytic continuation of $Z_F(s)$: the Eisenstein series $E_{0t}(Z,s)$ is reduced to a sum of so-called Epstein zeta-functions (see [K], [Kr]). Let us introduce the following positive definite quadratic form corresponding to the variable $Z = X + iY \in \mathbb{H}_2$

$$P_Z = \begin{pmatrix} Y & 0 \\ 0 & Y^{-1} \end{pmatrix} \left[\begin{pmatrix} E & 0 \\ X & E \end{pmatrix} \right] \qquad (M[N] = {}^tNMN).$$

Then

$$P_{\gamma<Z>}=P_Z[{}^t\gamma] \text{ and } Y(\gamma < Z >)^{-1}=P_Z\left[\begin{pmatrix} {}^tC \\ {}^tD \end{pmatrix}\right] \text{ for } \gamma = \begin{pmatrix} A & B \\ C & D \end{pmatrix} \in Sp_4(\mathbb{R}).$$

The quotient $v(\gamma < Z >)/|Y(\gamma < Z >)|$ is equal to the $(2,2)$-entry of the matrix $Y(\gamma < Z >)^{-1}$ and one can rewrite the series $L(2s,\chi_t)E_{0t}(Z,s)$ as a sum of Epstein zeta-functions of the quadratic form P_Z :

$$L(2s,\chi_t)E_{0t}(Z,s) = \sum_{\substack{N={}^t(n_1,n_2,n_3,n_4)\in\mathbb{Z}^4 \\ n_1,n_2,n_3\equiv 0 \bmod t,\ (n_4,t)=1}} P_Z[N]^{-s} = \sum_{\substack{g=(0,0,0,g_4) \\ g_4\in t^{-1}\mathbb{Z}\backslash\mathbb{Z} \\ (tg_4,t)=1}} t^{-2s}\zeta(s,g,0,P_Z),$$

where for $g, h \in \mathbb{R}^4$ (see [Ep], [T])

$$\zeta(s, g, h, P_Z) = \sum_{\substack{N \in \mathbb{Z}^4 \\ N+g \neq 0}} \exp(2\pi i\, {}^t N h)\, P_Z[N+g]^{-s}.$$

It is known (see [Ep]) that the function $\zeta^*(s, g, h, P_Z) = \pi^{-s}\Gamma(s)\zeta(s, g, h, P_Z)$ has the meromorphic continuation to the s–plane, satisfies the functional equation

$$(13) \qquad \zeta^*(s, g, h, P_Z) = \exp(-2\pi i < g, h >)\zeta^*(2-s, h, -g, P_Z^{-1}),$$

where $< g, h >= g_1h_1 + \cdots + g_4h_4$, and has simple pole with residue 1 at $s = 2$, if h is integral, and simple pole with residue -1 at $s = 0$, if g is integral.

As a corollary of the integral representation of Lemma 2, we have the meromorphic continuation of the function $Z_F^*(s)$ of Theorem 2.

Moreover, if $t > 1$ then the vectors g are not integral and $Z_F^*(s)$ could have a pole only at $s = k$.

If $t = 1$ (that could be possible only for even weight k), then the Eisenstein series $\pi^{-s}\Gamma(s)\zeta(2s)E_{01}(Z, s)$ is equal to the Epstein zeta-function $\zeta^*(s, 0, 0, P_Z)$, satisfies the functional equation (13) and has two poles at $s = 0,\ 2$ with residues ∓ 1 respectively. It gives us the functional equation of Theorem 2 in the case $t = 1$.

Moreover the residue $\mathrm{Res}_{s=k} L_F^*(s)$ is proportional to the scalar product $< F, G_1 >$ of the Siegel modular form F and the form $G_1 = J(f_1)$ containing in the Maass subspace, which is invariant with respect to the action of Hecke operators. This scalar product is zero if F orthogonal to this subspace, thus $Z_F^*(s)$ is entire for such F. Otherwise $F = G_1$ and the residue of $Z_F^*(s)$ at $s = k$ is equal to $\pi^{2-k} < F, F > / < f_1, f_1 >$.

We consider the case $t > 1$ below.

5. — Functional equation of the Spin L-functions for Siegel modular forms with the first Fourier-Jacobi coefficient $f_1 \equiv 0$

The integral representation of the Spin L-function obtained above gives us its meromorphic continuation, but the Eisenstein series, which is the kernel of the integral, has no good functional equation. As was shown in the proof of Theorem 3, the lifting $J(f_t)$ is "nearly" invariant with respect to a normal extension $\Gamma^*[t]$ of the paramodular group $\Gamma[t]$ generated by the group $\Gamma[t]$ and the element W_t (see (12)). To get the functional equation for $Z_F(s)$ one has to construct the second variant of the Rankin-Selberg convolution for this new group, since in that case the Eisenstein series satisfies a good functional equation. Without loss of generality we may restrict ourselves to the case of a prime number t.

LEMMA 3. — *Let F be a cusp form which is an eigenfunction of all Hecke operators. Let us assume that its first Fourier-Jacobi coefficients $f_1(\tau, z)$ vanishes. Then there exists a prime number p such that $f_p(\tau, z)$ is not identically equal to zero.*

Proof : let us consider the Fourier expansion $F(Z) = \sum_{N\in\mathfrak{B}} a(N)$ $\exp(2\pi i\,\mathrm{tr}(NZ))$, where the sum is taken over the set $\mathfrak{B}_2$ of all positive definite semi-integral symmetric matrices $N = \begin{pmatrix} m & l/2 \\ l/2 & n \end{pmatrix}$. The Fourier coefficient $a(N)$ depends only on the class of the quadratic form $N : a(N) = a({}^tXNX)$ for any $X \in SL_2(\mathbb{Z})$. If there is a primitive N ($(m,l,n) = 1$) such that $a(N) \neq 0$, then we may take any prime p represented by the quadratic form N. For such prime numbers $f_p(\tau, z) \not\equiv 0$.

Let us suppose that $a(N) = 0$ for all primitive matrices N. Consider the Fourier–Jacobi expansion of F

$$F(\tau, z, \omega) = \sum_{m\geq r>1} f_m(\tau, z)\exp(2\pi i\, m\omega).$$

The form F is an eigenfunction, therefore we have the following relation between the Fourier-Jacobi coefficients of F and $F|_kT(e)$ for any divisor e of the index r ($ed = r$) :

$$f_d\{F|_kT(e)\}(\tau, z) = \big(f_r\{F\}\,|_k\, T_+(e)\big)(\tau, z) = \lambda_F(T(e))f_d\{F\}(\tau, z) \equiv 0,$$

where $T(e)$ is the $Sp_4(\mathbb{Z})$–Hecke operator with index e, $\lambda_F(T(e))$ is its eigenvalue and $T_+(e)$ is the j_+-image of the $SL_2(\mathbb{Z})$–Hecke operator (see (7), (9), (10) and the proof of Lemma 1). In the Fourier expansion of the Jacobi form f_r

$$f_r(\tau, z)\exp(2\pi i\, r\omega) = \sum_{N=\begin{pmatrix} * & * \\ * & r \end{pmatrix}\in\mathfrak{B}_2} a(N)\exp\left(2\pi i\,(\mathrm{tr}(NZ))\right),$$

there are no $a(N)$ with a primitive N. If $a(\begin{pmatrix} em & el/2 \\ el/2 & ed \end{pmatrix}) \neq 0$, where $ed = r$, $e > 1$ and $(m, l, d) = 1$, then it is easy to see, that the function $\big(f_r|_k T_+(e)\big)(\tau, z)$, that is identically equal to 0 in accordance with the previous considerations, has at least one non–zero Fourier coefficient?! The lemma is proved.

In the next lemma we consider two "trace" operators for Siegel modular forms on $Sp_4(\mathbb{Z})$ which send them to modular forms on the paramodular group and on the group $\Gamma^*[t]$ respectively.

LEMMA 4. — *Let*

$$F(Z)=\sum_{m\geq 1} f_m(\tau,z)\exp(2\pi i\, m\omega)\in \mathfrak{M}_k(Sp_4(\mathbb{Z})),$$

then the functions

$$F_p=F|_k\,\nabla_p+F|_k\,J_p \qquad \text{and} \qquad F_p^*=F_p+F_p|_k\,W_p,$$

(see (1), (12) and Lemma 1) are modular forms of weight k with respect to the paramodular group $\Gamma[p]$ *and its (maximal) normal extension*

$$\Gamma^*[p]=\Gamma[p]\cup\Gamma[p]W_p,$$

respectively. Moreover, the function $F_p^*(Z)$ *has the Fourier–Jacobi expansion*

$$F_p^*(Z)=\sum_{m\geq 1} f_{pm}^*(\tau,z)\exp(2\pi i\, pm\omega)$$

where

$$f_{pm}^*=pf_{pm}+p^{-(2k-6)}f_{\frac{m}{p}}|_k\,\Lambda_-(p)+p^{-(k-3)}f_m|_k\,T_-(p).$$

Proof : the first part of the lemma is evident. To calculate the Fourier-Jacobi expansion we can rewrite the trace operator as follows

$$F_p^*=\sum_{x\in p^{-1}\mathbb{Z}/\mathbb{Z}} F|_k\,\nabla(x)+F|_k\,{}^tJ_1J_p$$
$$+\sum_{x\in p^{-1}\mathbb{Z}/\mathbb{Z}} F|_k\begin{pmatrix}0&1&0&0\\1&0&0&0\\0&0&0&1\\0&0&1&0\end{pmatrix}\nabla(x)W_p+F|_k\begin{pmatrix}0&0&0&-1\\0&0&-1&0\\0&1&0&0\\1&0&0&0\end{pmatrix}J_pW_p.$$

The first sum gives us only coefficients with indices divisible by p, the second sum is equivalent to the action of the operators $\Lambda_-(p)\Delta_p^{-1}$ and the last two summands coincide with the action of the Hecke operator $(\Delta_{\sqrt{p}})^{-1}T_-(p)$.

LEMMA 5. — *Let p be prime. Then the Eisenstein series*

$$E_p^*(Z,s)=\pi^{-s}\Gamma(s)(1+p^{-s})\zeta(2s)\sum_{\gamma\in\Gamma_\infty[p]\backslash\Gamma^*[p]}|Y(\gamma<Z>)|^s v(\gamma<Z>)^{-s},$$

has a meromorphic continuation to the whole s-plane and is invariant with respect to the transformation $s \to 2 - s$.

Proof : as in the proof of Lemma 2, $E_t^*(Z,s)$ can be represented as a sum of Epstein zeta-functions. The last rows of representatives of $\Gamma_\infty[p] \setminus \Gamma^*[p]$ form the set of all $\mathbb{Z}[p^{-1}]$-primitive (primitive outside p) vectors of the following types :

$$(pa, pb, pc, d) \quad \text{with } (p,d) = 1; \qquad p(a,b,c,d) \quad \text{with } (b,p) = 1;$$
$$\sqrt{p}(a, pb, c, d) \quad \text{without } a \equiv c \equiv 0 \bmod p.$$

Therefore there is a representation of E_p^* as a sum of Epstein zeta-functions. A simple computation shows that

$$E_p^*(Z,s) = \pi^{-s}\Gamma(s)p^{\frac{3s}{2}}\Bigg(\sum_{(a,b,c,d)\in\mathbb{Z}^4\setminus\{0\}} P_Z[{}^t(pa,pb,pc,d)]^{-s} + p^{-s}P_Z[{}^t(a,pb,c,d)]\Bigg)$$
$$= p^{\frac{s}{2}-1}\sum_{\substack{g=(0,g_2,0,0)\\ g_2 \bmod p}} \zeta^*(s,\frac{g}{p},0,P_Z) + p^{-\frac{s}{2}}\sum_{\substack{h=(0,0,0,h_4)\\ h_4 \bmod p}} \zeta^*(s,0,\frac{h}{p},P_Z).$$

Using the functional equation(13) and the identity $P_Z^{-1} = P_Z[J_1]$ we get the functional equation for E^*.

In accordance with Lemma 2 and with the identity (12), the product

$$\left(F_p(Z) + (-1)^k F_p|_k\, W_p(Z)\right) \overline{G_p(Z)}\,(\det Y)^k$$

(where $G_p = J(f_p)$) is invariant with respect to the action of $\Gamma^*[p]$ and we can construct an integral analogue to the integral of Lemma 2 for the group $\Gamma^*[p]$.

LEMMA 6. — *Let $F(Z)$ be a cusp form of weight k on $Sp_4(\mathbb{Z})$. Let F be an eigenfunction of all Hecke operators with the first Fourier–Jacobi coefficient $f_1(\tau,z) \equiv 0$ and let p be a prime number for which $f_p(\tau,z) \neq 0$ identically. For Res $> k+1$ the following identity holds*

$$p^{3-2k}\pi^{k-2}(p^{\frac{s}{2}-k+1} + (-1)^k p^{-\frac{s}{2}})) < f_p, f_p > Z_F^*(s) =$$
$$\int_{\Gamma^*[p]\setminus\mathbb{H}_2} \left(F_p(Z) + (-1)^k F_p|_k W_p\,(Z)\right) \overline{G_p(Z)}\, E_p^*(Z, s-k+2)|\,Y|^{k-3}dXdY,$$

where $< f_p, f_p > \neq 0$ is the scalar square and $G_p = J(f_p)$ is the lifting of the Jacobi form f_p.

We note that the functional equation for L-function $Z_F^*(s)$ of Theorem 2 follows from the above representation and the functional equation for the Eisenstein series $E_p^*(Z,s)$ from Lemma 5.

Proof of Lemma 6 : applying the same unfolding arguments with E_p^* as in Lemma 2 we find that the integral equals the following Dirichlet series

$$\pi^{k-2}(2\pi)^{-2s}p^{\frac{s}{2}-\frac{3k}{2}+2}\Gamma(s)\Gamma(s-k+2)(1-p^{-(s-k+2)})^{-1}$$
$$L(2s-2k+4,\chi_p)\sum_{m\geq 1}\frac{< f_{mp}^*|_kT_+(m), f_p >}{m^s},$$

where f_{mp}^* are the Fourier–Jacobi coefficients of the function $F_p+(-1)^kF_p|_k W_p$. These coefficients contain three summands, as has been shown in Lemma 5. Thus we have

$$\sum_{m\geq 1}\frac{f_{mp}^*|_kT_+(m)}{m^s}=$$
$$\sum_{m\geq 1}\frac{\left[pf_{pm}+p^{-(2k-6)}f_{\frac{m}{p}}|_k\,\Lambda_-(p)+p^{-(k-3)}f_m|_k\,T_-(p)\right]|_k\,T_+(m)}{m^s},\tag{14}$$

which is reduced to three Dirichlet series. The first series one can calculate using Corollary 2. For the third sum one gets

$$L(2s-2k+4,\chi_p)\sum_{m\geq 1}\frac{f_m|_k\,T_-(p)T_+(m)}{m^s}=p^{-s}(f_p|_kT_-(p)T_+(p))\,Z_F(s).\tag{15}$$

To prove this we may use the identity

$$\lambda_pf_{mp}=f_{mp}||_k\,T(p)=f_m|_k\,T_-(p)+f_{mp^2}|_k\,T_+(p),\tag{16}$$

where λ_p is the F-eigenvalue of the Hecke operator $T(p)$. Thus the series (15) is equal to

$$L(2s-2k+4,\chi_p)\sum_{m\geq 1}\left(\lambda_pf_{mp}-f_{mp^2}|_k\,T_+(p)\right)|_kT_+(m)\,m^{-s}.$$

The operators $T_+(p)$ and $T_+(m)$ commute, thus using Corollaries 1 and 2 of Lemma 1 we see that the last sum is equal to

$$\left(\lambda_pf_p-\left(f_{p^2}-f_p|_kT_-(p)p^{-s}\right)|_kT_+(p)\right)Z_F(s).$$

Applying (16) again for $mp = p^2$ and taking into account the assumption that $f_1 \equiv 0$ we get (15).

The calculation of the second sum in (14) can be done as follows. The standard identity between elements in the $SL_2(\mathbb{Z})$–Hecke ring $T(p)T(\frac{m}{p}) = T(m) + pT(p,p)T(\frac{m}{p^2})$ gives us after the "plus" j_+–embedding : $T_+(p)T_+(\frac{m}{p}) = T_+(m) + p\Lambda_+(p)T_+(\frac{m}{p^2})$. In Lemma 1 we have seen that $\Lambda_-(p)T_+(p) = p^2\Delta_p T_-(p)$ and $\Lambda_-(p)\Lambda_+(p) = p^4\Delta_p^2$. Therefore

$$\sum_{m\geq 1} f_m|_k \left(\Lambda_-(p)\Delta_p^{-1}T_+(mp)\right)(mp)^{-s}$$

$$= p^{2-s}\sum_{m\geq 1} f_m|_k T_-(p)T_+(m)\, m^{-s} - p^{5-2s}\sum_{l\geq 1} f_{pl}|_k T_+(l)\Delta_p\, l^{-s}$$

$$= \left(p^{2-s} f_p|_k T_-(p)T_+(p) - p^{2k-1-2s}\right) Z_F(s) L(2s-2k+4,\chi_p)^{-1},$$

where we have used (15) and Corollary 2. In the second and third sums there is a summand of type $f_p|_k T_-(p)T_+(p)$, which we shall calculate in the next lemma.

LEMMA 7. — *Let $f_p(\tau,z)$ be a Jacobi form of index p and weight k such that $f_p|_k T_+(p) \equiv 0$. Then*

$$f_p|_k T_-(p)T_+(p) = p^{2k-6}(p^3 - (-1)^k p^2) f_p.$$

Proof : one can prove the lemma almost entirely "inside" the formal Hecke ring $\mathcal{H}(\Gamma_\infty)$. As in the proof of Lemma 1

$$T_-(p)T_+(p) = (pT^J(p) + p^3 + p^2)\,\Delta_p, \text{ where } T^J(p) = \Gamma_\infty \mathrm{diag}\,(p^{-1},1,p,1)\Gamma_\infty.$$

We note that the operator $|_k T^J(p)$ coincides up to a constant with the Hecke-Jacobi operator $T^J(p)$ defined in §4 of [EZ]. On the other side it is easy to check that

$$T_+(p)T_-(p) = (T^J(p)\nabla_p + p\nabla_p + \Xi_p)\Delta_p, \tag{17}$$

where

$$\Xi_p = \sum_{x,y,r\in p^{-1}\mathbb{Z}/\mathbb{Z}} \Gamma_\infty \begin{pmatrix} 1 & 0 & 0 & y \\ -x & 1 & y & r \\ 0 & 0 & 1 & x \\ 0 & 0 & 0 & 1 \end{pmatrix}.$$

Let us take the standard expansion of $f_p(\tau, z)$ with respect to the basic Jacobi functions

$$\theta_{p,\mu}(\tau, z) = \sum_{l\in\mathbb{Z}} \exp\Big(2\pi i p\Big(l + \frac{\mu}{2p}\Big)^2 \tau + 2\pi i(2pl + \mu)z\Big)$$

(see [EZ], §5). If

$$f_p(\tau, z) = \sum_{\mu \bmod 2p} \phi_\mu(\tau)\theta_{p,\mu}(\tau, z),$$

then after obvious calculations with Gauss sums one gets

$$f_p(\tau, z)|_k\, \Xi_p = p^2 \sum_{\mu \bmod 2p} \phi_\mu(\tau)\theta_{p,-\mu}(\tau, z)$$

(this is the only place in the proof of the lemma in which we have to use Jacobi forms themselves). The invariance of $f_p(\tau, z)$ with respect to the minus identity matrix $-E_4$ is equivalent to the identity $\phi_\mu = (-1)^k\phi_{-\mu}(\tau, z)$, thus $f_p|_k\, \Xi_p = (-1)^k p^2 f_p$. Moreover, from (17) we get

$$f_p|_k\, T^J(p) = \begin{cases} -2p^2 f_p, & \text{if } k \text{ is even,} \\ 0 & \text{if } k \text{ is odd,} \end{cases}$$

since by our assumption $f_p|_k\, T_+(p) \equiv 0$. This prove the lemma.

To finish the prove of Lemma 6 and to get the functional equation of Theorem 2 one has to collect the three sums in (14) together and to take into account the result of Lemma 7.

We have proved in §4, that the function $Z_F^*(s)$ could have only one pole for $t > 1$. Together with the functional equation stated above it gives us that the Spin L-function is entire function on the whole s-plane if the first Fourier-Jacobi coefficients of the cusp form F vanishes. Thus Theorem 2 is proved completely.

Manuscrit reçu le 26 octobre 1993

REFERENCES

[A] A. N. ANDRIANOV. — *Euler products corresponding to Siegel modular forms of genus* 2, Russian Math. Survey **29** (1974), 45–116.

[EZ] M. EICHLER, D. ZAGIER. — *The theory of Jacobi forms*, Progress in Math. **55**, Birkhäuser, Boston, Basel, Stuttgart, 1985.

[Ep] P. EPSTEIN. — *Zur Theorie allgemeiner Zetafunctionen* Math. Ann. **56**, (1903), 614-644.

[Ev] S. A. EVDOKIMOV. — *A characterization of the Maass space of Siegel cusp forms of degree* 2, Matem. Sbornik **112** (1980), 133-142 (Russian); English transl. in Math. USSR Sbornik **40** (1981), 125–133.

[F] E. FREITAG. — *Siegelsche Modulfunktionen*, Grundlehren der math. Wissensch., 254, Springer, Berlin, Heidelberg, New York, 1983.

[G1] V. A. GRITSENKO. — *The action of modular operators on the Fourier-Jacobi coefficients of modular forms*, Matem. Sbornik **119** 1982, 248-277 (Russian); English transl. in Math. USSR Sbornik **47** (1984), 237–268.

[G2] V. A. GRITSENKO. — *Jacobi functions and Euler products for Hermitian modular forms*, Zap. Nauk. Sem. LOMI **183** (1990), 77–123 (Russian); English transl. in J. Soviet Math. **62** (1992), 2883–2914.

[G3] V. A. GRITSENKO. — *Jacobi functions of n–variables*,Zap. Nauk. Sem. LOMI **168** (1988), 32–45 (Russian); English transl. in J. Soviet Math. **53** (1991), 243-252.

[G4] V. A. GRITSENKO. — *Modular forms and moduli spaces of abelian and $K3$ surfaces*, Mathematica Gottingensis Schrift. des SFB "Geometrie und Analysis", Helt 26, 1993, p. 32; appears in St.Petersburg Math. Jour. **5** (1994).

[G5] V. A. GRITSENKO. — *Induction in the theory of zeta-functions*, Preprint 91–097 University Bielefeld, 1991 p. 76; appears in St.Petersburg Math. Jour. **5** (1994).

[G6] V. A. GRITSENKO. — *Moduli spaces of abelian surfaces* (in preparation).

[HKW] K. HULEK, C. KAHN, S. H. WEINTRAUB. — *Theta functions and compactification of moduli spaces of polarized abelian surfaces*, 1993.

[I] J. IGUSA. — *Theta function*, Grundlehren der math. Wissensch., 254, Springer Verlag, Berlin, Heidelberg, New York, 1972.

[K] W. KOHNEN. — *On character twists of certain Dirichlet series*, Mem. of the Fac. of Science Kyushu University, series A, Mathematics **47** (1993), 103–119.

[KS] W. KOHNEN, N.-P. SKORUPPA. — *A certain Dirichlet series attached to Siegel modular forms of degree two*, Invent. Math. **95** (1989), 449–476.

[Ko] H. KOJIMA. — *On construction of Siegel modular forms of degree two*, J. Math. Soc. Japan **34** (1982), 393–411.

[Kr] A. KRIEG. — *A Dirichlet series for modular forms of degree n*, Acta Arith. **59** (1991), 243–259.

[Ku] S. KUDLA. — *Seesaw dual reductive pairs*, Automorphic forms of several variables, Progress in Math. 46, Birkhäuser, Boston, Basel, Stuttgart, 1983, 244-268.

[L] R.P. LANGLANDS. — *Euler products*, Yale Univ. Press, 1971.

[O1] T. ODA. — *On modular forms associated with indefinite quadratic forms of signature* $(2, n-2)$, Math. Ann. **231** (1977), 97–144.

[O2] T. ODA. — *On the poles of Andrianov L-functions*, Math. Ann. **256** (1981), 323-340.

[P-S] I. I. PYATETSKII-SHAPIRO. — *Automorphic functions and the geometry of classical domains*, Gordon and Breach, New York, 1969.

[RS] S. RALLIS, G. SCHIFFMANN. — *On a relation between SL_2 cusp forms and cusp forms on tube domain associated to orthogonal groups*, Trans. Amer. Math. Soc **263** (1981), 1–58.

[SZ] N-P. SKORUPPA, D. ZAGIER. — *A trace form for Jacobi forms*, J. reine und angew. Math. **393** (1989), 168–198.

[T] A. TERRAS. — *Harmonic Analysis on Symmetric Spaces ans Applications, I*, Springer Verlag, Berlin, Heidelberg, New York, 1983.

Valeri GRITSENKO
Department Steklov Mathematical Institute
St. Petersburg
FONTANKA 27
191011 ST. PETERSBURG
RUSSIA

Number Theory
Paris 1992–93

Towards an arithmetical analysis of the continuum

Glyn Harman

1. — Introduction

Since the set of real numbers is uncountable, almost all (in a variety of senses) real numbers are effectively indescribable. Our curiousity forces us, however, to attempt to describe the irrationals in terms of their relation to the "known" set of rationals. An elementary theorem given by Dirichlet (1842) provides the simplest such relation.

For every real α, and any given $N \geq 1$, there exist coprime integers m, n such that

$$\left|\alpha - \frac{m}{n}\right| \leq \frac{1}{n(N+1)} \quad \text{with } 1 \leq n \leq N. \tag{1}$$

This result is best possible, even for almost all α (in the sense of Lebesgue measure), although if we are only interested in what is true for infinitely many N then better results are possible.
Hurwitz (1891) : *For every irrational α there are infinitely many fractions m/n in lowest terms such that*

$$\left|\alpha - \frac{m}{n}\right| < \frac{1}{5^{\frac{1}{2}} n^2}. \tag{2}$$

Khintchine (1926) : *Let $f(n)$ be a positive function defined on the integers which decreases to zero monotonically with increasing n, and such that*

$$\sum_{n-1}^{\infty} \frac{f(n)}{n} \tag{3}$$

diverges. Then, for almost all α there are infinitely many solutions to

$$\left|\alpha - \frac{m}{n}\right| < \frac{f(n)}{n^2}, \quad \text{with } (m,n) = 1. \tag{4}$$

These results are best possible since $5^{\frac{1}{2}}$ cannot be replaced by a larger number in (2), while if (3) converges then there are only finitely many solutions to (4) for almost all α. Of course, each irrational α can be expressed as a non-terminating continued fraction, and the approximations in (2) and (4) (as soon as $f(n) < \frac{1}{2}$) will come from convergents to the continued fraction. These results naturally lead to a more general question : what types of fractions are near irrationals? Here, by "types", we mean fractions whose numerator and/or denominator are restricted in various ways, for example to prime values. By "near" we mean we would like to get as close as possible to

$$\left|\alpha - \frac{m}{n}\right| < n^{-2},$$

but if we obtained

$$\left|\alpha - \frac{m}{n}\right| < n^{-1-\theta} \tag{5}$$

then the size of θ could be regarded as a measure of our "success". We will henceforth consider the question from two different perspectives corresponding to (2) and (4) above : what is true for *every* irrational? or : what is true for *almost all* irrationals? One could ask the question for different subsets of the irrationals, for example algebraic numbers, but we shall not pursue such topics here. Neither shall we deal with inhomogeneous approximation. The answers we obtain form "une contribution à l'analyse arithmétique du continu", which subject was begun in earnest in 1903 in Paris by E. Borel [4].

Before reviewing our current state of knowledge on these questions, it is amusing to observe that these, and related questions, are of interest in fields considerably removed from number theory. To make a clockwork model of the solar system requires approximations m/n to a number α (usually the ratio of a planet's "year" to the earth's year) with m and n composed of numbers with small prime factors (to make the gears used feasible). The Dutch scientist Christiaan Huygens used continued fractions to make such a model in the 17th century, a project which we note was "partiellement exécuté à Paris" [22]. For details of this problem see Chapter 4 of [26]. In the theory of music, and in particular in the construction of musical instruments, it is often regarded as desirable to have n equally spaced notes per octave such that

$$\frac{\log(3/2)}{\log 2} \approx \frac{m_1}{n} \quad \text{and} \quad \frac{\log(5/4)}{\log 2} \approx \frac{m_2}{n}, \tag{6}$$

where m_1 and m_2 are integers. The first approximation in (6) is to make the interval of a fifth correct, the second gives a major third. It would be useful to have other approximations with the same denominator also. The well-known value of 12 for n gives $2^{7/12} = 1.4983\ldots$ (very nearly 3/2), while $2^{4/12} = 1.2599\ldots$ (about 0.8% sharp). To improve on these approximations (and so possibly obtain sweeter music?) would require $n = 53$ (31/53, 17/53) or $n = 118$ (69/118, 38/118). The problem now becomes one of genetic engineering : breed people with lots of small fingers and specially developed brains to play the resulting instruments!

2. — Question One : restrict only one of numerator/denominator

Without loss of generality we restrict the denominator only in the following. In 1941 Duffin and Schaeffer [8] generalized Khintchine's theorem (4) as follows.

Suppose that $f(n)$ is a non-negative function such that

$$\sum_{n=1}^{N} \frac{\phi(n) f(n)}{n} > c \sum_{n=1}^{N} f(n) \tag{7}$$

for all N and some positive constant c, and the right hand side of (7) tends to infinity with N. Then, for almost all real α there are infinitely many solutions to

$$|n\alpha - m| < f(n) \quad \text{with} \quad (m, n) = 1 \tag{8}$$

(in (7) $\phi(n)$ denotes Euler's totient function).

As examples we may take $f(n) = \chi(n)g(n)$ where χ is the characteristic function of a set with number–theoretic interest and $g(n)$ is suitably behaved. We thus obtain :

$$|p\alpha - m| < 1/(2p)\,, \ (m, p) = 1, \ p \text{ a prime,} \tag{9}$$

$$|n^2\alpha - m| < 1/(n \ \log n)\,, \ (m, n) = 1\,, \tag{10}$$

$$|10^n\alpha - m| < 1/n\,, \ (m, 10) = 1\,, \tag{11}$$

all have infinitely many solutions for almost all α. The first inequality (9) shows that almost all real numbers have infinitely many convergents in their continued fraction expansion having prime denominator. The final inequality (11) actually says something about the decimal expansion of almost all α : for almost all α there are infinitely many n such that the n–th and $[\log_{10} n]$ following decimal places are all zero.

The major unsolved problem in this field is the Duffin–Schaeffer conjecture which states that in place of (7) we require only the divergence of the left hand side of (7). Although many cases of this conjecture have been settled (see [9], [17], [27] for example) it remains an open question whether the conjecture is true in its full generality (although it is known to be true in higher dimensions [25], and is true if "almost all α" is replaced by "a set of dimension 1"). Settling this problem would be a major advance in the arithmetical analysis of the continuum.

When we turn from what is almost always true to what is true for every irrational α we enter more difficult territory. Now (9), (10) and (11) are no longer always true. There are uncountably many α such that

$$|p\alpha - m| > \frac{\log p}{4000\, p \log\log p}, \tag{12}$$

for all large primes p [18]. There are uncountably many α such that

$$|10^n\alpha - m| > 1 \quad \text{if } (m, 10) = 1. \tag{13}$$

This is very easy. Take $\alpha = 0 \cdot a_1 a_2 \cdots$ where each a_j is either 4 or 5, but the decimal does not recur. The best approximations $m/10^n$ with $(m, 10) = 1$ must then have m ending in a 3 or a 7 which gives (13).

Inequalities like (9) were first investigated by Vinogradov [30] who showed that for every irrational α there are infinitely many solutions in primes p to

$$\|p\alpha\| < p^{-\theta+\varepsilon} \tag{14}$$

where $\|\ \|$ denotes distance to the nearest integer and $\theta = 1/5$. The value of θ was improved to $1/4$ by R.C. Vaughan [29] and subsequently improved further to $3/10$ by the present author [12] (the method presented there shows that (14) holds for a value of θ between $3/10$ and $1/3$: numerical calculation is required to arrive at the best value). The idea here is to approximate $\|x\|$ by a Fourier series and then convert a sum over primes to double sums, so that we require estimates for

$$\sum_{\ell \le L} \Big| \sum_{m \le M} a_m \sum_{n \le N} b_n\, e(\alpha m n \ell) \Big| . \tag{15}$$

Here $L \approx (MN)^{\theta-\varepsilon}$, a_m, $b_n \ll n^\varepsilon$ for any $\varepsilon > 0$, and $b_n = 1$ or $\log n$ if M is much smaller than N. The conversion to double sums may be done by Vaughan's identity or arise from a sieve method.

Hardy and Littlewood were the first to investigate (10). They were only able to show that $\min_{n\le N} \|\alpha n^2\| \to 0$ as $N \to \infty$ for all α [11]. Vinogradov gave the first quantitative formulation of this result and this was improved to

$$\min_{n\le N} \|\alpha n^2\| < N^{-\frac{1}{2}+\varepsilon}$$

by Heilbronn [20]. This result was generalized to αn^k by Danicic [7], also improving Vinogradov's earlier work. Here estimates are required for the Weyl sums :

$$\sum_{\ell\le L}\Big|\sum_{n\le N} e(\alpha n^k \ell)\Big|. \tag{16}$$

The two problems above suggest the further problem of $\|\alpha p^k\|$, for prime p. This requires information on double Weyl sums [2]. It would be very interesting to have a new means of attacking these problems. The current methods give not only one solution of the desired inequality, but the "correct number" with a smaller error (or lower bound with the expected order of magnitude). This sets a limit on how well the methods could be expected to work since we know, for example, that the sequence $\{\alpha n^2\}$ does not have small discrepancy [3].

3. — Question 2 : restrict both numerator and denominator, what happens for almost all α?

Let $\mathcal{A}$ and $\mathcal{B}$ be two sets of positive integers, and let $\rho(n)$ denote the probability that $n \in \mathcal{B}$. That is, we suppose

$$\sum_{\substack{n\le N\\ n\in\mathcal{B}}} 1 = \sum_{n\le N} \rho(n)(1+o(n)) \tag{17}$$

where $\rho(n)$ is continuous and non-increasing. For example we have :

$\mathcal{B}$:	$\rho(n) =$
prime numbers	$(\log n)^{-1}$
square-frees	$6/\pi^2$
$n \equiv a (\mathrm{mod}\, q)$	$1/q$
$n = r^2 + s^2$	$K/(\log n)^{\frac{1}{2}}$.

Now suppose that $f(n)$ is a decreasing positive function and consider the sum

$$\sum_{n\in\mathcal{A}} \rho(n) f(n). \tag{18}$$

If this sum diverges it would be reasonable to suppose that there were infinitely many solutions to

$$|n\alpha - m| < f(n) \quad \text{with} \quad n \in \mathcal{A},\ m \in \mathcal{B} \tag{19}$$

for almost all $\alpha > 0$. On the other hand, if (18) converges then it is easy to show that there are only finitely many solutions to (19) for almost all $\alpha > 0$. If $\mathcal{B}$ is just the set of positive integers (the question considered in the previous section) then it is known that a zero–one law operates ([5] and [10]) : the inequality has infinitely many solutions either for almost all α or for a set with measure zero. In the current situation it is possible to produce counter–examples to show that this is no longer the case [15], although it is likely that such a law will operate in all cases of number–theoretic interest.

As an example of the convergence of (18) we have the following result :

For almost all α there are only a finite number of solutions to

$$|p\alpha - q| < 1/p \quad \text{with } p, q \text{ primes}.$$

It is possible to show that the exceptional set here has Hausdorff dimension one.

To study the case where (18) diverges it is useful to impose the additional condition (compare (7))

$$\sum_{\substack{n \in \mathcal{E} \\ n \le X}} \frac{\phi(n)}{n} > c \sum_{\substack{n \in \mathcal{E} \\ n \le X}} 1 \quad \text{for} \quad \mathcal{E} = \mathcal{A} \text{ or } \mathcal{B}. \tag{20}$$

We then only count coprime solutions to (19). Some condition in addition to the divergence of (18) is necessary because if $\mathcal{A}$ and $\mathcal{B}$ both consist of numbers with many small prime factors then there will be a large number of solutions to

$$\frac{m}{n} = \frac{a}{b} \quad \text{with} \quad a \in \mathcal{A} \quad \text{and} \quad b \in \mathcal{B}.$$

This means the fractions "fall on top of each other" rather than spread out evenly. The condition (20) is certainly satisfied for sets such as the primes, square–frees, integers in arithmetic progressions or sums of two squares. In these cases (and some others) the present author has shown that there are infinitely many solutions to (19) given the divergence of (18) for almost all $\alpha > 0$. It is possible to give explicit examples of numbers in the exceptional sets for these problems, even though no numbers are known in the set of almost all α. For example, although there are infinitely many solutions to $|m\alpha - n| < 1/(2n)$ with m, n both sums of two squares for almost all $\alpha > 0$, the number $3 + 5^{-\frac{1}{2}}$ $(= [3, 2, \ddot{4}])$ has no such approximation because each numerator in its continued fraction expansion is congruent to 3 (mod 4) (they are $3, 7/2, 31/9, 131/38\cdots$). In like manner the number $(45 - 10^{\frac{1}{2}})/186$ ($= [0, 4, 2, \dot{4}, \dot{9}]$) is an example for the set of measure zero for approximation by fractions with square–free denominator. The convergents to the continued fraction expansion of this number are $1/4, 2/9, 9/40, 83/369, 341/1516, 3152/14013, \ldots$ and the denominators are divisible alternately by 4 and 9.

We now indicate how such results can be proved. The following lemma is fundamental to the start of the proof (see [16]). The lemma itself is a consequence of Cauchy's inequality and the Lebesgue density theorem.

LEMMA. — *Let $\mathcal{I}$ be a sub-interval of $\mathbb{R}$, and $\mathcal{D}_n$ a sequence of subsets of $\mathcal{I}$. For each open interval $\mathcal{J} \subset \mathcal{I}$ write $\mathcal{B}_n = \mathcal{D}_n \cap \mathcal{J}$ and suppose that*

$$\sum_n \lambda(\mathcal{B}_n) = \infty \tag{21}$$

and

$$\limsup_{N\to\infty} \Big(\sum_{n \le N} \lambda(\mathcal{B}_n) \Big)^2 \Big(\sum_{m,n \le N} \lambda(\mathcal{B}_n \cap \mathcal{B}_m) \Big)^{-1} \ge \delta \lambda(\mathcal{J}), \tag{22}$$

where $\lambda(\ )$ denotes Lebesgue measure, and δ is a positive constant independent of $\mathcal{J}$. Then almost all $\alpha \in \mathcal{I}$ belong to infinitely many $\mathcal{D}_n$.

Now suppose we are dealing with (19) with $\mathcal{A}$ and $\mathcal{B}$ the set of primes. We then take

$$\mathcal{D}_p = \mathcal{I} \cap \bigcup_{\substack{s \neq p \\ s \text{ a prime}}} \left(\frac{s - f(p)}{p}, \frac{s + f(p)}{p} \right),$$

which leads us to require an upper bound for

$$\sum_{p,q \le N} \lambda(\mathcal{B}_p \cap \mathcal{B}_q) \tag{22}$$

and a lower bound for

$$\sum_{p \le N} \lambda(\mathcal{B}_p) . \tag{23}$$

It is easy to give a lower bound for (23). If $\mathcal{I} = (A, B)$ and $\mathcal{J} = (a, b)$ then we obtain

$$2 \sum_{p \le N} \frac{f(p)}{\log Bp} (b - a)$$

for all large N. We therefore need to demonstrate that (22) is bounded by

$$K(b-a)\Big(\sum_{p\leq N}\frac{f(p)}{\log p}\Big)^2. \tag{24}$$

where K is independent of $\mathcal{J}$.

We note that for $p < q$ we have

$$\lambda(\mathcal{B}_p \cap \mathcal{B}_q) \leq \frac{2f(q)}{q} \sum_{\substack{r\sim q \ s\sim p \\ 0<|rp-sq|<2qf(p)}} 1.$$

Here $r \sim q$ signifies $r/q \in \mathcal{J}$. The problem has now been reduced to giving an upper bound for the number of solutions to

$$|rp - sq| < A \quad p, q, r, s \text{ all primes}. \tag{25}$$

One can use the Brun–Titchmarsh theorem to tackle (25) for p and q in certain ranges. When this result is inadequate the author fixed p and applied the three dimensional sieve to give an upper bound for the number of solutions to (25) in q, r and s. To deal with the remainder terms it is necessary to use exponential sums and the standard bound for the simplest incomplete Kloostermann sum. In this way (24) is established and the result proved.

4. — Question 3 : restrict both numerator and denominator, what happens for every irrational α?

In general our answers to this question are very much worse than for the previous questions. The most successful theorem in this area is the following result given by Heath–Brown [19] :

For every irrational α there are infinitely many fractions m/n with square–free numerator and denominator such that

$$|\alpha n - m| < n^{-2/3+\varepsilon}.$$

The proof uses a lattice argument for counting solutions to certain congruences. The problem of approximating any given irrational by a fraction whose numerator and denominator are both primes appears to be exceptionally difficult (like Goldbach's problem, another binary problem in primes). As yet no–one has shown that

$$|\alpha p - q| < 1 \tag{26}$$

il existe une constante effective K, dépendant uniquement de k, S et de Σ, telle que pour toute sous-variété X de Σ définie sur k, on ait

$$\deg \overline{(X(\overline{k}) \cap \Sigma_S)} \leq K(\deg X)^{2^{\dim X}(\dim X+1)(2g|S|+1)}.$$

Démonstration : pendant la démonstration, une *constante* signifira toujours un nombre ne dépendant que de k, S et de Σ. Fixons d'abord k et S, un entier $g \geq 1$ et désignons par $\mathcal{E}_g$ l'ensemble des variétés semi-abéliennes de dimension au plus g et définies sur k. Soit $\Sigma' \in \mathcal{E}_g$, soit T est le tore maximal de Σ' et soit $A = \Sigma'/T$, alors T et A sont définis sur k : si donc k' est une extension finie de k on a (d'après les "conjectures" de Weil) : $|T(k')| \leq (|k'|+1)^{\dim T}$, $|A(k')| \leq (\sqrt{|k'|}+1)^{2\dim A}$, d'où : $|\Sigma'(k')| \leq |T(k')|\,|A(k')| \leq (|k'|+1)^{\dim T})(\sqrt{|k'|}+1)^{2\dim A}$. Il s'ensuit que $|\Sigma_S(k')|$ est majoré uniquement en fonction de k et de g et donc que μ et N peuvent être majorés effectivement en fonction de (k, g, S).

Reprenons les notations utilisées lors de la démonstration du théorème A. D'après (5) et les propriétés du degré on obtient :

$$\text{(6)} \qquad \deg \overline{(V(\overline{k}) \cap \Sigma_S)} \leq |\Sigma[\mu_V n]| \deg B_V + \sum_W \deg \overline{(W(\overline{k}) \cap \Sigma_S)},$$

où n est divisible par $N_V \epsilon(H_{V,S})$ et les W parcourent une partie des $k_{\mu_V n}$-composantes irréductibles des intersections $V \cap T_Q V$ avec $Q \in \Sigma[n]$ tel que $\dim(V \cap T_Q) \leq \dim V - 1$.

Avant de continuer, nous aurons besoin d'un lemme. Rappelons que si V est une sous-variété de Σ, $\tilde{B}_V$ désigne le stabilisateur de V et B_V la composante connexe de $\tilde{B}_V$ contenant l'identité. Dans ce lemme, qui généralise le lemme 4 de [Bo1], k désigne momentairement un corps quelconque.

LEMME 3. — *On a* : $\deg \tilde{B}_V \leq (\deg V)^{\dim V+1}$.

Démonstration : écrivons $\tilde{B}$ (resp. B) à la place de $\tilde{B}_V$ (resp. B_V). On sait que $\tilde{B} = \cap_{Q \in V(\overline{k})} T_Q V$. Soit $\{V_\alpha\}$ l'ensemble des composantes connexes de V de dimension maximale. Si $\dim B = \dim V$, alors $\bigcup_\alpha V_\alpha$ est une réunion finie de translatés de $\tilde{B}$: comme le degré est stable par translation, on en tire que $\deg \tilde{B} \leq \deg V$ est le résultat est vrai. Si $\dim B \leq \dim V - 1$, comme $\overline{k}$ est infini et donc $V(\overline{k})$ ne peut pas être une réunion finie de points de sous-variétés propres, il existe Q_1, $Q_2 \in V(\overline{k})$ tels que $\tilde{B} \subseteq T_{Q_1} V \cap T_{Q_2} V$ et $\dim(T_{Q_1} V \cap T_{Q_2} V) < \dim V$. Soit alors $\{C_\beta\}$ l'ensemble des composantes irréductibles de $T_{Q_1} V \cap T_{Q_2} V$ de dimension $\dim V - 1$. Si $\dim B = \dim V - 1$ on conclut que les C_β sont des réunions finies de translatés de B et que $\bigcup_\beta C_\beta$ est une réunion de translatés de

is solvable in primes p and q. One can obtain p^λ on the right side of (26) where "almost all" intervals $[x, x + x^\lambda)$ contain primes (so $1/12 + \varepsilon$ is a possible value for λ). This result is true for rational α as well though!

Just as progress has been made on Goldbach's problem by using almost-primes instead of primes so the same approach can be made here. The first such result was given by Vaughan in 1976 [28] :

$$|p\alpha - P_4| < p^{-\delta}$$

where P_4 denotes a number with at most 4 prime factors and the value 10^{-6} is given for δ (the important fact being that $\delta > 0$). The present author improved this by replacing P_4 with P_3 [13] (also giving a less significant improvement on δ to 1/300). This problem thus appears "harder" than Goldbach where a prime and a P_2 suffices [6]. It has been shown by Iwaniec [23] (see also [14]) that one can approximate irrationals in this way with fractions whose numerator and denominator are sums of two squares.

We finish this article by giving an indication of how such results can be established. We replace α with a convergent a/q, with error $1/q^2$ and choose the size of our possible numerator and denominator in relation to q. To approximate with p/P_3 we pick $X = q^{8/5}$, and write

$$\mathcal{A} = \{[\![pa/q]\!] : p \le X\,,\, \|pa/q\| < X^{-\delta}/2\}\,.$$

Here $[\![\;]\!]$ indicates *nearest* integer. We then want to show that $\mathcal{A}$ contains P_3 numbers. To do this we need to consider how well $\mathcal{A}$ is distributed in arithmetic progressions. We write

$$\mathcal{A}_d = \#\{n \in \mathcal{A} : n \equiv 0 \pmod d\}\,, \quad R_d = \mathcal{A}_d - \frac{\mathcal{A}_1}{d}\,.$$

We then wish to show that

$$\sum_{d \le D} |R_d| = o\left[\frac{\mathcal{A}_1}{\log X}\right] \tag{27}$$

for as large a value of D as possible. Using a familiar argument (see Chapter 2 of [1]) we obtain

$$R - d \ll \frac{X^{1-\delta-\varepsilon}}{d} + \frac{X^{-\delta}}{d} \sum_{\ell \le L} \left| \sum_{p \le X} e\left[\frac{p\ell a}{qd}\right] \right|,$$

where $L = L(d) = dX^{\varepsilon+\delta}$. We can convert the sum over primes above into a double sum which leads us to seek estimates for

$$\sum_d \sum_\ell \left| \sum_m \sum_n a_n b_m\, e\left[\frac{mn\ell a}{qd}\right] \right|.$$

Using the large sieve and other devices we obtain a suitable estimate when $D < X^{1/3-\delta-\varepsilon}$. This establishes the result after employing Chen's rôle reversal technique [6]. The reader can see from these results that there is much work to be done in this area to give a more satisfactory "contribution à l'analyse arithmétique du continu".

Additional note : since delivering the above talk (November 1992) there has been further progress on two problems. The most important is the announcement by A. Zaharescu that the exponent $\frac{1}{2}$ can be improved to 4/7 in Heilbronn's Theorem on $\|\alpha n^2\|$, with a further improvement to 2/3 if one is only looking for infinitely many solutions. His proof uses character sums not exponential sums. As mentioned above, numerical calculation is required to obtain the best exponent ρ for $\|\alpha p\| < p^{-\rho}$ and Jia Chao–Hua (J. Number Theory 45 (1993), 241–253) has shown that one can take $\rho = 4/13$ $(0.308\ldots)$. The present author will show elsewhere how the method can be improved to yield $7/22$ $(0.318\ldots)$.

The referee in his comments pertinently remarked that I should have mentioned in section 4 the important work of Margulis (Discrete subgroups and ergodic theory in *Number Theory, trace formulas and discrete groups, Oslo 1987*, pages 377–398, Academic Press, Boston 1989) whereby every irrational α has infinitely many approximations

$$\left|\alpha - \frac{n^2}{u^2+v^2}\right| < \frac{\varepsilon}{u^2+v^2} \quad \text{for any} \quad \varepsilon > 0\,.$$

Manuscrit reçu le 2 septembre 1993

References

[1] R.C. BAKER. — *Diophantine Inequalities*, Clarendon Press, Oxford 1986.

[2] R.C. BAKER and G. HARMAN. — *On the distribution of αp^k modulo one*, Mathematika, **38** (1991), 170–184.

[3] H. BEHNKE. — *Über die Verteilung von Irrationalitäten* mod 1, Abh. Math. Sem. Hamburg, **1** (1922), 252–267.

[4] E. BOREL. — *Une contribution à l'analyse arithmétique du continu*, Journal de Mathématiques Pures et Appliquées, (5ème série), **9** (1903), 329–375.

[5] J.W.S. CASSELS. — *Some metrical theorems in Diophantine approximation I*, Proc. Cambridge Phil. Soc., **46** (1950), 209–218.

[6] J.-R. CHEN. — *On the representation of a large even integer as the sum of a prime and the product of at most two primes*, Sci. Sinica, **16** (1973), 157–176.

[7] I. DANICIC. — *Contributions to Number Theory*, Ph. D. Thesis, London 1957.

[8] R.J. DUFFIN and A.C. SCHAEFFER. — *Khintchine's problem in Metric Diophantine approximation*, Duke Math. J., **8** (1941), 243–255.

[9] P. ERDÖS. — *On the distribution of the convergents of almost all real numbers*, J. Number Theory, **2** (1970), 425–441.

[10] P.X. GALLAGHER. — *Approximation by reduced fractions*, J. Math. Soc. of Japan, **13** (1961), 342–345.

[11] G.H. HARDY and J.E. LITTLEWOOD. — *The fractional part of $n^k\theta$*, Acta Math., **37** (1914), 155–191.

[12] G. HARMAN. — *On the distribution of αp modulo one*, J. London Math. Soc., (2) **27** (1983), 9–18.

[13] G. HARMAN. — *Diophantine approximation with a prime and an almost prime*, J. London Math. Soc., (2) **29** (1984), 13–22.

[14] G. HARMAN. — *Diophantine approximation with almost primes and two squares*, Mathematika, **32** (1985), 301–310.

[15] G. HARMAN. — *Metric diophantine approximation with two restricted variables I*, Math. Proc. Cambridge Phil. Soc., **103** (1988), 197–206.

[16] G. HARMAN. — *Metric diophantine approximation with two restricted variables III*, J. Number Theory, **29** (1988), 364–375.

[17] G. HARMAN. — *Some cases of the Duffin and Schaeffer conjecture*, Quart. J. Math. Oxford, (2) **41** (1990), 395–404.

[18] G. HARMAN. — *Numbers badly approximable by fractions with prime denominator*, preprint Cardiff, 1993.

[19] D.R. HEATH-BROWN. — *Diophantine approximation with square-free integers*, Math. Zeit., **187** (1984), 335–344.

[20] H. HEILBRONN. — *On the distribution of the sequence* $n^2\theta$ (mod 1), Quart. J. Math. Oxford, (1) **19** (1948), 249–256.

[21] A. HURWITZ. — *Über die angenäherte Darstellung der Irrationalzahlen durch rationale Brüche*, Math. Ann., **39** (1891), 279–284.

[22] C. HUYGENS. — *Projet de 1680–81, partiellement exécuté à Paris, d'un planétaire tenant compte de la variation des vitesses des planètes dans leurs orbites supposées elliptiques ou circulaires, et considération de diverses hypothèses sur cette variation*, in Œuvres Complètes de Christian Huygens, **21**, 109–163, Martinus Nijhoff, La Haye (1944).

[23] H. IWANIEC. — *On indefinite quadratic forms in four variables*, Acta Arithmetica, **33** (1977), 209–229.

[24] A. KHINTCHINE. — *Zür metrischen Theorie der diophantischen Approximationen*, Math. Zeit., **24** (1926), 706–714.

[25] A.D. POLLINGTON and R.C. VAUGHAN. — *The k-dimensional Duffin and Schaeffer conjecture*, Mathematika, **37** (1990), 190–200.

[26] A.M. ROCKETT and P. SZÜSZ. — *Continued Fractions*, World Scientific, Singapore–New Jersey–London–Hong Kong 1992.

[27] J.D. VAALER. — *On the metric theory of Diophantine approximations*, Pacific J. Math., **76** (1978), 527–539.

[28] R.C. VAUGHAN. — *Diophantine approximation by prime numbers III*, Proc. London Math. Soc., (3) **33** (1976), 177–192.

[29] R.C. VAUGHAN. — *On the distribution of* αp *modulo* 1, Mathematika, **24** (1977), 135–141.

[30] I.M. VINOGRADOV. — *The method of trigonometric sums in the theory of numbers* (translated from the Russian by K.F. Roth and A. Davenport), Wiley–Interscience, London 1954.

Glyn HARMAN
School of Mathematics
University of Wales College of Cardiff,
23 Senghenydd Road,
P.O. Box 926,
CARDIFF CF2 4YH
United Kingdom

Number Theory
Paris 1992–93

On Λ-adic forms of half integral weight for $SL(2)_{/\mathbb{Q}}$

Haruzo HIDA[*]

1. — Let $\widetilde{S}$ be the two-fold metaplectic cover of $S = SL(2)_{/\mathbb{Z}}$ and fix a prime $p \geq 5$. In this short note[**], we want to describe a technique of lifting a family of complex automorphic representations of $\widetilde{S}(\mathbb{A})$ to a "Λ-adic automorphic" representation Π of $\widetilde{S}(\mathbb{A}^{(p\infty)})$, where Λ is a one variable power series ring over an appropriate p-adically complete discrete valuation ring, and $\mathbb{A}^{(p\infty)}$ is the adele ring $\mathbb{A}$ of $\mathbb{Q}$ the p and ∞-components removed. Then we will have a Λ-adic version of a result of Waldspurger [Wa2]. We begin with the study of p-adic cusp forms of half integral weight and prove in Section 3 that the classical cusp forms of weight $k + \frac{1}{2}$ is dense in the space of p-adic cusp forms of half integral weight if $k \geq 2$ (Theorem 1). Then we study Λ-adic forms of half integral weight in Section 4 by combining the techniques of Wiles [Wi] (introduced for integral weights) and the representation theoretic technique of Waldspurger [Wa1,2]. Taking the limit shrinking the congruence subgroup, we get the desired Λ-adic representation of $\widetilde{S}(\mathbb{A}^{(p\infty)})$ (Proposition 1). Then we prove the weak multiplicity one theorem for p-ordinary Λ-adic automorphic representations (Theorem 2 in Section 4). Although our construction is just the combination of these two existing techniques, we get a fairly strong result on p-adic standard L-functions of $G = GL(2)_{/\mathbb{Q}}$. That is, a certain ratio of the restriction of 2-variable p-adic standard L-functions [K] to the line interpolating

---

[*] The author is partially supported by an NSF grant. The final touch to the paper was given while the author was visiting the Isaac Newton Institute for Mathematical Sciences, Cambridge, England. The author acknowledges the support from the Institute for the month of April in 1993.

[**] Some part of the work presented in this note was actually done in 1988 in order to construct a p-adic standard L-functions for $GL(2)$ restricted at the center critical line (Theorem 4). The construction of two variable p-adic standard L-functions was later done by K. Kitagawa [K] using a different method.

the central critical values is shown to be square in the field of fractions of the Iwasawa algebra Λ (Theorems 3 and 4), which is the Λ-adic version of a result of Waldspurger ([Wa2] Corollary 2) we alluded to. A further scrutinizing of the representation we constructed might bring us a sharpening of this result giving a Λ-adic version of the result in [V]. However to make our presentation short, we will not touch this subject in the present account. Another interesting point which awaits further study is the behavior of the specialization $\pi_{wt=2}$ of irreducible factors π of Π at weight 2. In [GS], Greenberg and Stevens gave an interesting limit formula of the derivative of the p-adic standard L-function at the center critical point, when the L-function has an exceptional zero at this point. This is the unique case where the specialized automorphic representation $\pi_{wt=2}$ of $\widetilde{S}(\mathbb{Z}^{p\infty)})$ (supplemented with the p-component) becomes super cuspidal at p although the integral image of $\pi_{wt=2}$ under the Shimura correspondence is special and p-ordinary. Thus the study of the behavior of the other local components of $\pi_{wt=2}$ might cast some new insight upon the p-adic analog of the conjecture of Birch–Swinnerton Dyer formulated in [MTT]. Although I have only worked out here the result for $SL(2)$ defined over $\mathbb{Q}$, our idea works fine for $SL(2)$ over general number fields. However, in the general case, the many variable standard p-adic L-functions defined on the spectrum of the p-adic Hecke algebra are not yet constructed.

2. — Let Δ be a congruence subgroup of level prime to p. When we consider modular forms of half integral weight, we assume that Δ is contained in $\Gamma_0(4)$. We write $\Delta_1(p^\alpha) = \Delta \cap \Gamma_1(p^\alpha)$ and $\Delta(p^\alpha) = \Delta_1(p) \cap \Gamma_0(p^\alpha)$. We use the same notation introduced in [H1] Sections 1 and 2 for classical modular forms. In particular, for each integer k and an algebra A, $\mathcal{P}_{k+(1/2)}(\Delta_1(p^\alpha); A))$ stands for the space of A-integral cusp forms of half integral weight $k + \frac{1}{2}$ with respect to $\Delta_1(p^\alpha)$, while for each integer κ, $\mathcal{S}_\kappa(\Delta_1(p^\alpha); A))$ stands for the space of A-integral cusp forms of integral weight κ. Here the A-integrality is given by the q-expansion at the cusp ∞. For each Dirichlet character χ modulo Np^α, $\mathcal{P}_{k+(1/2)}(\Gamma_0(Np^\alpha); \chi; A)$ consists of cusp forms g in $\mathcal{P}_{k+(1/2)}(\Gamma_1(Np^\alpha); A)$ with $g|_{k+(1/2)}\sigma = \chi(d)g$ for each $\sigma = \begin{pmatrix} a & b \\ c & d \end{pmatrix} \in \Gamma_0(N)$, where $g|_{k+(1/2)}\sigma$ is the action of σ defined in [H1] (2.2a) which is a little different from the normalization of [Sh1] p. 447. Our normalization is :

$$g|_{k+(1/2)}\sigma(z) = g(\sigma(z))j(\sigma, z)^{-1}J(\sigma, z)^{-k} \text{ for } \sigma = \begin{pmatrix} a & b \\ c & d \end{pmatrix},$$

where $J(\sigma,z)=(cz+d)$ and $j(\sigma,z)=\theta(\sigma(z))/\theta(z)$ for $\theta(z)=\sum_{n=-\infty}^{\infty}\exp(2\pi in^2z)$. By [H1] Theorem 2.2 or its proof, $\mathcal{P}_{k+(1/2)}(\Gamma_1(Np^\alpha);A)$ is stable under this action of $\sigma\in\Gamma_0(Np^\alpha)$.

We now give an interpretation in adelic language following [Wa2] III. We write S for the algebraic group $SL(2)_{/\mathbb{Z}}$. We write $\widetilde{S}$ for the two fold metaplectic cover of $SL(2)$ defined in [Wa2] II.4. Thus $\widetilde{S}(\mathbb{Q}_p)$, $\widetilde{S}(\mathbb{A})$ and $\widetilde{S}(\mathbb{R})$ have meaning, where $\mathbb{A}$ is the adele ring of $\mathbb{Q}$. In other words, we have a non-splitting exact sequence of groups :

$$1\longrightarrow\{\pm1\}\longrightarrow\widetilde{S}(A)\longrightarrow S(A)\longrightarrow 1,$$

where A is either $\mathbb{A}$, $\mathbb{Q}_p$ or $\mathbb{R}$. Now let us describe the 2-cocycle β giving the extension $\widetilde{S}(\mathbb{Q}_v)$ for a place v of $\mathbb{Q}$. For each $\sigma=\begin{pmatrix}a&b\\c&d\end{pmatrix}$, we put $x(\sigma)=d$ or c according as $c=0$ or $c\neq0$. We also put

$$s_v(\sigma)=\begin{cases}(c,d)_v & \text{if } cd\neq0,\ v \text{ is finite},\quad v_p(c)\text{ is odd},\\ 1 & \text{otherwise},\end{cases}$$

where $(c,d)_v$ is the Hilbert symbol at v (that is, $\sqrt{c}^{(d,\mathbb{Q}_v)}=(c,d)_v\sqrt{c}$ for the Artin symbol $(d,\mathbb{Q}_v)$ of d). Then we have

$$\beta_v(\sigma,\sigma')=(x(\sigma),\ x(\sigma'))_v(-x(\sigma)x(\sigma'),\ x(\sigma\sigma'))_v s_v(\sigma)s_v(\sigma')s_v(\sigma\sigma').$$

For $\sigma\in S(\mathbb{Q})$, the product $s(\sigma)=\Pi_v s_v(\sigma_v)$ is well defined. Similarly we may define $\beta(\sigma,\sigma')=\Pi_v\beta(\sigma_v,\sigma'_v)$ for σ, $\sigma'\in S(\mathbb{A})$. Then we identify $\widetilde{S}(\mathbb{A})$ with $S(\mathbb{A})\times\{\pm1\}$ under the multiplication law given by $(g,\varepsilon)(h,\varepsilon')=(gh,\beta(g,h)\varepsilon\varepsilon')$. By the product formula of the Hilbert symbol, $\beta(\sigma,\sigma')=s(\sigma)s(\sigma')s(\sigma\sigma')$ for σ, $\sigma'\in S(\mathbb{Q})$. Thus $\sigma\longmapsto(\sigma,s(\sigma))$ gives a section : $S(\mathbb{Q})\to\widetilde{S}(\mathbb{A})$. We identify $S(\mathbb{Q})$ with its image in $\widetilde{S}(\mathbb{A})$. We also identify the standard maximal compact subgroup $SO_2(\mathbb{R})$ with $\mathbb{R}/\mathbb{Z}$ by $\theta\longmapsto\begin{pmatrix}\cos2\pi\theta&\sin2\pi\theta\\-\sin2\pi\theta&\cos2\pi\theta\end{pmatrix}$. Then the pull back image of $SO_2(\mathbb{R})$ in $\widetilde{S}(\mathbb{R})$ can be identified with $\mathbb{R}/2\mathbb{Z}$. We write the corresponding element $r(\theta)$ in $\widetilde{S}(\mathbb{R})$ and $C_\infty=\{r(\theta)\mid\theta\in\mathbb{R}/2\mathbb{Z}\}$. Then $r(\theta)\longmapsto e((k+\frac12)\theta)$ for an integer k and $e(\theta)=\exp(2\pi i\theta)$ is a character of C_∞. Via $(g,\varepsilon)\longmapsto g(i)\in H$, we have $\widetilde{S}(\mathbb{R})/C_\infty\cong H$ for the upper half complex plane H. Let $\mathbf{e}:\mathbb{A}/\mathbb{Q}\to\mathbb{C}$ be the standard additive character such that $e(x_\infty)=\exp(2\pi ix_\infty)$. We write $\mathbf{e}_v$ for the restriction of $\mathbf{e}$ to $\mathbb{Q}_v$ for each place v, and we define $\gamma_v(t)$ to be the Weil's constant with respect to $\mathbf{e}_v$ and the quadratic form tx^2 on $\mathbb{Q}_v$ [W] p. 161. We put, following [Wa2], $\tilde{\gamma}(t)=(t,t)_v\gamma_v(t)\gamma_v(1)^{-1}$. Then we have $\tilde{\gamma}_v(tt') = (t,t')_v\tilde{\gamma}_v(t)\tilde{\gamma}_v(t')$, $\tilde{\gamma}_v(t^2) = 1$ for arbitrary v, and $\tilde{\gamma}_\ell(t) = 1$ if $t \in \mathbb{Z}_\ell^\times$ and $\ell \neq 2$, $\tilde{\gamma}_2(1) = \tilde{\gamma}_2(5) = 1$ and $\tilde{\gamma}_2(3) = \tilde{\gamma}_2(7) = -i$. Let

$$U_0(N) = \left\{\begin{pmatrix} a & b \\ c & d \end{pmatrix} \in S(\widehat{\mathbb{Z}}) \mid c \in N\widehat{\mathbb{Z}}\right\} \quad (\widehat{\mathbb{Z}} = \Pi_{\ell\ \mathrm{prime}}\mathbb{Z}_\ell)$$

and write $U_0(N)_\ell$ for the ℓ-component of $U_0(N)$. Defining for $\sigma = \begin{pmatrix} a & b \\ c & d \end{pmatrix} \in S(\mathbb{Q}_2)$

$$\tilde{\varepsilon}_2(\sigma) = \begin{cases} \tilde{\gamma}_2(d)^{-1}(c,d)_2 s_2(\sigma) & \text{if } c \neq 0, \\ \tilde{\gamma}_2(d) & \text{if } c = 0, \end{cases}$$

we can check that $\tilde{\varepsilon}$ extends to a character of $U_0(4)_2 \times \{\pm 1\}$ in $\widetilde{S}(\mathbb{Q}_2)$ non-trivial on $\{\pm 1\}$. Let χ be a character of $(\mathbb{Z}/N\mathbb{Z})^\times$ with $\chi(-1) = 1$. For $(u,\varepsilon) \in U_0(N) \times \{\pm 1\}$, we define $\chi(u) = \chi(d)$ if $u = \begin{pmatrix} a & b \\ c & d \end{pmatrix}$. We then consider the space of functions f satisfying :

$$\text{(m1)} \qquad f(\alpha x(u,\varepsilon)r(\theta)) = \tilde{\varepsilon}_2(u_2)\varepsilon\chi(u)f(x)e((k+\frac{1}{2})\theta)$$

for $\alpha \in S(\mathbb{Q})$, $(u,\varepsilon) \in U_0(N) \times \{\pm 1\}$ and $r(\theta) \in C_\infty$. We impose another condition at ∞ :

(m2) $Df = \left(\frac{k'(k'-2)}{2}\right)f$ for $k' = k+\frac{1}{2}$ for the Casimir operator D at ∞.

We write $P_{k+(1/2)}(N,\chi;\mathbb{C})$ for the space of functions satisfying $(m1-2)$ which are cusp forms. Writing $J(g,z) = cz+d$ for $g = \begin{pmatrix} a & b \\ c & d \end{pmatrix} \in SL_2(\mathbb{R})$ and $z \in H$, we can identify

$$\widetilde{S}(\mathbb{R}) = \{(g,t(g,z)) | g \in SL_2(\mathbb{R}),\ t(g,z) : \text{ holomorphic on } H \text{ with } t(g,z)^2 = J(g,z)\}.$$

The product is then given by $(g,t(g,z))(h,t(h,z)) = (gh,t(g,h(z))t(h,z))$. We have a natural inclusion map $\widetilde{S}(\mathbb{R}) \to \widetilde{S}(\mathbb{A})$ and $\widetilde{S}(\mathbb{Q}_p) \to \widetilde{S}(\mathbb{A})$. We have the theta series : $\theta(z) : \sum_{n=-\infty}^{\infty} \exp(2\pi i n^2 z)$ defined on H. As is well known, putting $j(\gamma, z) = \theta(\gamma(z))/\theta(z)$, $j(\gamma, z)^2 = \left(\frac{-1}{d}\right) J(\gamma, z)$ if $\gamma \in \Gamma_0(4)$. Thus $\gamma \longmapsto (\gamma, \varepsilon_2(\gamma) j(\gamma, z))$ defines an inclusion of $\Gamma_0(4)$ into $\widetilde{S}(\mathbb{R})$. It is known that the extension splits over $U_1(4) = \left\{ \begin{pmatrix} a & b \\ c & d \end{pmatrix} \in S(\widehat{\mathbb{Z}}) \mid c \in 4\widehat{\mathbb{Z}} \text{ and } a \equiv d \equiv 1 \mod 4\widehat{\mathbb{Z}} \right\}$. Thus we have by the strong approximation theorem that $\widetilde{S}(\mathbb{A}) = S(\mathbb{Q}) U_1(4) \widetilde{S}(\mathbb{R})$. We can identify these two realizations by $\widetilde{S}(\mathbb{A}) \ni (g, t(g, z)) \longmapsto (g, t(g_\infty, z) J(g_\infty, z)^{-1/2})$ where the square root is taken so that $-\pi/2 < \arg(cz + d)^{1/2} \leq \pi/2$.

For each cusp form $f \in P_{k+(1/2)}(N, \chi; \mathbb{C})$, we define $F : H \to \mathbb{C}$ by $F(z) = f((g, 1)) J(g, i)^{k+(1/2)}$ for $z = g(i)$ $(g \in S(\mathbb{R}))$. Then as shown in [Wa2] Proposition 3, $f \longmapsto F$ induces an isomorphism :

$$P_{k+(1/2)}(N, \chi; \mathbb{C}) \cong \mathcal{P}_{k+(1/2)}(\Gamma_0(N), \chi; \mathbb{C}). \tag{2.1}$$

When f is cuspidal, the holomorphy of F follows from $(m1 - 2)$. Let us prove the above isomorphism. We have put $F(z) = f((g, 1)) J(g, i)^{k+(1/2)}$ for $g \in S(\mathbb{R})$. Then

$$F(\gamma(z)) = f((\gamma_\infty g, 1)) J(\gamma g, i)^{k+(1/2)}.$$

Suppose $\gamma \in \Gamma_0(N)$. Then note that

$$\begin{aligned}(\gamma_\infty g, 1) &= (\gamma g \gamma_f^{-1}, 1) = (\gamma, s(\gamma))(g\gamma_f^{-1}, 1)(1, s(\gamma^{-1})\beta(\gamma, g\gamma_f^{-1})) \\ &= (\gamma, s(\gamma))(g, 1)(\gamma_f^{-1}, 1)(1, s(\gamma^{-1})\beta(g, \gamma_f^{-1})\beta(\gamma, g\gamma_f^{-1})).\end{aligned}$$

Since β is a 2–cocycle, $\beta(h, k)\beta(g, hk) = \beta(gh, k)\beta(g, h)$. This shows

$$(\gamma_\infty g, 1) = (\gamma, s(\gamma))(g, 1)(\gamma_f^{-1}, 1)(1, s(\gamma^{-1})\beta(\gamma g, \gamma_f^{-1})\beta(\gamma, g)).$$

Thus :

$$\begin{aligned}F(\gamma(g)) &= f((\gamma, s(\gamma))(g, 1)(\gamma_f^{-1}, 1)(1, s(\gamma^{-1})\beta(\gamma g, \gamma_f^{-1})\beta(\gamma, g))) J(\gamma g, i)^{k+(1/2)} \\ &= f((g, 1)(\gamma_f^{-1}, 1)(1, s(\gamma^{-1})\beta(\gamma g, \gamma_f^{-1})\beta(\gamma, g))) J(\gamma g, i)^{k+(1/2)} \\ &= s(\gamma^{-1})\beta(\gamma g, \gamma_f^{-1})\beta(\gamma, g)\tilde{\varepsilon}_2(\gamma_f^{-1})\chi(\gamma_f^{-1}) f((g, 1)) J(\gamma g, i)^{k+(1/2)}.\end{aligned}$$

Since $J(\gamma g,i)^{1/2} = \beta(\gamma,g)J(\gamma,z)^{1/2}J(g,i)^{1/2}$ and $\chi(\gamma_f^{-1}) = \chi(d)$ if $\gamma = \begin{pmatrix} a & b \\ c & d \end{pmatrix}$, we see :

$$\begin{aligned} F(\gamma(z)) &= s(\gamma^{-1})\beta(\gamma g,\gamma_f^{-1})\tilde{\varepsilon}_2(\gamma_f^{-1})\chi(d)J(\gamma,z)^{k+(1/2)}f(g,1)J(g,i)^{k+(1/2)} \\ &= s(\gamma^{-1})\beta(\gamma g,\gamma_f^{-1})\tilde{\varepsilon}_2(\gamma_f^{-1})\chi(d)F(z)J(\gamma,z)^{k+(1/2)}. \end{aligned}$$

Thus we need to prove $s(\gamma^{-1})\beta(\gamma g,\gamma_f^{-1})\tilde{\varepsilon}_2(\gamma_f^{-1}) = \left(\frac{c}{d}\right)\tilde{\gamma}_2(d)$. If $c = 0$, the both sides are trivial. Thus we may assume that $c \neq 0$. The case $c \neq 0$ is treated in [Wa2] p. 388.

For any open subgroup U of $U_0(4)$, we write $\Gamma_U = S(\mathbb{Q}) \cap US(\mathbb{R})$. We write $P_{k+(1/2)}(U;\mathbb{C})$ for the space of holomorphic cusp forms on $\widetilde{S}(\mathbb{A})$ satisfying $(m2)$ and

(m′1)
$$f(\alpha x(u,\varepsilon)r(\theta)) = \tilde{\varepsilon}_2((u_2,\varepsilon))f(x)e((k+\frac{1}{2})\theta) \text{ for } u \in U \text{ and } \alpha \in S(\mathbb{Q}).$$

Then $\mathcal{P}_{k+(1/2)}(\Gamma_U;\mathbb{C}) \cong P_{k+(1/2)}(U;\mathbb{C})$. Thus we can transfer the rational structure from the classical side to the adelic side to have the spaces $P_{k+(1/2)}(U;A)$ for any subalgebra A of $\mathbb{C}$.

3. — In this section, we first prove the density theorem of low weight classical cusp forms in the space of p-adic cusp forms of half integral weight. Using this fact, we describe another way, much closer to Weil's original definition in [W] and due to Shimura [Sh2], to define $\widetilde{S}(\mathbb{A})$. By the strong approximation theorem, we have a bijection :

$$\{\text{congruence subgroups of } S(\mathbb{Z}) \text{ of level prime to } p\} \longleftrightarrow \mathcal{Z} = \{\text{open subgroups of } S(\mathbb{Z}^{(p)})\}$$

$$\Delta = \widehat{\Delta} \cap S(\mathbb{Z}) \leftrightarrow \widehat{\Delta} : \text{the closure of } \Delta \text{ in } S(\mathbb{Z}^{(p)}),$$

where $\mathbb{Z}^{(p)} = \prod_{\ell \neq p} \mathbb{Z}_\ell$. We put

$$\mathcal{S}_\kappa(\widehat{\Delta};A) = \bigcup_\alpha \mathcal{S}_\kappa(\Delta_1(p^\alpha);A) \text{ and } \mathcal{P}_{k+(1/2)}(\widehat{\Delta};A) = \bigcup_\alpha \mathcal{P}_{k+(1/2)}(\Delta_1(p^\alpha);A).$$

Let O be the ring of Witt vectors with coefficients in an algebraic closure $\overline{\mathbb{F}}_p$ of $\mathbb{F}_p$ and K be the field of fractions of O. Let Ω_p be the completion of an algebraic closure $\overline{\mathbb{Q}}_p$ of $\mathbb{Q}_p$ under its standard p-adic norm $|\,|_p$. We take an embedding : $K \to \Omega_p = \widehat{\overline{\mathbb{Q}}}_p$ and fix two embeddings $\overline{\mathbb{Q}} \to \mathbb{C}$ and $\overline{\mathbb{Q}} \to \Omega_p$ for an algebraic closure $\overline{\mathbb{Q}}$ of $\mathbb{Q}$. Put $A = O \cap \overline{\mathbb{Q}}$ and $\mathcal{S}_\kappa(\widehat{\Delta}; O) = \mathcal{S}_\kappa(\widehat{\Delta}; A) \otimes_A O$, $\mathcal{P}_{k+(1/2)}(\widehat{\Delta}; O) = \mathcal{P}_{k+(1/2)}(\widehat{\Delta}; A) \otimes_A O$. We write $\widehat{\mathcal{S}}(\widehat{\Delta}; O)$ (resp. $\widehat{\mathcal{P}}(\widehat{\Delta}; O)$) for the p-adic completion of $\mathcal{S}_\kappa(\widehat{\Delta}; O)$ (resp. $\mathcal{P}_{k+(1/2)}(\widehat{\Delta}; O)$), which is independent of κ (resp. k) if $\kappa \geq 2$ (resp. $k \geq 2$). This fact is proven in [H2] and [H6] for integral weight and is conjectured in [H1] for half integral weight. Now we can give a proof of this fact for half integral weight.

THEOREM 1. — *If $k \geq 2$ and $p > 3$, we have an isomorphism preserving q-expansions :*

$$\widehat{\mathcal{P}}_{k+(1/2)}(\widehat{\Delta}; O) \cong \widehat{\mathcal{P}}_{k+(3/2)}(\widehat{\Delta}; O).$$

Proof : let U be an open subgroup of $G(\widehat{\mathbb{Z}})$ ($G = GL(2)_{/\mathbb{Z}}$) and $Y(U)$ be the corresponding open modular curve. Suppose that $U \supset G(\mathbb{Z}_p)$ and we put

$$U(p^\alpha) = \left\{ s \in \mathbb{S} \mid s_p \equiv \begin{pmatrix} * & * \\ 0 & 1 \end{pmatrix} \bmod p^\alpha \right\}.$$

For each positive integer N, we put $\zeta_N = \exp\left(\frac{2\pi i}{N}\right)$. Then $Y_\alpha = Y(U(p^\alpha))$ has a model over $A = \mathbb{Z}[1/6N, \zeta_N]$ for the level N of U which is the moduli space parametrizing an elliptic curve E with U–structure and a Drinfeld style level structure at p; that is, a morphism $\phi : \mathbb{Z}/p^\alpha\mathbb{Z} \to E$ of group schemes such that $\sum_{P \in \mathbb{Z}/p^\alpha\mathbb{Z}} [\phi(P)]$ is of degree p^α as a relative Cartier divisor (see [KM] Chapter 1 or [H7]). Suppose that $U \subset U_0(4)$. We can compactify Y_α adding cusps to get the proper curve X_α, which is regular proper over $\mathbb{Z}_p$ [KM]. Let $\omega_{/Y_\alpha}$ be the invertible sheaf corresponding to weight 1 modular forms studied in [KM]. Let I_α be the Igusa curve containing the cusp ∞ which is the irreducible component of $X_\alpha \bmod p^\alpha$. If we consider the p–ordinary moduli problem $\phi : \mu_{p^\alpha} \subset E$ of generalized semi–stable elliptic curves, it gives an open subscheme U_α of X_α whose fiber at p is I_α–{super singular points}. Then there exists a unique invertible sheaf $\omega_{1/2}$ on U_α such that $\omega_{1/2}^{\otimes 2} = \omega$ and $\theta \in \Gamma(U_{\alpha/\mathbb{C}}, \omega_{1/2})$. By the q–expansion principle and $p > 3$, θ is a section defined over $\mathbb{Z}_p$. We first suppose that U is contained in the principal congruence subgroup of level 24. Then the Dedekind η function is a section of $H^0(U_\alpha, \omega_{1/2})$. Writing $\omega(2 + (k/2))$ for $\omega_{1/2}^{\otimes k} \otimes \omega^\circ\ |_{U_\alpha}$, we consider the following commutative diagram :

$$\begin{array}{ccc}
 & & 0 \\
 & & \downarrow \\
H^0(U_\alpha,\omega(k+\frac{1}{2}))\otimes\mathbb{Z}/p^\alpha\mathbb{Z} & \longrightarrow & H^0(U_\alpha,\omega(k+\frac{1}{2})\otimes\mathbb{Z}/p^\alpha\mathbb{Z}) \\
\downarrow \eta & & \downarrow \eta \\
H^0(U_\alpha,\omega_1(k+1))\otimes\mathbb{Z}/p^\alpha\mathbb{Z} & \longrightarrow & H^0(U_\alpha,\omega_1(k+1)\otimes\mathbb{Z}/p^\alpha\mathbb{Z}) \\
\downarrow & & \downarrow \rho \\
H^0(U_\alpha,O(D))\otimes\mathbb{Z}/p^\alpha\mathbb{Z} & \longrightarrow & H^0(U_\alpha,O(D)\otimes\mathbb{Z}/p^\alpha\mathbb{Z}) \\
\downarrow & & \\
0, & &
\end{array}$$

where D is a cuspidal divisor given by $\operatorname{div}(\eta) = \sum_{s\in U_\alpha,\operatorname{ord}_s(\eta)>0} (\operatorname{ord}_s(\eta))s$ and the first horizontal maps are given by the multiplication by η. Here we regard D as a closed subscheme of U_α in a natural way, and $O(D)$ is its structure sheaf. The first row is exact. When $k \geq 2$, $\deg\left(\omega\left(k+\frac{1}{2}\right)\otimes_A\mathbb{Q}\right) > \deg(\Omega_{X_\alpha/A}\otimes_A\mathbb{Q})$. Thus the Riemann–Roch theorem tells us the vanishing of $H^1\left(X_\gamma,\omega\left(k+\frac{1}{2}\right)\right)\otimes_A\mathbb{Q} = H^1\left(U_\gamma,\omega\left(k+\frac{1}{2}\right)\right)\otimes_A\mathbb{Q}$. Since $\omega\left(k+\frac{1}{2}\right)$ is A-flat, this shows the vanishing of $H^1\left(U_\gamma,\omega\left(k+\frac{1}{2}\right)\right)$ and the exactness of the second row. Since the vertical maps are injective, we have a commutative diagram whose rows are exact if $k \geq 2$

$$\begin{array}{ccc}
0 & & 0 \\
\downarrow & & \downarrow \\
H^0(U_\gamma,\omega(k+\frac{3}{2}))\otimes\mathbb{Z}/p^\beta\mathbb{Z} & \xrightarrow{E_\alpha} & H^0(U_\gamma,\omega(k+\frac{1}{2}))\otimes\mathbb{Z}/p^\beta\mathbb{Z} \\
\downarrow & & \downarrow \\
H^0(U_\gamma,\omega(k+2))\otimes\mathbb{Z}/p^\beta\mathbb{Z} & \xrightarrow{E_\alpha} & H^0(U_\gamma,\omega(k+1))\otimes\mathbb{Z}/p^\beta\mathbb{Z} \\
\downarrow & & \downarrow \\
H^0(U_\gamma,O(D))\otimes\mathbb{Z}/p^\beta\mathbb{Z} & = & H^0(U_\gamma,O(D))\otimes\mathbb{Z}/p^\beta\mathbb{Z}
\end{array}$$

where $0 < \beta \leq \alpha \leq \gamma$ and E_α is the modular form on U_α of weight 1 with $E_\alpha \equiv 1 \bmod p^\alpha$. Taking injective limit with respect to γ, we write

$$H^0(U_\infty, \omega(\ell)) = \varinjlim_{\gamma} H^0(U_\gamma, \omega(\ell)).$$

Then we have by the p-adic density theorem of integral weight modular forms, if $k \geq 2$

$$\begin{array}{ccc} 0 & & 0 \\ \downarrow & & \downarrow \\ H^0(U_\infty, \omega(k+\frac{3}{2})) \otimes \mathbb{Z}/p^\beta\mathbb{Z} & \xrightarrow{E_\alpha} & H^0(U_\infty, \omega(k+\frac{1}{2})) \otimes \mathbb{Z}/p^\beta\mathbb{Z} \\ \downarrow & & \downarrow \\ H^0(U_\infty, \omega(k+2)) \otimes \mathbb{Z}/p^\beta\mathbb{Z} & \overset{\sim}{=\!=\!=} & H^0(U_\infty, \omega(k+1)) \otimes \mathbb{Z}/p^\beta\mathbb{Z} \\ \downarrow & & \downarrow \\ H^0(U_\infty, O(D)) \otimes \mathbb{Z}/p^\beta\mathbb{Z} & =\!=\!= & H^0(U_\infty, O(D)) \otimes \mathbb{Z}/p^\beta\mathbb{Z}. \end{array}$$

This shows the p-adic density theorem for half integral weight if $24 \mid N$. If not, we just use restriction and transfer maps and recover the result in general if $p > 3$.

Put

$$\begin{aligned} \mathcal{S}_\kappa(A) &= \bigcup_{\widehat{\Delta} \in \mathcal{Z}} \mathcal{S}_\kappa(\widehat{\Delta}; A) \quad \text{and } \mathcal{P}_\kappa(A) = \bigcup_{\widehat{\Delta} \in \mathcal{Z}} \mathcal{P}_\kappa(\widehat{\Delta}; A), \\ \widehat{\mathcal{S}}(O) &= \bigcup_{\widehat{\Delta} \in \mathcal{Z}} \widehat{\mathcal{S}}(\widehat{\Delta}; O) \quad \text{and } \widehat{\mathcal{P}}(O) = \bigcup_{\widehat{\Delta} \in \mathcal{Z}} \widehat{\mathcal{P}}(\widehat{\Delta}; O). \end{aligned}$$

If $f \in \mathcal{S}_\kappa(A)$, one can find Γ such that $f \in \mathcal{S}_\kappa(\Gamma; A)$. Then for each $x \in S(\mathbb{A}^{(p\infty)})$ ($\mathbb{A}^{(p\infty)} = \{x \in \mathbb{A} \mid x_p = x_\infty = 0\}$), one can find $u \in \widehat{\Gamma} \subset S(\mathbb{A}^{(\infty)})$ and $\gamma \in S(\mathbb{Q})$ such that $x = u\gamma$, where $\widehat{\Gamma}$ is the closure of Γ in $S(\mathbb{A}^{(\infty)})$ ($\mathbb{A}^{(\infty)} = \{x \in \mathbb{A} \mid x_\infty = 0\}$). Some time ago, Shimura defined the action of $x \in S(\mathbb{A}^{(p\infty)})$ on f by $f^x = f \mid \gamma$ [Sh2]. Then he showed that the action is a smooth action of $S(\mathbb{A}^{(p\infty)})$ on $\mathcal{S}_k(\mathbb{Q}_{ab}^{(p)})$, where $\mathbb{Q}_{ab}^{(p)} = \mathbb{Q}[\zeta_N \mid (p, N) = 1]$ is the maximal abelian extension of $\mathbb{Q}$ unramified at p.

Using Katz's theory of p-adic modular forms (see [H7] Chapter 2), it is easy to check that the action of $S(\mathbb{A}^{(p\infty)})$ preserves $\mathcal{S}_\kappa(A)$ and extends to $\widehat{\mathcal{S}}(O)$ by p-adic continuity. Note that the representation of $S(\mathbb{A}^{(p\infty)})$ we obtained is smooth, but not of finite type. I like to call this representation the p-adic automorphic representation of $S(\mathbb{A}^{(p\infty)})$.

According to Shimura [Sh2], we can give a definition of $\widetilde{S}(\mathbb{A}^{(p\infty)})$ as follows :

$$\widetilde{S}(\mathbb{A}^{(p\infty)})=\{(x,v)\in S(\mathbb{A}^{(p\infty)})\times GL(\widehat{\mathcal{P}}(O)) \mid (f^v)^2=(f^2)^x \text{ for all } f\in\widehat{\mathcal{P}}(O)\}.$$

Then we have an exact sequence : $1\to\{\pm 1\}\to\widetilde{S}(\mathbb{A}^{(p\infty)})\xrightarrow{\pi}S(\mathbb{A}^{(p\infty)})\to 1$. It is basically shown in [Sh2] that any $x\in S(\mathbb{A}^{(p\infty)})$ is liftable to an automorphism v of $\mathcal{P}_{k+(1/2)}(\mathbb{Q}_{ab}^{(p)})$. Since x preserves A-integrality, v keeps A-integrality and hence gives an automorphism of $\widehat{\mathcal{P}}(O)$. This shows the surjectivity of π. There is an alternative way of showing the surjectivity of π. One can check that the action of $S(\mathbb{Z}^{(p)})$ is liftable to half integral weight by multiplying half integral weight cusp forms by η (or θ), because the action of $S(\mathbb{Z}^{(p)})$ preserves A-integral structure of integral weight cusp forms. It is easy to check the liftability of the action of upper triangular matrices. Thus by the Iwasawa decomposition, every $x\in S(\mathbb{A}^{(p\infty)})$ is liftable. By definition, we have a smooth p-adic "automorphic" representation of $\widetilde{S}(\mathbb{A}^{(p\infty)})$ on $\widehat{\mathcal{P}}(O)$.

Although we do not have a good action of $\widetilde{S}(\mathbb{Q}_p)$ on $\widehat{\mathcal{P}}(O)$, we can at least define an action of the maximal split torus $T(\mathbb{Z}_p)=\mathbb{Z}_p^{\times}$ in $S(\mathbb{Z}_p)$. Take a subgroup Δ corresponding to $\widehat{\Delta}\in\mathcal{Z}$. Thus its level N is prime to p. We assume that $\Gamma_0(4)\supset\Delta$. When A is a $\mathbb{Z}_p$-algebra, we can show multiplying by θ as done in [H1] §3 that $\mathcal{P}_{k+(1/2)}(\Delta_1(p^r);A)$ is stable under the action of $Z_N=\mathbb{Z}_p^{\times}\times(\mathbb{Z}/N\mathbb{Z})^{\times}$ for the level N of Δ, which is given for $f\in\mathcal{P}_{k+(1/2)}(\Delta_1(p^r);A)$ by

$$f\mid z=z_p^k f\mid\sigma_z \text{ for } \sigma_z\in SL_2(\mathbb{Z}) \text{ with } \sigma_z\equiv\begin{pmatrix} z^{-1} & 0\\ 0 & z\end{pmatrix} \bmod Np^r. \tag{3.1}$$

This action of $\mathbb{Z}_p^{\times}$ extends by continuity to $\widehat{\mathcal{P}}(O)$.

4. — We put $W=1+p\mathbb{Z}_p$ in $\mathbb{Z}_p^{\times}$. Then $W\cong\mathbb{Z}_p$ as topological groups, and $\mathbb{Z}_p^{\times}=W\times\mu$ for the subgroup μ of $(p-1)$-th roots of unity. Simplifying the notation, we write $\mathcal{P}_{k+(1/2)}(Np^\alpha;A)$ for $\mathcal{P}_{k+(1/2)}(\Gamma_1(Np^\alpha);A)$. We put, for $\Delta\leftrightarrow\widehat{\Delta}\in\mathcal{Z}$ and a character ε of W modulo p^α,

$$\mathcal{P}_{k+(1/2)}(\Delta(p^\alpha);\varepsilon;A)=\{f\in\mathcal{P}_{k+(1/2)}(\Delta_1(p^\alpha);A)\;\big|\; f\mid z=\varepsilon(z)z_p^k f \text{ for } z\in W\},$$

where $\Delta(p^\alpha)=\Delta_1(p)\cap\Gamma_0(p^\alpha)$ and A is a ring either in Ω_p or in $\mathbb{C}$ containing all the values of ε on W. We now consider the action of the

Hecke operator $T(q^2)$ for each prime q on $\mathcal{P}_{k+(1/2)}(\Delta_1(p^\alpha);\mathbb{C})$. As shown in [Sh1] Theorem 1.7, we know

$$\begin{aligned} &a(n,f|T(q^2)) = a(p^2n,f) + q^{-1}\left(\tfrac{n}{p}\right)a(n,f|q) + q^{-1}a(n/q^2,f|q^2) \text{ if } q\nmid Np^\alpha, \\ &a(n,f|T(q^2)) = a(p^2n,f) \text{ if } q|Np^\alpha, \end{aligned} \tag{4.1}$$

where N is the level of Δ and $q \in Z_N$ $(= \mathbb{Z}_p{}^\times \times (\mathbb{Z}/N\mathbb{Z})^\times)$ acts on f as in (3.1). This combined with [H1] Theorem 2.2 shows that $\mathcal{P}_{k+(1/2)}(\Delta_1(p^\alpha);O)$ is stable under $T(q^2)$. In particular, we can define the idempotent e in $\mathrm{End}_O(\mathcal{P}_{k+(1/2)}(\Delta_1(p^\alpha);O))$ by taking the limit :

$$e = \lim_{n\to\infty} T(p^2)^{n!}. \tag{4.2}$$

We write M^{ord} for eM for any module M with an action of e.

Hereafter we allow as a base ring a finite extension of the ring of Witt vectors with coefficients in $\overline{\mathbb{F}}_p$ and write the ring as O and its field of fractions as K. All the definitions we have given for the ring of Witt vectors carry over to this slightly general situation by extending scalar to O from the ring of Witt vectors. Write $\Lambda = O[[W]]$ for the completed group algebra of W. Then Λ is isomorphic to the one variable power series ring $O[[X]]$ via $u \longmapsto 1+X$ if we fix a generator $u \in W$. We fix an algebraic closure $\overline{\mathbb{L}}$ of the quotient field $\mathbb{L}$ of Λ and consider the algebraic closure of K in Ω_p as a subfield of $\overline{\mathbb{L}}$. For each normal integral domain $\mathbb{I}$ in $\overline{\mathbb{L}}$ finite over Λ, let $\mathcal{X}(\mathbb{I}) = \mathrm{Hom}_{O\text{-alg}}(\mathbb{I},\Omega_p)$ be the space of all Ω_p-valued points of $\mathrm{Spec}(\mathbb{I})$ and $\mathcal{A}(\mathbb{I})$ be the subset of arithmetic points, that is, those O-algebra homomorphisms $P:\mathbb{I}\to\Omega_p$ such that $P(\gamma) = \gamma^{k(P)}$ for an integer $k(P)\geq 0$ on a neighborhood of the identity of W. Thus $\varepsilon_P(\gamma) = P(\gamma)\gamma^{-k(P)}$ defines a finite order character of W, whose order will be denoted by $P^{r(P)-1}$. We write $\mathcal{A}(\mathbb{I};O) = \{P \in \mathcal{A}(\mathbb{I}) \mid O \supset P(\mathbb{I})\}$. For each congruence subgroup Δ (with level N) associated with $\widehat{\Delta} \in \mathcal{Z}$, let $\mathbb{P}(\Delta;\mathbb{I})$ be the space of $\mathbb{I}$-adic cusp forms. Thus $\mathbf{f} \in \mathbb{P}(\Delta;\mathbb{I})$ is a formal q-expansion :

$$\mathbf{f} = \sum_{n=0}^{\infty} \mathbf{a}(n/N,\mathbf{f})q^{n/N} \in \mathbb{I}[[q^{1/N}]]$$

whose specialization $\mathbf{f}(P) = \sum_{n=0}^{\infty} P(\mathbf{a}(n/N,\mathbf{f}))q^{n/N} \in P(\mathbb{I})[[q^{1/N}]]$ at $P \in \mathcal{A}(\mathbb{I})$ is a classical cusp form in $\mathcal{P}_{k(P)+(1/2)}(\Delta(p^{r(P)}),\varepsilon_p;\Omega_p)$ for all $P \in \mathcal{A}(\mathbb{I})$ with sufficiently large $k(P)\geq 0$. When $\Delta = \Gamma_1(N)$ $(4 \mid N)$, we write $\mathbb{P}(N;\mathbb{I})$ for $\mathbb{P}(\Delta;\mathbb{I})$. Since Λ is a regular local ring of dimension 2, $\mathbb{I}$ is Λ-free. Fixing a base $\{\mathbf{i}_j\}$ of $\mathbb{I}$ over Λ, we can write formally that $\mathbf{f} = \Sigma_j \mathbf{f}_j \mathbf{i}_j$. Then it is easy to see that $\mathbf{f}_j$ is a Λ–adic form. Thus $\mathbb{P}(\Delta; \mathbb{I}) = \mathbb{P}(\Delta; \Lambda) \otimes_\Lambda \mathbb{I}$.

There is another interpretation of the above space of Λ–adic forms. We first identify Λ with the measure algebra on W having values in O. Then to each $\mathbf{f} \in \mathbb{P}(\Delta; \Lambda)$, we associate a p–adic measure $\phi \longmapsto \int_W \phi d\mathbf{f}$ on W having values in $O[[q^{1/N}]]$ by

$$\int_W \phi d\mathbf{f} = \sum_{n=1}^{\infty} \int_W \phi d\mathbf{a}(n/N, \mathbf{f}) q^{n/N} \in O[[q^{1/N}]]. \tag{4.3}$$

Writing $\chi_P(w) = \varepsilon_P(w) w^{k(P)}$ for each arithmetic point P (that is, the character of W corresponding to P), we have $\int_W \chi_P d\mathbf{f} = \mathbf{f}(P) \in \mathcal{P}_{k(P)+(1/2)}(\Delta(p^{r(P)}), \varepsilon_P; \Omega_p)$ for sufficiently large $k(P)$. Since $\{\chi_P \mid k(P) \gg 0\}$ spans a dense subspace of continuous functions on W having values in K, as a measure, $d\mathbf{f}$ has values in $\widehat{\mathcal{P}}(O)$. In particular, the new measure $\phi \longmapsto \int_W \phi d\mathbf{f} \mid s$ for $s \in \widetilde{S}(\mathbb{A}^{(p\infty)})$ again comes from a Λ–adic form $\mathbf{f} \mid s \in \mathbb{P}(\Delta_s; \Lambda)$ for a suitable congruence subgroup Δ_s corresponding to $\widehat{\Delta}_s \in \mathcal{Z}$. Thus, we have a natural action of $\widetilde{S}(\mathbb{A}^{(p\infty)})$ on $\mathbb{P}(\mathbb{I}) = \bigcup_{\widehat{\Delta} \in \mathcal{Z}} \mathbb{P}(\Delta; \mathbb{I})$. Similarly, we have an action of Hecke operators $T(q^2)$ and the group $\mathbb{Z}_p^\times$ on $\mathbb{P}(N; \mathbb{I})$. Writing $\iota : w \longmapsto [w]$ for the tautological character of W into Λ, we know that $w \in W$ acts on $\mathbb{P}(N; \mathbb{I})$ via ι, that is, $\mathbf{f} \mid w = [w]\mathbf{f}$. Since the projector e naturally acts on $\mathcal{P}_{k+(1/2)}(O)$ and hence on $\widehat{\mathcal{P}}(O)$, e again acts on $\mathbb{P}(\Delta; \mathbb{I})$ and $\mathbb{P}(\mathbb{I})$. We note this fact as

PROPOSITION 1. — *As long as q is prime to the level N of Δ, we have Hecke operators $T(q^2)$ given by* (4.1) *and the ordinary projector e on $\mathbb{P}(\Delta; \mathbb{I})$, and the metaplectic group $\widetilde{S}(\mathbb{A}^{(p\infty)})$ naturally acts on $\mathbb{P}(\mathbb{I})$ through a smooth representation.* Here the smoothness means that the stabilizer of each vector in the representation space is open in $\widetilde{S}(\mathbb{A}^{(p\infty)})$.

We can think of the corresponding notion of $\mathbb{I}$–adic cusp forms for integral weight modular forms (cf. [H5] Chapter 7). We briefly recall the definition. For $\widehat{\Delta} \in \mathcal{Z}$, a formal q–expansion $\mathbf{f} \in \mathbb{I}[[q^{1/N}]]$ is called an $\mathbb{I}$–adic cusp form of integral weight if $\mathbf{f}(P) \in \mathcal{S}_{k(P)}(\Delta(p^{r(P)}), \varepsilon_P; \Omega_p)$ whenever P is arithmetic and $k(P)$ is sufficiently large. We write $\mathbb{S}(\Delta; \mathbb{I})$ for the space of $\mathbb{I}$–adic cusp forms (of integral weight). Then similar to Proposition 1, we have Hecke operators $T(n)$ (cf. [H5] Chapter 7) and the ordinary projector e on $\mathbb{S}(\Delta; \mathbb{I})$. In this case, e is given on the space of p–adic cusp forms by $e = \lim_{n\to\infty} T(p)^{n!}$. The group $S(\mathbb{A}^{(p\infty)})$ naturally acts on $\bigcup_{\widehat{\Delta} \in \mathcal{Z}} \mathbb{S}(\Delta; \mathbb{I})$. We actually need to have $G(\mathbb{A}^{(p\infty)})$–action (recall $G = GL(2)_{/\mathbb{Z}}$). Note that

$G(\mathbb{A}) = G(\mathbb{Q})G(\widehat{\mathbb{Z}})G_+(\mathbb{R})$ for the identity connected component $G_+(\mathbb{R})$ of $G(\mathbb{R})$. For each open subgroup U of $G(\widehat{\mathbb{Z}})$, we consider cusp forms $f : G(\mathbb{A}) \to \mathbb{C}$ satisfying :

(M1) $$f(\alpha x u) = f(x)\det(u_\infty)J(u_\infty, i)^{-k} \text{ for } u \in UC_\infty\mathbb{R}^\times ;$$

(M2) $$Df = \left(\frac{k(k-2)}{2}\right)f ;$$

(M3) $$\int_{\mathbb{Q}\backslash\mathbb{A}} f\left(\begin{pmatrix} 1 & u \\ 0 & 1 \end{pmatrix} x\right) du = 0 \text{ for all } x \in G(\mathbb{A}) .$$

We write $S_k(U;\mathbb{C})$ for the space of functions f satisfying (M1–3). Choosing a complete representative set $R = R(U)$ for $G(\mathbb{Q})\backslash G(\mathbb{A})/UG_+(\mathbb{R})$ in $G(\widehat{\mathbb{Z}})$, we can define $F_t \in S_k(\Gamma_{tUt^{-1}};\mathbb{C})$ $(\Gamma_{tUt^{-1}} = S(\mathbb{Q}) \cap tUt^{-1}S(\mathbb{R}))$ for each $t \in R$ by $F_t(z) = f(tg)\det(g)^{-1}J(g,i)^k$, where $g \in G_+(\mathbb{R})$ such that $g(i) = z$. Then it is easy to see $S_k(U;\mathbb{C}) \cong \oplus_{t\in R}S_k(\Gamma_{tUt^{-1}};\mathbb{C})$. We then define $S_k(U;A)$ by the image of $\oplus_{t\in R}S_k(\Gamma_{tUt^{-1}};A)$. We can take R inside $\mathcal{R} = \left\{\begin{pmatrix} a & 0 \\ 0 & 1 \end{pmatrix} \mid a \in \mathbb{Z}^{(p)}\right\}$. We always choose R in this way. Then we have e and $T(p)$ well defined on $S_k(U;\Omega_p)$. Let

$$\mathcal{U} = \{U : \text{ open subgroup of } G(\mathbb{Z}^{(p)})\} .$$

Write $U_0 = U \times GL_2(\mathbb{Z}_p)$ for $U \in \mathcal{U}$. Taking $R(U_0)$ in $\mathcal{R}$ so that $R(U_0) \supset R(V_0)$ if $V \supset U$ for all $U, V \in \mathcal{U}$, we define $\mathbb{S}(U;\mathbb{I}) = \oplus_{t\in R(U_0)}\mathbb{S}(\Gamma_{tU_0t^{-1}};\mathbb{I})$ and $\mathbb{S}(\mathbb{I}) = \bigcup_{U\in\mathcal{U}} \mathbb{S}(U;\mathbb{I})$. Using the stability of $\bigcup_{\widehat{\Delta}\in\mathcal{Z}} \mathbb{S}(\Delta;\mathbb{I})$ under $S(\mathbb{A}^{(p\infty)})$, it is easy to check that $\mathbb{S}(\mathbb{I})$ is stable under $S(\mathbb{A}^{(p\infty)})$. Since $\begin{pmatrix} a & 0 \\ 0 & 1 \end{pmatrix}$ with $a \in \mathbb{A}^{(p\infty)}$ basically permutes the direct summands $\mathbb{S}(\Gamma_{tU_0t^{-1}};\mathbb{I})$ of $\mathbb{S}(U;\mathbb{I})$, $\mathbb{S}(\mathbb{I})$ is stable under $G(\mathbb{A}^{(p\infty)})$. We thus have

PROPOSITION 2. — *The space $\mathbb{S}(U;\mathbb{I})$ has, as $\mathbb{I}$–linear endomorphisms, the ordinary projector e and the Hecke operators $T(q)$ for primes q prime to the level of U. The group $G(\mathbb{A}^{(p\infty)})$ acts on $\mathbb{S}(\mathbb{I})$ smoothly.*

5. — Before going into a hard work, we like to give a sketch of the theory. The first main result is

THEOREM 2. — *The automorphic representation of* $\widetilde{S}(\mathbb{A}^{(p\infty)})$ *on* $\mathbb{P}^{\mathrm{ord}}(\mathbb{I})$ *is smooth and, after having extended scalar to the field of fractions of* $\mathbb{I}$, *is a discrete direct sum of irreducible admissible representations with multiplicity at most* 1.

Putting off all the details to the end of this paper for attentive readers, we here give a sketch of the proof. It is well known that $\mathbb{S}^{\mathrm{ord}}(N;\Lambda) = \mathbb{S}^{\mathrm{ord}}(\Gamma_1(N);\Lambda)$ is free of finite rank over Λ (see [H5] Chapter 7), and if $k(P) \geq 2$ for $P \in \mathcal{A}(\Lambda;O)$, then

$$(*) \qquad \mathbb{S}^{\mathrm{ord}}(\Delta;\Lambda)/P\mathbb{S}^{\mathrm{ord}}(\Delta;\Lambda) \cong S_k^{\mathrm{ord}}(\Delta(p^{r(P)}),\varepsilon_P;O)\,.$$

This implies that there are only finitely many, bounded independently of weights, of complex irreducible automorphic representations of $G(\mathbb{A})$ which is p-ordinary and of conductor dividing Np. On the other hand, one has the Shimura correspondence :

$$Sh : \{\text{irreducible holomorphic automorphic representations of } \widetilde{S}(\mathbb{A}) \text{ of weight } k+\frac{1}{2}\} \to \{\text{irreducible holomorphic automorphic representation of } G(\mathbb{A}) \text{ of weight } 2k\}\,.$$

By a result of Waldspurger, there exists a bound $M > 0$ such that

$$\text{(i)} \qquad \#Sh^{-1}(\pi) \leq M \text{ for all } k\,, \text{ if } C(\pi) \mid Np\,,$$

where $C(\pi)$ is the conductor of π. If $\widetilde{\pi}$ is p-ordinary (that is, the eigenvalue of $T(p^2)$ in $\widetilde{\pi}$ is a p-adic unit), $Sh(\widetilde{\pi})$ is p-ordinary. Moreover, if we write V for the space of $\widetilde{\pi}$, we have a positive bound M' independently of weights (but depending on $\widehat{\Delta}$) such that

$$\text{(ii)} \qquad \dim_C H^0(\widehat{\Delta}(p),V) < M'\,.$$

Then (i) + (ii) $\Rightarrow$ $\mathrm{rank}_O\, \mathcal{P}^{\mathrm{ord}}_{k+(1/2)}(\Delta(p);O) < M''(\widehat{\Delta})$ independently of k for a positive bound $M''(\widehat{\Delta})$. Take a subset $\{\phi_1,\dots,\phi_m\}$ in $\mathbb{P}^{\mathrm{ord}}(\Delta;\Lambda)$ which is linearly independent over Λ. Then we can find m rational numbers $n_1,\dots,n_m$ such that $D = \det(a(n_i,\phi_j)) \neq 0$. Therefore for arithmetic P with $k(P)$ sufficiently large and $\varepsilon_P = id$, $\phi_i(P)$ is and element of $\mathcal{P}^{\mathrm{ord}}_{k(P)+(1/2)}(\Delta(p);O)$ and $D(P) \neq 0$. In other words, $\{\phi_i(P)\}_i$ is linearly independent over O. Therefore $m < M''$. This implies that $\mathrm{rank}_\Lambda\, \mathbb{P}^{\mathrm{ord}}(\Delta;\Lambda) < M''$. As we will see later, $\mathbb{P}^{\mathrm{ord}}(\Delta;\Lambda)$ is actually free of finite rank over Λ. Then all the assertion follows from the weak multiplicity one theorem of Waldspurger by reducing the Λ-adic reprensentation modulo P.

Thus we have the Λ–adic Shimura correspondence :

$$Sh : \{\text{irreducible } \Lambda\text{–adic ordinary automorphic representations of } \widetilde{S}(\mathbb{A})\} \to \{\text{irreducible } \Lambda\text{–adic ordinary automorphic representations of } G(\mathbb{A})\}.$$

Suppose $\Pi = Sh(\widetilde{\Pi})$. We write $\widetilde{\pi}_P = \widetilde{\Pi} \bmod P$ and $\pi_P = Sh(\widetilde{\pi}_P)$. Then π_P for an arithmetic P is a scalar extension of classical representation if $k(P) \geq 2$. This means that one can supplement a (unique) local representation at p with π_P to get a complex automorphic representation if $k(P) \geq 2$, which we again write π_P. Similarly $\widetilde{\pi}_P$ is associated with a complex automorphic representation of the metaplectic group if $k(P)$ is sufficiently large, because we can only prove the metaplectic version of (*) under the assumption that $k(P)$ is sufficiently large. Here note that $\pi_P \neq \Pi \bmod P$ but $\pi_P = Sh(\widetilde{\pi}_P) = \Pi \bmod P^2$, because representations of weight $k + \frac{1}{2}$ correspond to those of weight $2k$. Here we used the group structure of $\mathrm{Spec}(\Lambda)(O) = \mathrm{Hom}_{gr}(W, O^\times)$ to define P^2. The above fact characterizes the Λ–adic Shimura correspondence. By (*), the prime to p–part $C(\Pi)$ of the conductor of π_P is independent of P. Moreover the central character of Π can be written as $\iota\psi^2$ for a finite order even character ψ modulo $4pC(\Pi)$, where ι is the tautological character of W into $\Lambda^\times$ composed with the "norm" character : $(\mathbb{A}^{(p\infty)})^\times \ni x \longmapsto |x|_{\mathbb{A}}^{-1}\omega^{-1}(x) \in \mathbb{Z}_p{}^\times$ for the Teichmüller character ω. We put $\psi_P = \varepsilon_P\psi\omega^{-k(P)}$ for each arithmetic P. As a striking consequence of his theory, Waldspurger expressed the square of a certain ratio of two Fourier coefficients of a cusp form of half integral weight by a ratio of L–values attached to the image under the Shimura correspondence. Applying this result, we get a Λ–adic version of his result :

THEOREM 3. — *For each pair (m, n) of positive square free integers with $m/n \in \Pi_{l|4Np}\mathbb{Q}_l^2$, we find two elements Φ and Ψ in $\mathbb{I}$ such that if $k(P) > 1$ or $\psi_P^2 \neq 1$, we have :*

$$\frac{\Phi(P)^2}{\Psi(P)^2} = \frac{L(\frac{1}{2}, \pi_P \otimes \psi_P^{-1}\chi_n)\psi_P(n+m)(n/m)^{k-(1/2)}}{L(\frac{1}{2}, \pi_P \otimes \psi_P^{-1}\chi_m)}$$

as long as

$$L(\tfrac{1}{2}, \pi_P \otimes \psi_P^{-1}\chi_m) \neq 0,$$

where χ_t is the quadratic character associated with $\mathbb{Q}(\sqrt{t})$.

6. — We now start filling the details with the argument in Section 5. Fix a character ψ of $(\mathbb{Z}/Np\mathbb{Z})^\times$. For each arithmetic point $P \in \mathcal{A}(\Lambda)$, we define a character ψ_P of Z_N by $\psi_P(z) = \psi(z)\chi_P(<z>)z^{-k(P)} = \psi\varepsilon_P\omega^{-k(P)}(z)$, where $z \longmapsto <z>$ is the projection to W and ω is the Teichmüller character. We now prove

PROPOSITION 3. — *The dimension of $\mathcal{P}^{\text{ord}}_{k+(1/2)}(\Delta_0(p^{r(P)}), \psi_P; \Omega_p))$ is bounded independent of $P \in \mathcal{A}(\Lambda)$ if $k(P) \geq 1$ (the dimension depends on $\widehat{\Delta} \in \mathcal{Z}$).*

To prove the proposition, we prepare several lemmas. Let ℓ be a prime and put

$$U_r = U_{r,\ell} = \left\{ \begin{pmatrix} a & b \\ c & d \end{pmatrix} \in SL_2(\mathbb{Z}_\ell) \mid c \equiv 0 \mod \ell^r \right\},$$

For each character χ of $\mathbb{Z}_\ell^\times$ modulo ℓ^r and a U_r–module M, we write $M(\chi)$ for the χ–eigenspace. That is, $M(\chi) = \left\{ m \in M \mid \begin{pmatrix} a & b \\ c & d \end{pmatrix} m = \chi(d)m \text{ for } \begin{pmatrix} a & b \\ c & d \end{pmatrix} \in U_r \right\}$. When the reference to the level ℓ^r is necessary, we write $M(\ell^r, \chi)$ in place of $M(\chi)$.

LEMMA 1. — *Let π be an irreducible admissible representation of the metaplectic covering group $\widetilde{S}(\mathbb{Q}_\ell)$ of $SL_2(\mathbb{Q}_\ell)$ and V denote its representation space. Let χ be a character of $\mathbb{Q}_\ell^\times$ modulo ℓ^r. Suppose that π appears as a local factor of a holomorphic automorphic representation of weight $k + (1/2)$ $(k \geq 2)$. Then the dimension of $V(\ell^r, \chi)$ is bounded independent of V and χ (but it depends on r).*

Proof : when π is special or principal, then we can realize π as a subquotient of the induced representation space $\mathcal{B}_\mu = \mathcal{B}_{\mu,\mathbf{e}_\ell}$ of a character μ of the standard Borel subgroup, as in [Wa1, II.2] and [Wa2, II], for the ℓ–part $\mathbf{e}_\ell$ of the standard additive character $\mathbf{e}$ of $\mathbb{A}/\mathbb{Q}$ and a quasi character μ of the standard Borel subgroup of $\widetilde{S}(\mathbb{Q}_\ell)$. Since the left translation by the upper triangular matrices of $\widetilde{S}(\mathbb{Q}_\ell)$ is already prescribed on $\mathcal{B}_\mu$, any function in $\mathcal{B}_\mu$ is determined by its restriction to $SL_2(\mathbb{Z}_\ell) \times \{\pm 1\}$. Then for each given open compact subgroup U of $SL_2(\mathbb{Z}_\ell) \times \{\pm 1\}$, the dimension of $H^0(U, \mathcal{B}_\mu)$ is bounded by the index $2(SL_2(\mathbb{Z}_\ell) : U)$. A more effective bound can be obtained using the explicit calculation of the space $\mathcal{B}_\mu(\ell^r, \chi)$ done in [Wa2] Proposition 9 (p. 417) (see also Lemma 3 in the text). We then have $\dim(\mathcal{B}_\mu(\ell^r, \chi)) \leq 2(r + 1)$. This settles the problem in the case of non–super cuspidal representations. Let $\widetilde{\Pi}$ be a holomorphic automorphic representation of $\widetilde{S}(\mathbb{A})$ of weight $k + \frac{1}{2}$ $(k \geq 2)$ having π as its factor at ℓ. Let W be the space of the ℓ–component of the automorphic representation of $GL_2(\mathbb{A})$ corresponding to $\widetilde{\Pi}$ by the Shimura correspondence. Using the notation of [Wa1] V.4 (p. 99), we mean by W the ℓ–component of $\mathcal{V}'(\mathbf{e}, V) \otimes \chi$. We know from [C] that $\dim W(\ell^r, \chi^2) \leq r+1$. By [Wa2] V, Proposition 5 (p. 404), $V(\ell^r, \chi)$ is a subspace of the space spanned by, with the notation in [Wa2], $i_{\nu,\ell} \circ j_{\nu,\ell}(w)(f_{r,\nu})$ for r sufficiently large (if $r \geq \max(2v_\ell(2)+1, v_\ell(C(\chi)))$ for the conductor $C(\chi)$ of χ), where $w \in W(\ell^r, \chi)$ and $\nu \in (\mathbb{Q}_ell^\times(V)/(\mathbb{Q}_ell^\times)^2)$. Here $f_{r,\nu}$ is a Schwartz–Bruhat function on $H_\ell = \{x \in M_2(\mathbb{Q}_\ell) \mid Tr(x) = 0\}$ determined by (r, ν) as specified in [Wa2] Chapter V. The choice of $\nu \in \mathbb{Q}_ell^\times$ is bounded by $\#(\mathbb{Q}_ell^\times/(\mathbb{Q}_ell^\times)^2)$ which is 4 if $\ell > 2$ and 8 if $\ell = 2$. Thus we have, for general V,

$$\dim(V(\ell^r, \chi)) \leq 8(r+1) \quad \text{for } r \text{ sufficiently large}.$$

This finishes the proof.

LEMMA 2. — *Let π be an irreducible admissible representation of $\widetilde{S}(\mathbb{Q}_\ell)$ with representation space V. Suppose that π is super cuspidal. Then, for sufficiently large m, $T(\ell^m)$ annihilates $V(\ell^r, \chi)$ if $r > 0$.*

Proof : note that

$$U_0(\ell^r)\left(\begin{pmatrix} \ell^m & 0 \\ 0 & \ell^{-m} \end{pmatrix}, 1\right)U_0(\ell^r) = \bigcup_{u \in \mathbb{Z}_\ell/\ell^m\mathbb{Z}_\ell} \left(\begin{pmatrix} \ell^m & u\ell^{-m} \\ 0 & \ell^{-m} \end{pmatrix}, 1\right)U_0(\ell^r)$$

and $$\left(\begin{pmatrix} \ell^m & u\ell^{-m} \\ 0 & \ell^{-m} \end{pmatrix}, 1\right) = \left(\begin{pmatrix} \ell^m & 0 \\ 0 & \ell^{-m} \end{pmatrix}, 1\right)\left(\begin{pmatrix} 1 & \ell^{-2m}u \\ 0 & 1 \end{pmatrix}, 1\right).$$

Thus for $v \in H^0(U_0(\ell^r), V)$, we define an operator $\widetilde{T}(\ell^m)$ by

$$v \mid \widetilde{T}(\ell^m) = \sum_{u \in \mathbb{Z}_\ell/\ell^m\mathbb{Z}_\ell} \left(\begin{pmatrix} \ell^m & u\ell^{-m} \\ 0 & \ell^{-m} \end{pmatrix}, 1\right)v.$$

The operator $\widetilde{T}(\ell^m)$ coincides with the Hecke operator $(\widetilde{T}'_\ell)^m$ acting on $V(\ell^r, \chi)$ defined in [Wa2] III.3, pp. 388–389. Then we have

$$v \mid \widetilde{T}(\ell^m) = \pi\left(\left(\begin{pmatrix} \ell^m & 0 \\ 0 & \ell^{-m} \end{pmatrix}, 1\right)\right) \int_{\ell^{-2m}\mathbb{Z}_\ell} \pi\left(\begin{pmatrix} 1 & u \\ 0 & 1 \end{pmatrix}, 1\right)v\,du = 0$$

for sufficiently large m by the definition of super cuspidality. As shown in [Wa2] Lemma 4, p. 389, we know that $\widetilde{T}'_\ell = \ell^{(3-2k)/2}\tilde{\gamma}_\ell(\ell\chi(\ell)^{-1}T(\ell^2)$ for $T(\ell^2)$ defined in [Sh1], where $\tilde{\gamma}_\ell(t) = (t,t)_\ell\gamma_\ell(t)\gamma_\ell(1)^{-1}$. Thus we know the lemma from the above result. Here we should note that the definition of our space of modular forms of half integral weight is different by the character $\left(\frac{-1}{\cdot}\right)^k$ from that of [Wa2], and thus we do not replace χ by χ_0 as was done in [Wa2] for these formulas.

LEMMA 3 ([V]). — *Suppose that $\ell > 2$. Let $V = \mathcal{B}_\mu$ and let χ be a continuous character of $\mathbb{Q}_\ell^\times$ into $\mathbb{C}^\times$. We consider the Hecke operator $T(\ell^r) = \chi(\ell)\tilde{\gamma}_\ell(\ell)^{-1}\ell^{(2k-3)/2}\tilde{T}'_\ell$. Suppose that $r \geq \mathrm{Sup}(v_\ell(C(\chi)), v_\ell(C(\mu\chi)C(\mu\chi^{-1})))$, where $C(\chi)$ is the conductor of χ. Then we have the following assertions :*

(i) *If both $\mu\chi$ and $\mu\chi^{-1}$ are non-trivial on $\mathbb{Z}_e ll^\times$, then $T(\ell^2)$ is nilpotent on $V(\ell^r;\chi)$ for $r > 0$;*

(ii) *Suppose that $\mu\chi^{-1}$ is unramified but $\mu\chi$ is ramified. Then we can decompose $V(\ell^r;\chi) = N \oplus V(C(\chi);\chi)$ so that $T(\ell^2)$ is nilpotent on N and $V(C(\chi);\chi)$ is one dimensional on which $T(\ell^2)$ acts by scalar multiplication of $\chi(\ell)\ell^{(2k-1)/2}\mu(\ell^{-1})$;*

(iii) *Suppose that $\mu\chi$ is unramified but $\mu\chi^{-1}$ is ramified. Then we can decompose $V(\ell^r;\chi) = N \oplus V(C(\chi);\chi)$ so that $T(\ell^2)$ is nilpotent on N and $V(C(\chi);\chi)$ is one dimensional on which $T(\ell^2)$ acts by scalar multiplication of $\chi(\ell)\ell^{(2k-1)/2}\mu(\ell)$;*

(iv) *Suppose that both $\mu\chi$ and $\mu\chi^{-1}$ are unramified. Then we can decompose $V(\ell^r;\chi) = N \oplus V(\ell;\chi)$ so that* (i) *$T(\ell^2)$ is nilpotent on N,* (ii) *$V(\ell;\chi)$ is two dimensional, and* (iii) *we have a base $\{v_1, v_2\}$ of $V(\ell;\chi)$ such that $v_1|T(\ell^2) = \chi(\ell)\ell^{(2k-1)/2}\mu(\ell^{-1})v_1$ and $v_2|T(\ell^2) = \chi_\ell(\ell)\ell^{(2k-1)/2}\mu(\ell)v_2 + cv_1$ with some constant c.*

Proof : write $\nu(\varepsilon)$ for the exponent of ℓ in $C(\varepsilon)$ for any character of ε of $\mathbb{Z}_e ll^\times$. As shown in [Wa2] Proposition 9, p. 417, under the assumption of $r \geq \mathrm{Sup}(\nu(\chi), 1)$, $V(\ell^r;\chi) \neq 0$ if and only if $r \geq \nu(\mu\chi) + \nu(\mu\chi^{-1})$. As long as $r > \nu(\chi)$ and $r > 1$, $T(\ell^2)$ sends $V(\ell^r;\chi)$ to $V(\ell^{r-1};\chi)$ (cf. [Wa2] Lemma 7 or [H2] (8.6)). This shows that for sufficiently large m, $V(\ell^r;\chi)|T(\ell^{2m})$ is contained in $V(C(\chi);\chi)$ or $V(\ell;\chi)$ if $\nu(\chi) \leq 1$. Unless χ is quadratic, $\nu(\chi) = \nu(\chi^2)$ since $\ell > 2$. Thus if $\chi^2 \neq id$, then $\nu(\mu\chi) + \nu(\mu\chi^{-1}) \geq \mathrm{Max}(\nu(\mu\chi), \nu(\mu\chi^{-1})) \geq \nu(\chi)$. If moreover both $\nu(\mu\chi)$ and $\nu(\mu\chi^{-1})$ are positive, then $\nu(\mu\chi) + \nu(\mu\chi^{-1}) > \nu(\chi)$ and thus $V(C(\chi);\chi) = 0$. Therefore $T(\ell^2)$ is nilpotent on $V(\ell^r;\chi)$ if $\chi^2 \neq id$ and if both $\nu(\mu\chi)$ and $\nu(\mu\chi^{-1})$ are positive. Now suppose that $\chi^2 = id$ and both $\nu(\mu\chi)$ and $\nu(\mu\chi^{-1})$ are positive. Then if $\chi \neq id$, then $C(\chi) = \ell$ and $V(C(\chi);\chi) = 0$ because $\nu(\mu\chi) + \nu(\mu\chi^{-1}) > 1$. If $\chi = id$, then again $V(\ell;\chi) = 0$ because $\nu(\mu) + \nu(\mu) > 1$. Thus $T(\ell^2)$ is nilpotent if both $\nu(\mu\chi)$ and $\nu(\mu\chi^{-1})$ are positive. Now suppose that $\nu(\mu\chi^{-1}) = 0$ but $\nu(\mu\chi) > 0$. Then $\chi^2 \neq id$ because $\nu(\mu\chi) = \nu(\chi^2) > 0$ (and hence $\nu(\chi^2) = \nu(\chi)$), and $V(C(\chi);\chi)$ is one dimensional by [Wa2] Proposition 9. Moreover by [Wa2] Proposition 10, (ii), we know that $T(\ell^2)$ acts on $V(C(\chi);\chi)$ by the scalar multiplication of $\chi(\ell)\ell^{(k/2)-1}\mu(\ell^{-1})$. Thus we can decompose $V(\ell^r;\chi) = N \oplus V(C(\chi);\chi)$ such that on N, $T(\ell^2)$ is nilpotent, and on the one dimensional space $V(C(\chi);\chi)$, $T(\ell^2)$ acts via the multiplication of $\chi(\ell)\ell^{(k/2)-1}\mu(\ell^{-1})$. Suppose $\nu(\mu\chi^{-1}) = \nu(\mu\chi) = 0$ and $\chi \neq id$. Then $\chi^2 = id$ because $\nu(\mu\chi) = \nu(\chi^2) = 0$. By [Wa2] Proposition 2, $V(\ell;\chi)$ is 2–dimensional, and there is a base $\{v_1, v_2\}$ of $V(\ell;\chi)$ such that

$$v_1 \mid T(\ell^2) = \chi(\ell)\ell^{(k/2)-1}\mu(\ell^{-1})v_1 \text{ and}$$
$$v_2 \mid T(\ell^2) = \chi(\ell)\ell^{(k/2)-1}\mu(\ell)v_2 + \ell^{(k/2)-2}\tilde{\gamma}_\ell(\ell)^{-1}\chi(\ell)(\ell-1)v_1 .$$

Thus we can decompose $V(\ell^r;\chi) = N \oplus V(\ell;\chi)$ such that on N, $T(\ell^2)$ is nilpotent and on the 2–dimensional space $V(\ell;\chi)$, it acts by the above formula. Next suppose that $\nu(\mu\chi^{-1}) = \nu(\mu\chi) = 0$ and $\chi = id$. Then $V(\ell;\chi)$ is 2–dimensional, and we can find a base $\{v_1, v_2\}$ by [Wa2] Proposition 10 such that $(v_1 + v_2) \in V(1;\chi)$ and

$$v_1 \mid T(\ell^2) = \chi(\ell)\ell^{(k/2)-1}\mu(\ell^{-1})v_1 \text{ and } v_2 \mid T(\ell^2) = \chi(\ell)\ell^{(k/2)-1}\mu(\ell)v_2 + cv_1$$

with some constant c. The value of c is given by [Wa2] p. 420. This shows that $V(\ell^r;\chi) = N \oplus V(\ell;\chi)$ such that on N, $T(\ell^2)$ is nilpotent, and on the 2–dimensional space $V(\ell;\chi)$, $T(\ell^2)$ is an automorphism described as above. Finally we assume that $\nu(\mu\chi) = 0$ but $\nu(\mu\chi^{-1}) > 0$. Then $\nu(\mu\chi^{-1}) = \nu(\chi^{-2}) > 0$ and hence $\nu(\chi^2) = \nu(\chi)$. Thus again by [Wa2] Propositions 9 and 10, $V(C(\chi);\chi)$ is one dimensional and $T(\ell^2)$ acts on it by the multiplication of $\chi(\ell)\ell^{(k/2)-1}\mu(\ell)$. Therefore, we can decompose $V(\ell^r;\chi)$ into $V(C(\chi);\chi) \oplus N$, where on N, $T(\ell^2)$ is nilpotent and on the one–dimensional space $V(C(\chi);\chi)$, it acts by the scalar $\chi(\ell)\ell^{(k/2)-1}\mu(\ell)$.

LEMMA 4 ([Wa1] Proposition 18, p. 68). — *Let ρ^* be an irreducible admissible representation of $PGL_2(\mathbb{Q}_\ell)$ and let ρ be the corresponding irreducible admissible representation of $\widetilde{S}(\mathbb{Q}_\ell)$ via Weil representation with respect to the additive character $\mathbf{e}_\ell^\xi$ ($\xi \in \mathbb{Q}_e ll^\times$). Then we have*

Equivalence class of ρ^*	*Equivalence class of ρ*
$\pi(\mu,\mu^{-1})$ ($\mu^2 \neq \alpha$)	$\tilde{\pi}_{\mu\chi_\xi}$
$\sigma(\mu,\mu^{-1})$ ($\mu^2 \neq \alpha$, $\mu \neq \alpha^{1/2}$)	$\tilde{\sigma}_{\mu\chi_\xi}$
$\sigma(\alpha^{1/2},\alpha^{-1/2})$	*Supercuspidal*
Supercuspidal	*Supercuspidal*

where $\mathbf{e}_\ell$ is the standard additive character of $\mathbb{Q}_\ell$ and $\mathbf{e}_\ell^\xi(x) = \mathbf{e}_\ell(\xi x)$ and we have used the notation of [Wa1] *Propositions* 1 *and* 2.

Here note that $\tilde{\pi}_{\mu\chi_\xi}$ (resp. $\tilde{\sigma}_{\mu\chi_\xi}$) with respect to $\mathbf{e}_\ell^\xi$ is isomorphic to $\tilde{\pi}_{\mu\chi_{\xi\eta}}$ (resp. $\tilde{\sigma}_{\mu\chi_{\xi\eta}}$) with respect to $\mathbf{e}_\ell^{\xi\eta}$, and hence the right-hand side is well defined independent of the additive character.

A cusp form $f \in \mathcal{S}_\kappa(\Delta_1(p^r); \mathbb{C})$ is called *ordinary* at p if $f \mid T(p) = \lambda f$ and $|\lambda|_p = 1$. An automorphic representation π of $GL_2(\mathbb{A})$ spanned by a holomorphic primitive form f is called *ordinary* at p if f is ordinary at p.

LEMMA 5 (e.g. [H3] §2). — *Let π be a unitary holomorphic automorphic representation of $GL_2(\mathbb{A})$. Suppose that π is irreducible and ordinary at p. Then the local component π_p of π is either a principal series representation $\pi(\alpha, \beta)$ with unramified α or a special representation $\sigma(\alpha, \beta)$ with unramified α. Let f be the primitive form of weight k on $GL_2(\mathbb{A})$ belonging to π and write μ for the central character of π and $\lambda(T(p))$ for the eigenvalue of $T(p)$ on f. Then if $\pi_p = \pi(\alpha, \beta)$ and α and β are both unramified, then $\alpha(p) + \beta(p) = p^{(1-k)/2}\lambda(T(p))$, $\alpha(p)\beta(p) = \mu(p)$ and $|\lambda(T(p))|_p = 1$. If $\pi_p = \pi(\alpha, \beta)$ and β is ramified, then $\alpha(p) = p^{(1-k)/2}\lambda(T(p))$, $\alpha(p)\beta(p) = \mu(p)$ and $|\lambda(T(p))|_p = 1$. If $\pi_p = \sigma(\alpha, \beta)$, then π_∞ is of weight 2 and $\lambda(T(p)) = \alpha(p)$.*

LEMMA 6. — *Let F be a number field of finite degree. Let ρ be a cuspidal automorphic representation of $PGL_2(F_A)$ and let R be the set of all cuspidal automorphic representations of $\widetilde{S}(F_\mathbb{A})$. Define for each integral ideal N of F,*

$$R(\rho; N) = \{\pi \in R \mid \pi_v^* \cong \rho_v \text{ for all } v \text{ outside } N\},$$

where π_v^* denotes the corresponding representation of $PGL_2(F_v)$ via Weil representations defined in [Wa1, V.4] *(where it is written as : $T \longmapsto \mathcal{V}'(\mathbf{e}, T)$; see Lemma 4). Then we have*

$$\#R(\rho; N) \leq \#\{\Pi_{v|N} F_v^\times/(F_v^\times)^2\}.$$

Proof : we know from Lemma 4 (or the remark after the lemma) that if T_v is principal or special, then $\mathcal{V}'(\mathbf{e}_v^x, T_v) \cong \mathcal{V}'(\mathbf{e}_v, T_v)$. Moreover if $x/y \in (F_v^\times)^2$, then $\mathcal{V}'(\mathbf{e}_v^x, T_v) = \mathcal{V}'(\mathbf{e}_v^y, T_v)$ for all T_v by [Wa1] Theorem 2, p. 80, Proposition 28, p. 98 and [Wa2] Assertion 3, p. 394. Thus the number of isomorphism classes in $\{\mathcal{V}'(\mathbf{e}_v^\times, T_v) \mid x \in F_v^\times\}$ for all v outside N are at most $\#\{\Pi_{v|N} F_v^\times/(F_v^\times)^2\}$. Then the weak multiplicity one theorem [Wa1] VI shows the result.

Proof of Proposition 3 : we only prove the assertion when $\Delta = \Gamma_1(N)$. The general case follows from this special case because any Δ contains a conjugate of $\Gamma_1(N)$ for a suitable N. We shall prove the boundedness for $P \in \mathcal{A}(\Lambda)$ with $k(P) \geq 1$. We write $\chi = \psi_P$ for a given $P \in \mathcal{A}(\Lambda)$

and $\psi : (\mathbb{Z}/Np\mathbb{Z})^\times \to \overline{\mathbb{Q}}_p^\times$ and consider χ as an idele character so that $\chi(\varpi) = \chi(\ell)$ for a prime element ϖ at any prime ℓ outside Np. Let V be the subspace of functions on $\widetilde{S}(\mathbb{A})$ spanned by right translations of elements in

$$\mathcal{P}^{\text{ord}}_{k(P)+(1/2)}(Np^{r(P)}, \psi_P; \mathbb{C})$$

under the Hecke algebra of $\widetilde{S}(\mathbb{A})$. We decompose $V = \oplus_\rho V(\rho)$ into the sum of irreducible subspaces $V(\rho)$. Then by the weak multiplicity one theorem proven by [Wa1] p. 131, each irreducible representation ρ occurs at most once. Decompose $\rho = \otimes_\ell \rho_\ell$ into the tensor product of local representations. Then by Lemma 2, ρ_p is either $\tilde{\sigma}_\mu$ or $\tilde{\pi}_\mu$ for a quasi character $\mu : \mathbb{Q}_p{}^\times \to \mathbb{C}^\times$. By the Weil representation, $\tilde{\sigma}_\mu$ corresponds to $\sigma(\mu, \mu^{-1})$ and $\pi(\mu, \mu^{-1})$, which is a representation of $PGL_2(\mathbb{A})$ (Lemma 4 and [Wa1] Proposition 27 and Lemma 70). Then the Shimura correspondence is given locally by

$$\tilde{\sigma}_\mu \longmapsto \sigma(\mu\chi, \mu^{-1}\chi) \text{ and } \tilde{\pi}_\mu \longmapsto \pi(\mu\chi, \mu^{-1}\chi)$$

and globally by $\rho \longmapsto \rho^* \otimes \chi$, where $\rho \longmapsto \rho^*$ is given via the global Weil representation. The eigenvalue for $T(p^2)$ on $V(\rho_p)(p^r; \chi)$ $(r = r(P))$ is given as follows (Lemma 3) : if $\mu\chi$ is unramified, then it is $\mu\chi(p)p^{(2k-1)/2}$ $(k = k(P))$; if $\mu^{-1}\chi$ is unramified, then $\mu^{-1}\chi(p)p^{(2k-1)/2}$ and if both $\mu\chi$ and $\mu^{-1}\chi$ are ramified, it vanishes. On the other hand, these values are the eigenvalue of $T(p)$ on $V(\rho_P^* \otimes \chi)(p^r; \chi^2)$ by Lemma 5. Note that even if both $\mu\chi$ and $\mu^{-1}\chi$ are unramified, at most one eigenvalue in $\mu^{-1}\chi(p)p^{(2k-1)/2}$ and $\mu\chi(p)p^{(2k-1)/2}$ can be a p-adic unit in $\overline{\mathbb{Q}}_p$. Thus ρ corresponds to the ordinary ρ^* of character χ^2 and of level at most $Np^{r(P)}$. Then by [H5] Theorem 7.3.3, the number of such automorphic representations occurring in $\mathcal{S}^{\text{ord}}_{2k(P)}(Np^{r(P)}, \chi^2)$ is bounded independent of P if $k(P) \geq 1$. Then by Lemmas 2, 4 and 6, we know the assertion of the proposition.

We say an element $\mathbf{f} \in \mathbb{P}(\Delta; \mathbb{I})$ is ordinary, if for all $P \in \mathcal{A}(\mathbb{I})$ with sufficiently large $k(P)$, $\mathbf{f}_P \in \mathcal{P}^{\text{ord}}_{k(P)+(1/2)}(\Delta(p^{r(P)}), \varepsilon_P; \Omega_p)$. We denote the space of all $\mathbb{I}$-adic ordinary cusp forms as $\mathbb{P}^{\text{ord}}(\Delta; \mathbb{I})$. Then $\mathbb{P}^{\text{ord}}(\Delta; \mathbb{I}) = e\mathbb{P}(\Delta; \mathbb{I})$.

PROPOSITION 4. — *For each $\widetilde{\Delta} \in \mathcal{Z}$ with $\Delta \subset \Gamma_0(4)$, $\mathbb{P}^{\text{ord}}(\Delta; \mathbb{I})$ is free of finite rank over $\mathbb{I}$.*

Proof : we prove the assertion for $\mathbb{P}^{\text{ord}}(N; \mathbb{I})$ applying the argument of Wiles [Wi]. The other cases can be treated similarly. Let $\Delta = \Gamma_1(N)$. Let $\mathbb{K}$ be the quotient field of $\mathbb{I}$, which is a finite extension of $\mathbb{L}$. We put $\mathbb{P}^{\text{ord}}(N; \mathbb{K}) = \mathbb{P}^{\text{ord}}(N; \mathbb{I}) \otimes_{\mathbb{I}} \mathbb{K}$. Let $\mathbf{f}_1, \mathbf{f}_2, \ldots, \mathbf{f}_r$ be a finite set of linearly independent elements in $\mathbb{P}^{\text{ord}}(N; \mathbb{I})$ over $\mathbb{I}$. Then we can find positive integers

$n_1, \dots, n_r$ so that

$$D = \det(\mathbf{a}(n_i, \mathbf{f}_j)) \neq 0.$$

We now choose $P \in \mathcal{A}(\mathbb{I})$ so that for all $i = 1, \dots, r$,

$$\mathbf{f}_i(P) \in \mathcal{P}^{\mathrm{ord}}_{k(P)+(1/2)}(\Delta(p^{r(P)}), \varepsilon_P; \Omega_p) \text{ and } D(P) \neq 0.$$

Then $0 \neq D(P) = \det(a(n_i, \mathbf{f}_j(P))$ and thus $\mathbf{f}_i(P)$ are linearly independent. Namely, we have

$$r \leq \dim \mathcal{P}^{\mathrm{ord}}_{k(P)+(1/2)}(\Delta(p^{r(P)}), \varepsilon_P; \Omega_p),$$

which is bounded independently of P by Proposition 3. Thus there is a maximal set $\{\mathbf{f}_1, \mathbf{f}_2, \dots, \mathbf{f}_r\}$ of linearly independent elements in $\mathbb{P}^{\mathrm{ord}}(N, \mathbb{I})$. That is, $\dim_{\mathbb{K}} \mathbb{P}^{\mathrm{ord}}(N; \mathbb{K}) = r < \infty$. For any $\mathbf{f}$ in $\mathbb{P}^{\mathrm{ord}}(N; \mathbb{I})$, we can write $\mathbf{f} = \sum_{i=1}^{r} c_i(\mathbf{f})\mathbf{f}_i$ and $Dc_i(\mathbf{f}) \in \mathbb{I}$. Thus $D^{-1}(\mathbb{I}\mathbf{f}_1 + \cdots + \mathbb{I}\mathbf{f}_r) \supset \mathbb{P}^{\mathrm{ord}}(N; \mathbb{I})$ and hence $\mathbb{P}^{\mathrm{ord}}(N; \mathbb{I})$ is of finite type over $\mathbb{I}$ as $\mathbb{I}$–module, because $\mathbb{I}$ is noetherian. Now we see by definition that $\mathbb{P}^{\mathrm{ord}}(N; \mathbb{I}) = \cap_P \mathbb{P}^{\mathrm{ord}}(N; \mathbb{I}_P)$ where P runs over all prime ideals of height 1, $\mathbb{I}_P$ is the localizaton at prime P and $\mathbb{P}^{\mathrm{ord}}(N; \mathbb{I}_P) = \mathbb{P}^{\mathrm{ord}}(N; \mathbb{I}) \otimes_{\mathbb{I}} \mathbb{I}_P$. This shows that $\mathbb{P}^{\mathrm{ord}}(N; \mathbb{I})$ is $\mathbb{I}$–reflexive and hence if $\mathbb{I} = \Lambda$, then $\mathbb{P}^{\mathrm{ord}}(N; \Lambda)$ is Λ–free of finite rank. Since we already know that $\mathbb{P}^{\mathrm{ord}}(N; \mathbb{I}) = \mathbb{P}^{\mathrm{ord}}(N; \Lambda) \otimes_{\Lambda} \mathbb{I}$, we conclude that $\mathbb{P}^{\mathrm{ord}}(N; \mathbb{I})$ is $\mathbb{I}$–free of finite rank.

Proposition 5. — *Let $P \in \mathcal{A}(\mathbb{I})$. Then each $f \in \mathcal{P}^{\mathrm{ord}}_{k(P)+(1/2)}(\Delta(p^{r(P)}), \varepsilon_P; O)$ can be lifted to an ordinary Λ–adic form $\mathbf{f} \in \mathbb{P}^{\mathrm{ord}}(\Delta; \mathbb{I})$ such that $\mathbf{f}(P) = f$.*

Proof : it is sufficient to prove the assertion for $\mathbb{I} = \Lambda$. Let $E(X) \in \Lambda[[q]]$ be the Λ–adic Eisenstein series (cf. [H5] §7.1) such that for the generator $w = 1 + p$ of W

$$E(Q) = (Q(w) - 1)\Big\{L_p(1 - k(Q), \varepsilon_Q w^{-k(Q)})/2 + \sum_{n=1}^{\infty}\Big(\sum_{0<d|n} Q(<d>)d^{-1}\Big)q^n\Big\}$$

in $\mathcal{M}_{k(Q)}(\Delta(p^{r(Q)}), \varepsilon_Q; \overline{\mathbb{Q}}_p)$ for all $Q \in \mathcal{A}(\Lambda)$. Then we see for the point P_0 of $\mathcal{X}(\Lambda)$ corresponding to the trivial character of W, $E(P_0) = (1 - p)\log(w)/p$, which is a p-adic unit. We then put $F = E(P_0)^{-1}E$ and consider the product fF inside $\Lambda[[q]]$. Then $Ff(Q) = fF(Q) \in \mathcal{P}_{k(P)+k(Q)+(1/2)}(\Delta(p^{r(P)}), \varepsilon_P\varepsilon_Q; \Omega)$. We define a formal q–expansion $F * f(X)$ by $Ff(\varepsilon_P^{-1}(w)w^{-k}X + (\varepsilon_P^{-1}(w)w^{-k} - 1)$, which is a Λ–adic cusp form in $\mathbb{P}(N; \Lambda)$ ([H5] Lemma 7.1.1). Then we see that $F * f(P) = fF(P_0) = f$. Then $e(F * f)(P) = (F * f(P)) \mid e = f$ by Lemma 7 and the assertion of the theorem follows.

COROLLARY 1. — *For $P \in \mathcal{A}(\mathbb{I}; O)$ with sufficiently large $k(P)$ depending on Δ, we have*

$$\mathcal{P}^{\mathrm{ord}}_{k(P)+(1/2)}(\Delta(p^{r(P)}), \varepsilon_P; O) \cong \mathbb{P}^{\mathrm{ord}}(\Delta; \mathbb{I})/P\mathbb{P}^{\mathrm{ord}}(\Delta; \mathbb{I}).$$

Proof : Choose a base $\mathbf{f}_1, \ldots, \mathbf{f}_r$ of $\mathbb{P}^{\mathrm{ord}}(\Delta; \mathbb{I})$. We can find $a > 0$ so that

(i) $\mathbf{f}_i(P) \in \mathcal{P}^{\mathrm{ord}}_{k(P)+(1/2)}(\Delta(p^{r(P)}); \varepsilon_P; O)$ *for all i and all P with $k(P) > a$, and*

(ii) *there exist integers n_i such that* $\det(\mathbf{a}(n_i/N, \mathbf{f}_j))(P) \neq 0$ *if* $k(P) > a$.

Then $\mathbf{f}_i(P)$ are linearly independent over O. Thus $\mathbb{P}^{\mathrm{ord}}(\Delta; \mathbb{I})/P\mathbb{P}^{\mathrm{ord}}(\Delta, \mathbb{I})$ injects into $\mathcal{P}^{\mathrm{ord}}_{k(P)+(1/2)}(\Lambda(p^{r(P)}); \varepsilon_P; O)$. Surjectivity of the morphism follows from Proposition 5.

COROLLARY 2. — *Let $\mathbf{f}_1, \ldots, \mathbf{f}_r$ be a base of $\mathbb{P}^{\mathrm{ord}}(\Delta; \mathbb{I})$. Then we can find integers $n_1, \ldots, n_r$ so that* $\det(\mathbf{a}(n_i/N, \mathbf{f}_j)) \in \mathbb{I}^\times$.

Proof : let $f_1, \ldots, f_r$ be a base of $\mathcal{P}^{\mathrm{ord}}_{k(P)+(1/2)}(\Delta(p^{r(P)}), \varepsilon_P; O)$. let ϖ be a prime element of O. If $\det(a(n_i/N, f_j) \equiv 0 \bmod \varpi O$ for all choice of integers $n_, \ldots, n_r$, then $\{f_i \bmod \varpi\}$ are linearly dependent and hence we can find $\lambda_i \in O$ not all divisible by ϖ such that $\Sigma_i \lambda_i f_i \equiv 0 \bmod \varpi O$. Then

$$\varpi^{-1}\Sigma_i \lambda_i f_i \in \mathcal{P}^{\mathrm{ord}}_{k(P)+(1/2)}(\Delta(p^{r(P)}), \varepsilon_P; O)$$

but $\varpi^{-1}\lambda_i$ are not all in O. This contradicts to the fact that $\{f_i\}$ forms a base. Thus we can find the n_i's so that $\det(a(n_i, f_j)) \in O^\times$. Now applying this argument to a base $\{\mathbf{f}_i(P)\}$ by choosing P with sufficiently large $k(P)$, we find that $\det(\mathbf{a}(n_i/N, \mathbf{f}_j))(P) \in O^\times$ which implies that $\det(\mathbf{a}(n_i/N, \mathbf{f}_j)) \in \mathbb{I}^\times$.

Analogs of all the assertion so far we proved in this paragraph holds for $\mathbb{S}^{\mathrm{ord}}(U; \mathbb{I})$ in an obvious sense (see [H5] Chapter 7). In particular, the statement corresponding to Corollary 1 for $\mathbb{S}^{\mathrm{ord}}(\Delta; \mathbb{I})$ holds if $k(P) \geq 2$.

7. — We now restate Theorem 3 in the language of p-adic Hecke algebras. Let $\mathbf{h}^{\mathrm{ord}}(N; O)$ be the p-adic ordinary Hecke algebra defined in [H5] §7.3. Let us recall the definition. The algebra $\mathbf{h}^{\mathrm{ord}}(N; O)$ is the Λ–subalgebra of $\mathrm{End}_\Lambda(\mathbb{S}^{\mathrm{ord}}(N; \Lambda))$ generated by $T(n)$ for all n. There is another description of the algebra. Writing $\mathbf{h}^{\mathrm{ord}}_k(Np^r; O)$ for the O–subalgebra of $\mathrm{End}_0(\mathcal{S}^{\mathrm{ord}}_k(Np^r; O))$ generated by $T(n)$ for all n, we have a natural isomorphism : $\mathbf{h}^{\mathrm{ord}}(N; O) \cong \varprojlim_\alpha \mathbf{h}^{\mathrm{ord}}_k(Np^\alpha; O)$ if $k \geq 2$, which takes $T(n)$ to $T(n)$ [H2]. Under the natural pairing $< h, \mathbf{f} >= \mathbf{a}(1, \mathbf{f} \mid h)$, we know

$$\text{(7.1)} \qquad \begin{aligned} &\mathrm{Hom}_\Lambda(\mathbf{h}^{\mathrm{ord}}(N; O), \Lambda) \cong \mathbb{S}^{\mathrm{ord}}(N; \Lambda) \text{ and} \\ &\mathrm{Hom}_\Lambda(\mathbb{S}^{\mathrm{ord}}(N; \Lambda), \Lambda) \cong \mathbf{h}^{\mathrm{ord}}(N; O). \end{aligned}$$

We have a smooth representation of $G(\mathbb{A}^{(p\infty)})$ on $\mathbb{S}^{\mathrm{ord}}(\mathbb{I}) = \bigcup_{U \in \mathcal{U}} \mathbb{S}^{\mathrm{ord}}(U;\mathbb{I}) = e\mathbb{S}(\mathbb{I})$ and $\mathbb{S}(\mathbb{I})$. Thus compactly supported smooth functions on $G(\mathbb{A}^{(p\infty)})$ with values in $\mathbb{I}$ act on $\mathbb{S}(\mathbb{I})$. We fix an algebraic closure $\overline{\mathbb{L}}$ of $\mathbb{L}$. We then consider $\mathbb{S}^{\mathrm{ord}}(A) = \mathbb{S}^{\mathrm{ord}}(\Lambda) \otimes_\Lambda A$ as a $G(\mathbb{A}^{(p\infty)})$–module for any Λ–subalgebra A in $\overline{\mathbb{L}}$. Each irreducible factor of the representation on $\mathbb{S}^{\mathrm{ord}}(\overline{\mathbb{L}})$ of $G(\mathbb{A}^{(p\infty)})$ is admissible by the control theorem ([H5] Theorem 7.3.3, which is the integral weight counterpart of Corollary 1 and is valid for all arithmetic points of weight $k \geq 2$). Pick an arithmetic point P with $k(P) \geq 2$ and consider the localization Λ_P at P. Then by the control theorem, $\mathbb{S}^{\mathrm{ord}}(\Lambda_P) \otimes_\Lambda K(P)$ for $K(P) = \Lambda_P/P$ is a semi–simple $GL_2(\mathbb{A}^{(p\infty)})$–module. Since there are Zariski dense arithmetic points in $\mathrm{Spec}(\Lambda)$ at which the control theorem holds, we see that $\mathbb{S}^{\mathrm{ord}}(\overline{\mathbb{L}})$ is semi–simple as a $G(\mathbb{A}^{(p\infty)})$–module. Thus $\mathbb{S}^{\mathrm{ord}}(\overline{\mathbb{L}})$ is a sum of irreducible subspaces. The multiplicity is one by the control theorem combined with the multiplicity one theorem in classical situation. Since the proof of the factorization theorem in [JL] §9 is purely algebraic, it carries over to our situation, and each irreducible factor π of $\mathbb{S}^{\mathrm{ord}}(\overline{\mathbb{L}})$ is factored into the tensor product of local representations : $\pi = \otimes_{\ell \neq p} \pi_\ell$. Let $\lambda; \mathbf{h}^{\mathrm{ord}}(C;O) \to \mathbb{I}$ be a primitive Λ–algebra homomorphism. Then by the control theorem, we have a unique automorphic representation $\pi(P) = \otimes_\ell \pi_\ell(P)$ corresponding to $\lambda \bmod P$ for $P \in \mathcal{A}(\mathbb{I})$ with $k(P) \geq 2$. Thus λ corresponds a unique factor $\pi = \pi(\lambda)$ of $\mathbb{S}^{\mathrm{ord}}(\overline{\mathbb{L}})$, and $\pi_\ell(P) = \pi_\ell \bmod P$. We write $V(\pi)$ for the subspace of $\mathbb{S}^{\mathrm{ord}}(\overline{\mathbb{L}})$ on which $S(\mathbb{A}^{(p\infty)})$ acts via π. Thus for each arithmetic point P with $k(P) \geq 2$, $\lambda_P(T(n)) = \lambda(T(n))(P)$ is an algebraic number. Then for each Dirichlet character φ, we can define the complex L–function :

$$L(s, \lambda_P \otimes \varphi) = \sum_{n=1}^{\infty} \varphi(n) \lambda_P(T(n)) n^{-s}.$$

Note that $L(s, \pi(P)) = L(s + \frac{k(P)-1}{2}, \lambda_P)$ is the standard L–function of $\pi(P)$. As is well known, the L–function $L(s, \lambda_P \otimes \varphi)$ has a motivic interpretation. Since $\mathbb{I}$ is an integral domain, we see that $Z_C = \mathbb{Z}_p^\times \times (\mathbb{Z}/C\mathbb{Z})^\times \ni z \longmapsto \lambda(<z>) \in \mathbb{I}$ is a character, where $<z>$ is the operator induced by the central action of $z \in (\mathbb{Z}^{(p)})^\times \subset G(\mathbb{A}^{(p\infty)})$ on $\mathbb{S}(\mathbb{I})$ (see (2.1)). In particular, it restriction to $\mu_{p-1} \times (\mathbb{Z}/C\mathbb{Z})^\times$ gives a character $\psi_0 : \mu_{p-1} \times (\mathbb{Z}/C\mathbb{Z})^\times \to \overline{\mathbb{Q}}_p^\times$. We regard this character as a character of Z_C composing the projection : $Z_N \to \mu_{p-1} \times (\mathbb{Z}/C\mathbb{Z})^\times \cong (\mathbb{Z}/Cp\mathbb{Z})^\times$ and call it the character of λ.

We now consider the following conditions on λ :

(H_ℓ). *Writing* $\pi_\ell \cong \pi(\alpha, \beta)$ *when* π_ℓ *is principal* ($\ell \neq p$), *we have* $\alpha(-1) = \beta(-1) = 1$;

(H_p). *$\psi_0 = \psi^2$ for an even character ψ modulo N for N divisible by C and 4.*

Under this condition, the automorphic representation associated to λ is in the image of the Shimura correspondence (see [Wa2] Proposition 2). Now we consider the automorphism σ_m of Λ which takes w to w^m ($w \in W$) for m prime to p. This ring automorphism extends to an automorphism σ_m of $\mathbb{I}$ if $\mathbb{I}$ is sufficiently large. For each $P \in \mathcal{A}(\mathbb{I})$, we denote P^2 for $P \circ \sigma_2$. Then $k(P^2) = 2k(P)$ and $\varepsilon_{P^2} = \varepsilon_P^2$. As constructed in [K] and [GS], for each character φ of $(\mathbb{Z}/Np\mathbb{Z})^\times$, there is a two variable p-adic L-function $\mathcal{L}_p(P,Q;\lambda \otimes \varphi)$ defined on $\mathcal{X}(\mathbb{I}) \times \mathcal{X}(\Lambda)$ interpolating the value $L(k(Q), \lambda_P \otimes \varepsilon_Q^{-1}\omega^{k(Q)}\varphi)$ for $(P,Q) \in \mathcal{A}(\mathbb{I}) \times \mathcal{A}(\Lambda)$ with $0 < k(Q) < k(P)$. Here is a result slightly stronger than Theorem 3 :

THEOREM 4. — *Let $\lambda : \mathbf{h}^{\mathrm{ord}}(C;O) \to \mathbb{I}$ be a primitive Λ-algebra homomorphism. Suppose (H_ℓ) and (H_p). Then for any pair (m,n) of two square free positive integers with $m/n \in \prod_{l|Np} \mathbb{Q}_l^2$, there exists an element Φ in $\mathbb{K}$ such that for all $P \in \mathcal{A}(\mathbb{I})$ with $k(P) \geq 1$, if $\mathcal{L}_p(P^2,P;\lambda \otimes \psi^{-1}\chi_m) \neq 0$, we have*

$$\Phi(P)^2 = \frac{\mathcal{L}_p(P^2,P;\lambda \otimes \psi^{-1}\chi_n)\psi_P(n/m)(n/m)^{k(P)-(1/2)}}{\mathcal{L}_p(P^2,P;\lambda \otimes \psi^{-1}\chi_m)},$$

where χ_t is the quadratic character corresponding to $\mathbb{Q}(\sqrt{t})$. Here note that under our assumption on (m,n), (m/n) is prime to Np.

Proof : we take $P \in \mathcal{A}(\mathbb{I})$ with $k(P)$ sufficiently large. Let φ be a Dirichlet character. Then for the least common multiple N' of C and the conductor of φ, we find $\lambda \otimes \varphi : \mathbf{h}(N';O) \to \mathbb{I}$ such that $\lambda \otimes \varphi(T(n)) = \varphi(n)\lambda(T(n))$. Then the character of $\lambda \otimes \varphi$ is given by $\psi^2\varphi^2$. Taking even φ with sufficiently large 2-power conductor, we may assume that the conductor C' of $\lambda \otimes \varphi$ is divisible by 16. If we replace λ by $\lambda \otimes \varphi$, the role of ψ will be replaced by $\psi\varphi$. Since $\lambda \otimes \psi^{-1}\chi_n = (\lambda \otimes \varphi) \otimes (\varphi\psi)^{-1}\chi_n$, the L-value appearing in the assertion of the theorem is unchanged even if we replace λ by $\lambda \otimes \varphi$. Thus we may assume that $16 \mid C$ (hence π satisfies the condition (H2) in [Wa2] p. 378). Let f be the cusp form in $\mathcal{P}^{\mathrm{ord}}_{k(P)+1/2}(\Gamma_0(N^2p^{r(P)}), \psi_P; \Omega_p)$ which is a linear combination of the base defined in [Wa2] Theorem 1 for $\pi(P^2)$. Let us take $\mathbf{f} \in \mathbb{P}^{\mathrm{ord}}(N^2;\mathbb{I})$ such that $\mathbf{f} \mid T(q^2) = \sigma_2 \circ \lambda(T(q))\mathbf{f}$ for all prime q outside Np and $\mathbf{f}(P) = cf$ with $0 \neq c \in O$. Such $\mathbf{f}$ exists by Corollary 1. Then by [Wa2] Corollary 2, for any $Q \in \mathcal{A}(\mathbb{I})$ such that $\mathbf{f}(Q)$ is classical, we have :

$$\mathbf{a}(m,\mathbf{f})^2(Q)L(k(Q), \lambda_{Q^2} \otimes \psi_Q^{-1}\chi_n)\psi_Q(n/m)(n/m)^{k(Q)-(1/2)}$$
$$= \mathbf{a}(n,\mathbf{f})^2(Q)L(k(Q), \lambda_{Q^2} \otimes \psi_Q^{-1}\chi_m).$$

To get the p-adic interpolation, we need to remove certain Euler factor at p and divide the special value by a certain period. However the Euler factor and the period are the same for n and m under the condition of the theorem. Thus using two variable p-adic L-functions, the above identity can be stated as :

$$\mathbf{a}(m,\mathbf{f})(Q)^2\mathcal{L}_p(Q^2,Q;\lambda\otimes\psi^{-1}\chi_n)\psi_Q(n/m)(n/m)^{k(Q)-(1/2)} = \mathbf{a}(n,\mathbf{f})(Q)^2\mathcal{L}_p(Q^2,Q;\lambda\otimes\psi^{-1}\chi_m).$$

If $\mathcal{L}_p(Q^2,Q;\lambda\otimes\psi^{-1}\chi_n)=0$ for all Q as above, the p-adic $\mathcal{L}$-function $\mathcal{L}_p(\lambda\otimes\psi^{-1}\chi_n)$ vanishes. Hence there is nothing to prove. If $\mathcal{L}_p(\lambda\otimes\psi^{-1}\chi_n)\neq 0$, by the assumption of the theorem, $\mathcal{L}_p(\lambda\otimes\psi^{-1}\chi_m)\neq 0$. Then we may assume that

$$\mathcal{L}_p(P^2,P;\lambda\otimes\psi^{-1}\chi_m)\mathcal{L}_p(P^2,P;\lambda\otimes\psi^{-1}\chi_n)\neq 0$$

by moving around P. Then we may assume by Theorem 1 of [Wa2] that the m-th and n-th Fourier coefficients of f are both non-zero. Therefore $\mathbf{a}(m;\mathbf{f})\mathbf{a}(n;\mathbf{f})\neq 0$. Thus we can take $\Phi=\mathbf{a}(n,\mathbf{f})/\mathbf{a}(m,\mathbf{f})$. Now we have the evaluation property of Φ described in the theorem for almost all P. Note that $\mathcal{L}_p(P,Q;\lambda\otimes\psi^{-1}\chi_n)$ for a fixed n is a p-adic analytic function of (P,Q) (see [K] and [GS]). Thus as long as the removed Euler factor does not vanish, we get the result. The only case where the Euler factor vanishes is the case where $k(P)=1$ and the character of $\pi(P^2)$ is trivial. However this case is excluded because of the vanishing of the p-adic L-function in the denominator at (P^2,P).

Since $\mathcal{L}_p(P^2,P;\lambda\otimes\psi^{-1}\chi_n)=0\iff L(k(P),\lambda_{P^2}\otimes\psi_P^{-1}\chi_n)=0$ if either $k(P)>1$ or $\psi_P^2\neq 1$, Theorem 3 follows from Theorem 4.

Manuscrit reçu le 20 juin 1993

References

[C] W. CASSELMAN. — *On some results of Atkin and Lehner*, Math. Ann. **201** (1973), 301–314.

[GS] R. GREENBERG and G. STEVENS. — *p–adic L–functions and p–adic periods of modular forms*, Inventiones Math. **111** (1993), 407–447.

[H1] H. HIDA. — *p–adic L–functions for base change lifts of GL_2 to GL_3*, Perspective in Math. **11** (1990), 93–142.

[H2] H. HIDA. — *On p–adic Hecke algebras for GL_2 over totally real fields*, Ann. of Math. **128** (1988), 295–384.

[H3] H. HIDA. — *Nearly ordinary Hecke algebras and Galois representations of several variables*, JAMI inaugural conference proceedings, 1988 May, Supplement to Amer. J. Math. (1990), 115–134.

[H4] H. HIDA. — *A p–adic measure attached to the zeta functions associated with two elliptic modular forms II*, Ann. l'Institut Fourier **38** N° 3 (1988), 1–83.

[H5] H. HIDA. — *Elementary theory of L–functions and Eisenstein series*, LMS Student Texts tenbfbk 26, Cambridge University Press, 1993.

[H6] H. HIDA. — *On nearly ordinary Hecke algebras for $GL(2)$ over totally real fields*, Adv. Studies in Pure Math. **17** (1989), 139–169.

[H7] H. HIDA. — *Geometric modular forms*, Proc. CIMPA Summer School at Nice, 1992.

[JL] H. JACQUET and R.P. LANGLANDS. — *Automorphic forms on $GL(2)$*, Lecture notes in Math. **114**, 1970.

[KM] N.M. KATZ and B. MAZUR. — *Arithmetic moduli of elliptic curves*, Ann. of Math. Studies **108**, Princeton University Press, 1985.

[K] K. KITAGAWA. — *On standard p–adic L–functions of families of elliptic cusp forms*, preprint.

[MTT] B. MAZUR, J. TATE and J. TEITELBAUM. — *On p–adic analogues of the conjectures of Birch and Swinnerton–Dyer*, Inventiones Math. **81** (1986), 1–48.

[Sh1] G. SHIMURA. — *On modular forms of half integral weight*, Ann. of Math. **97** (1973), 440–481.

[Sh2] G. SHIMURA. — *On certain reciprocity laws for theta functions and modular forms*, Acta Math. **141** (1978), 35–71.

[V] M.–F. VIGNERAS. — *Valeurs au centre de symétrie des fonctions L associées aux formes modulaires*, Séminaire de Théorie des Nombres, Paris 1979–80, Progress in Math. **12**, Birkhäuser (1981), 331–356.

[Wa1] J.–L. WALDSPURGER. — *Correspondance de Shimura*, J. Math. pures et appl. **59** (1980), 1–133.

[Wa2] J.–L. WALDSPURGER. — *Sur les coefficients de Fourier des formes modulaires de poids demi–entier*, J. Math. pures et appl. **60** (1981), 375–484.

[W] A. WEIL. — *Sur certains groupes d'opérateurs unitaires*, Acta Math. **111**, 143–211.

[Wi] A. WILES. — *On ordinary λ–adic representations associated to modular forms*, Inventiones Math. **94** (1988), 529–573.

Haruzo HIDA
Department of Mathematics
UCLA
Los Angeles, Ca 90024
U.S.A.

Number Theory
Paris 1992–93

Structures algébriques sur les réseaux

Jacques Martinet[(*)]

PREMIÈRE PARTIE : **rappels sur les réseaux**

1. — On note E un espace euclidien de dimension n, souvent identifié à $\mathbb{R}^n$ par le choix d'une base orthonormée de E. La norme d'un vecteur $x \in E$ est $\mathrm{N}(x) = x.x$, le carré de la norme euclidienne $\|x\|$. Par *réseau*, on entend un sous-groupe discret Λ de E de rang n. La *norme de* Λ est $\mathrm{N}(\Lambda) = \min_{x\in\Lambda, x\neq 0} \mathrm{N}(x)$. On pose $S(\Lambda) = \{x \in \Lambda \mid \mathrm{N}(x) = \mathrm{N}(\Lambda)\}$ et $s(\Lambda) = \frac{1}{2}|S(\Lambda)|$. Le *déterminant de* Λ est le déterminant de la *matrice de Gram d'une base de* Λ (matrice des produits scalaires deux à deux des vecteurs de la base). L'*invariant d'Hermite de* Λ est $\gamma_n(\Lambda) = \mathrm{N}(\Lambda).\det(\Lambda)^{-1/n}$, et la *constante d'Hermite pour la dimension* n est $\gamma_n = \sup_\Lambda \gamma_n(\Lambda)$.

On dit qu'un réseau Λ est *entier* si le produit scalaire de E est à valeurs entières sur Λ, et qu'il est *pair* si ses vecteurs sont de norme paire. Le *réseau dual de* Λ est $\Lambda^* = \{x \in E \mid \forall y \in \Lambda, x.y \in \mathbb{Z}\}$. Les réseaux entiers sont les réseaux qui sont contenus dans leur dual. Ceux qui sont égaux à leur dual sont dits *unimodulaires*; ce sont les réseaux entiers de déterminant 1.

Les réseaux que nous rencontrerons seront tous proportionnels à des réseaux entiers. Dans ce cas, il existe une plus petite norme qui les rend entiers. On définit l'invariant de Smith d'un réseau Λ en considérant le réseau entier Λ' qui lui est ainsi associé; le couple (Λ'^*, Λ') de $\mathbb{Z}$-modules libres de rang n possède lui-même un invariant de Smith (suite des "facteurs invariants" ou "diviseurs élémentaires"), qui est l'*invariant de Smith* $\mathrm{Smith}(\Lambda)$ *de* Λ. Si $\mathrm{Smith}(\Lambda) = (a_1, \ldots, a_n)$, on a $a_n = 1$ et $\mathrm{Smith}(\Lambda^*) = (\frac{a_1}{a_n}, \frac{a_1}{a_{n-1}}, \ldots, \frac{a_1}{a_1})$.

2. — Soit E' un sous-espace de E de dimension r coupant Λ suivant un réseau Λ' de E'. Alors, $E'^{\perp}$ coupe Λ^* suivant un réseau $\Lambda'^{\perp}$ de $E'^{\perp}$, et

[(*)] Recherche effectuée au sein de l'unité mixte C.N.R.S.– Enseignement Supérieur U.R.M. 9936

l'on a entre déterminants la relation

$$\det(\Lambda') = \det(\Lambda).\det(\Lambda'^{\perp}).$$

Considérons le cas particulier dans lequel il existe une similitude u de Λ sur Λ^*, que nous prenons égale à l'identité dans le cas unimodulaire. On associe alors à tout réseau Λ' comme ci-dessus le réseau relatif $\Lambda'_u = u(E')^{\perp} \cap \Lambda \subset \Lambda$. En désignant par $\sqrt{\lambda}$ le rapport de similitude, on obtient la formule

$$\det(\Lambda'_u) = \lambda^r.\det(\Lambda).\det(\Lambda').$$

3. — Soit Λ un réseau. Nous appellerons *défaut de perfection de* Λ la différence entre la dimension ($\frac{n(n+1)}{2}$) de l'espace $\mathrm{End}^s(E)$ des endomorphismes symétriques de E et celle du sous-espace de $\mathrm{End}^s(E)$ engendré par les projections sur les directions des vecteurs minimaux de Λ; on appelle *relation d'eutaxie* toute expression de l'identité de E comme combinaison linéaire de ces projections. On dit que Λ est *parfait* s'il est de défaut nul et qu'il est *eutactique* s'il possède une relation d'eutaxie à coefficients positifs. On sait (théorème de Voronoï) que Λ est *extrême* (c'est-à-dire qu'il réalise un maximum local de son invariant d'Hermite) si et seulement s'il est parfait et eutactique. (En exprimant les endomorphismes de E dans un couple de bases $(\mathcal{B}, \mathcal{B}^*)$ où $\mathcal{B}$ est une base de Λ, on transforme ces définitions géométriques issues de [Be-M1] en les définitions classiques de la théorie des formes quadratiques.)

Une condition suffisante de perfection, due à Barnes, est l'existence d'une section hyperplane parfaite de même norme et de n vecteurs minimaux indépendant en-dehors de cette section (*perfection relative*). Soit Λ_0 un réseau de dimension n_0. Cette condition de perfection relative est vérifiée par les réseaux de dimension n_0+1 dont le déterminant est minimum parmi ceux possédant Λ_0 comme section hyperplane de même norme. Les réseaux faiblement laminés au-dessus de Λ_0 sont ceux que l'on obtient en itérant le procédé ci-dessus, et l'on parle de *réseaux fortement laminés* dans le cas de ceux qui sont de déterminant minimum dans chacune des dimensions $n_0, n_0+1, n_0+2, \ldots$ (ce vocabulaire est emprunté à Plesken et Pohst qui ont étudié les variantes des procédés de lamination dans lesquelle on considère des réseaux entiers de norme donnée). Les réseaux laminés sans autre précision sont ceux qui ont été obtenus par Conway et Sloane par "laminations fortes" au-dessus de $\Lambda_0 = \{0\}$ auquel est attribué la norme 4 ([C-S], ch. 6); pour $n \leq 8$, ces réseaux, notés Λ_n, sont les renormalisations à la norme 4 des réseaux $\{0\}, \mathbb{Z}, \mathbb{A}_2, \mathbb{A}_3, \mathbb{D}_4, \mathbb{D}_5, \mathbb{E}_6, \mathbb{E}_7, \mathbb{E}_8$, puisque la constante d'Hermite est atteinte dans ces dimensions sur les réseaux qui leurs sont semblables ("théorème de Blichfeldt-Vetchinkin"; Korkine et Zolotareff pour $n \leq 5$, Barnes pour $n = 6$).

Certains résultats de perfection et d'eutaxie que nous présentons dans cette note ont été obtenus en utilisant deux programmes de Batut, l'un calculant le rang des projections sur les vecteurs minimaux d'un réseau défini par une matrice de Gram et indiquant s'il existe une relation à coefficients d'eutaxie égaux, et l'autre donnant une base de l'espace des relations qui existent entre ces projections et l'identité, ainsi que divers programmes disponibles dans le système *PARI*.

L'invariant d'Hermite dual de Λ, introduit dans [Be-M1], est $\gamma'_n(\Lambda) = (\mathrm{N}(\Lambda)\mathrm{N}(\Lambda^*))^{1/2}$; sa borne supérieure sur les réseaux ("constante de Bergé-Martinet" de [C-S3]) est notée γ'_n. On dit que Λ est dual-extrême si son invariant γ'_n est un maximum local. Pour qu'il en soit ainsi, il suffit ([Be-M1], 3.20) que Λ soit extrême et que Λ^* soit eutactique.

4. — Rappelons les définitions de quelques réseaux classiques (cf [C–S], ch. 4). Soit (ε_i), $0 \leq i \leq n$ (resp. $1 \leq i \leq n$) la base canonique de $\mathbb{Z}^{n+1}$ (resp. de $\mathbb{Z}^n$). On pose

$$\mathbb{A}_n = \{x \in \mathbb{Z}^{n+1} \mid \sum x_i = 0\} \quad \text{et} \quad \mathbb{D}_n = \{x \in \mathbb{Z}^n \mid \sum x_i \equiv 0 \bmod 2\}.$$

Ce sont des réseaux pairs de norme 2. Le dual de $\mathbb{D}_n$ est le réseau cubique centré, de norme 1 lorsque n est ≥ 4. Il est isométrique au sous-réseau de $\mathbb{Z}^n$ muni de la forme $\frac{1}{4}\sum x_i y_i$ défini par les $n-1$ congruences $x_1 \equiv x_2 \cdots \equiv x_n \bmod 2$.

Pour n pair ≥ 8, soit $\mathbb{D}_n^+ = \mathbb{D}_n \cup \frac{1}{2}(\varepsilon_1 + \varepsilon_2 + \cdots + \varepsilon_n)\mathbb{D}_n$. On obtient un réseau isométrique en considérant le double système de congruences

$$x_1 \equiv x_2 \equiv \ldots \equiv x_n \bmod 2 \quad \text{et} \quad x_1 + x_2 + \cdots + x_n \equiv 0 \bmod 4$$

sur $\mathbb{Z}^n$ muni de la forme $\frac{1}{4}\sum x_i y_i$. Sous cette forme, on voit que $\mathbb{D}_n^+$ est isométrique à son dual, et même qu'il est unimodulaire pour $n \equiv 0 \bmod 4$, pair pour $n \equiv 0 \bmod 8$.

On pose $\mathbb{E}_8 = \mathbb{D}_8^+$, et l'on définit $\mathbb{E}_7$ (resp. $\mathbb{E}_6$) comme l'orthogonal dans $\mathbb{E}_8$ d'un vecteur minimal (resp. d'un sous-réseau isométrique à $\mathbb{A}_2$), cf. N$^\text{o}$ 2 (à isométrie près, les choix faits ci-dessus sont sans importance). Les *réseaux de racines* sont les sommes orthogonales de réseaux de racines irréductibles, isométriques à $\mathbb{Z}$, $\mathbb{A}_n$ $(n \geq 1)$, $\mathbb{D}_n$ $(n \geq 4)$ ou $\mathbb{E}_n$ $(n = 6, 7, 8)$. Ces derniers sont extrêmes, ont des duals eutactiques, et sont donc aussi dual-extrêmes.

DEUXIÈME PARTIE : **autour du réseau de Coxeter–Todd**

5. — Soit A l'anneau des entiers d'un corps de nombres K totalement réel ou de type C.M., de degré q. On note $x \mapsto \bar{x}$ l'involution de K (l'identité dans le premier cas), et l'on munit K de la forme bilinéaire $\mathrm{Tr}_{K/\mathbb{Q}}(\lambda\overline{\mu})$ (ou parfois d'une forme qui lui est proportionnelle), ce qui fait de A un réseau entier de $\mathbb{R}\otimes K$.

Soit $\mathfrak{a}$ un idéal de A stable par l'involution de K. On considère sur A^m les congruences suivantes :

$$\begin{array}{lll}(\mathrm{C1}) & \lambda_1 \equiv \lambda_2 \equiv \cdots \equiv \lambda_m & \mathrm{mod}\,\mathfrak{a}\\ (\mathrm{C2}) & \lambda_1+\lambda_2+\cdots+\lambda_m \equiv 0 & \mathrm{mod}\,\mathfrak{a}\\ (\mathrm{C'2}) & \lambda_1+\lambda_2+\cdots+\lambda_m \equiv 0 & \mathrm{mod}\,\mathfrak{a}^2,\end{array}$$

qui définissent des réseaux de dimension $n = qm$.

La congruence C'2 n'interviendra qu'en même temps que la congruence C'1 et seulement lorsque m est un multiple de la norme de $\mathfrak{a}$. En notant d le discriminant du corps K (i.e. le déterminant du réseau A), on trouve pour les déterminants des réseaux définis par les congruences C1 (resp. C2, resp. C1 et C'2) les valeurs $|d|^m \mathrm{N}_{K/\mathbb{Q}}(\mathfrak{a})^{2(m-1)}$ (resp. $|d|^m \mathrm{N}_{K/\mathbb{Q}}(\mathfrak{a})^2$, resp. $|d|^m \mathrm{N}_{K/\mathbb{Q}}(\mathfrak{a})^{2m}$).

On observe que, pour $m \in \mathfrak{a}$, le réseau défini par C1 ou par C1 et C'2 est encore entier lorsqu'on munit A de la forme $\frac{1}{m}\mathrm{Tr}(\lambda\overline{\mu})$: on a en effet

$$\sum_i \lambda_i\overline{\mu_i} \equiv \lambda_1 \sum_i \overline{\mu_i} \equiv m\lambda_1\overline{\mu_1} \;\mathrm{mod}\,\mathfrak{a}\,.$$

Les déterminants donnés ci-dessus sont alors à diviser par m^{qm}.

6. — Dans les numéros 6 à 9, sauf dans la remarque 8.3, A est l'anneau $\mathbb{Z}[\omega]$ ($\omega^2+\omega+1=0$) des entiers d'Eisenstein. Les congruences C1, C2, C'2 ont été considérées dans les années cinquante par Coxeter, Coxeter et Todd, et Barnes ([Cox], [Cox–T], [Bar]).

Soit $n = 2r + t$. Le réseau L_n^r de Barnes est formé des éléments de A^{r+t} qui vérifient la congruence C2 et dont les t dernières coordonnées sont réelles. Les réseaux L_n^r sont parfaits pour $n \geq 5$ et $r \geq 2$ ([Bar]; cela se voit par réduction à la dimension 5 en utilisant des arguments de perfection relative). Pour $n = 2r \geq 6$, ces réseaux sont extrêmes et dual extrêmes ([Bar], [Be-M1]). Dans le cas $n = 6, r = 3$, considéré initialement par Coxeter ([Cox]), on trouve un réseau semblable à E_6^*, et l'on obtient donc E_6 par la congruence C1 avec la forme $\frac{1}{3}\mathrm{Tr}$.

Le *réseau de Coxeter-Todd*, noté K_{12}, est défini par les congruences C1 et C3, avec la forme $\frac{1}{3}\mathrm{Tr}$. Cette définition par congruences, jointe au fait que $A \simeq \mathbb{A}_2$ est semblable à son dual, montre que $x \mapsto \frac{1}{1-\omega}x$ est un isomorphisme de K_{12} sur son dual, un résultat noté par Conway et Sloane, qui l'interprètent en faisant remarquer que K_{12} est $\mathbb{Z}[\omega]$-unimodulaire ([C-S], ch. 4, § 9). Une variante de cette construction, analogue à la définition de D_n^+ (cf. n° 3) consiste en l'adjonction à L_{12}^6 du vecteur $\frac{1}{1-\omega}(1,1,1,1,1,1)$. Sous cette forme, on voit immédiatement que K_{12} est extrême et dual-extrême (on a des résultats analogues dans toutes les dimensions multiples de 6 et

≥ 12).

Il est facile de vérifier que le réseau $\Lambda_6 \sim \mathbb{E}_6$ se plonge dans K_{12}. Comme les réseaux Λ_n réalisent la constante γ_n pour $n \leq 6$, ce sont les réseaux de plus petit déterminant contenus dans K_{12} pour les dimensions comprises entre 0 et 6; c'est la série K_n pour $0 \leq n \leq 6$. La méthode du n° 2, appliquée en prenant $u = (x \mapsto \frac{1}{1-\omega}x)$, permet de construire une suite descendante $K_{12} \supset K_{11} \supset \cdots \supset K_7 \supset K_6^{\perp}$ de réseaux dont les déterminants sont minimaux pour les dimensions comprises entre 12 et 6. On obtient une suite K_n, $0 \leq n \leq 12$ en raccordant les deux suites en dimension 6. Cela est bien connu depuis Leech (et également entre les dimensions 12 et 24 que nous examinerons plus loin), cf. [C–S], ch. 6, § 1.

Toutefois, comme on va le voir, ces plongements ne sont pas compatibles avec les $\mathbb{Z}[\omega]$-structures qui existent naturellement sur $\mathbb{D}_4$ et sur $\mathbb{E}_6$ (on a rencontré une telle structure dans le cas de $\mathbb{E}_6$, et l'on peut identifier $\mathbb{D}_4$ à l'ordre de Hurwitz $\mathfrak{M}$ des quaternions usuels sur $\mathbb{Q}$, puis plonger A dans $\mathfrak{M}$ par $\omega \mapsto \frac{-1+i+j+k}{2}$).

7. — Nous nous intéressons maintenant à des réseaux Λ pour lesquels le produit scalaire est de la forme $\mathrm{Tr} \circ h$ où $h : \Lambda \to A$ est une forme hermitienne (nous dirons simplement *$\mathbb{Z}[\omega]$-réseaux*), et nous considérons les plongements qui sont des isométries pour les structures hermitiennes, ce qui est plus restrictif que d'être seulement une isométrie pour la structure euclidienne qui s'en déduit.

Le théorème suivant sera démontré au n° 9.

7.1. THÉORÈME. — *Soit Λ un $\mathbb{Z}[\omega]$-réseau entier de norme* 4. *Si $n = 4$ et si $\det(\Lambda)$ est ≤ 81 (resp. si $n = 6$ et si $\det(\Lambda)$ est ≤ 243), alors Λ est $\mathbb{Z}[\omega]$-semblable à $\mathbb{D}_4$ ou à L_4^2 (resp. à $\mathbb{E}_6$ ou à E_6^*). Ces réseaux ont des A-bases (e_1, e_2) (resp. (e_1, e_2, e_3)) formées de vecteurs minimaux, et sont définis par les suites de produits scalaires $(e_1.e_2, e_1.\omega e_2)$ (resp. $(e_1.e_2, e_1.\omega e_2, e_1.e_3, e_1.\omega e_3, e_2.e_3, e_2.\omega e_3)$); des choix possibles pour ces quatre réseaux sont les suites $(0,2), (1,1)$ (resp. $(0,2,0,2,0,0), (1,1,1,1,1,1)$).*

[Le th. 7.1 prouve en particulier l'unicité à $\mathbb{Z}[\omega]$-isométrie près des réseaux $\mathbb{D}_4$ et $\mathbb{E}_6$. Feit a démontré un résultat analogue par une formule de masse pour le réseau K_{12} dans son article [Fe] consacré aux réseaux $\mathbb{Z}[\omega]$-unimodulaires. Des résultats d'unicité concernant en particulier $\mathbb{D}_4$ sur $\mathbb{Z}[\zeta_8]$ et sur $\mathbb{Z}[\zeta_{12}]$ et $\mathbb{E}_6$ sur $\mathbb{Z}[\zeta_9]$ figurent dans [Be-M2], th. 4.3 et 4.6].

En examinant les produits scalaires entre vecteurs minimaux de K_{12}, on s'aperçoit qu'il n'est pas possible de plonger $\Lambda_4 \sim \mathbb{D}_4$ dans K_{12} en tant que $\mathbb{Z}[\omega]$-réseau, et donc non plus $\Lambda_6 \sim \mathbb{E}_6$. En revanche, la définition de K_{12} montre que l'on peut plonger L_4^2 et $L_6^3 \sim \mathbb{E}_6^*$. En utilisant la méthode du n° 2, on construit une suite croissante de $\mathbb{Z}[\omega]$-réseaux K_n' (n pair) plongés dans

K_{12}, que l'on complète pour n impair en prenant le réseau de déterminant minimum parmi ceux qui sont contenus dans K'_{n+1} et contiennent K'_{n-1}. Ces réseaux K'_n, comme les K_n, sont bien définis à un automorphisme de K_{12} près.

On voit tout de suite que l'on a $K'_1 = \Lambda_1 \sim \mathbb{Z}$, $K'_2 = \Lambda_2 \sim \mathbb{A}_2$, $K'_{11} = K_{11}$, $K'_{12} = K_{12}$, et que, pour $4 \leq n \leq 8$, K'_n est isométrique à L^r_n avec $r = \lfloor \frac{n}{2} \rfloor$. Le réseau K'_3, connu des cristallographes (cf. [C–S3]) est caractérisé à similitude près comme le réseau d'invariant γ_3 minimum parmi ceux qui satisfont l'inégalité $s \geq 5$. Une vérification informatique à partir d'une matrice de Gram de K_{12} montre :

7.2. PROPOSITION. — *Les réseaux K_n (resp. K'_n) sont parfaits sauf pour $n = 7$ et $n = 8$ (resp. $n = 3$ et $n = 4$) où le défaut de perfection est égal à* 1.

Le tableau suivant décrit les principaux invariants des réseaux K'_n :

réseau	K'_3	K'_4	K'_5	K'_6	K'_7	K'_8	K'_9	K'_{10}
$\det(K'_n)$	36	81	162	243	486	729	972	972
$s(K'_n)$	5	9	15	27	36	54	81	135
$\mathrm{Smith}(K'_n)$	12.3	9.3^2	6.3^3	3^5	18.3^3	9.3^4	36.3^3	$6^2.3^3$

Signalons que les réseaux K'_{10} et K'^{*}_{10} (pour lequel on a $s = 120$) sont extrêmes et donc dual-extrêmes.

[Voici une construction explicite de ces deux réseaux. Par division par $1 - \omega$, on transforme le vecteur minimal $(0, 0, 0, 0, 1 - \omega, -(1 - \omega))$ de K_{12} en le vecteur $(0, 0, 0, 0, 1, -1)$ de K^*_{12}, dont l'orthogonal permet de définir K'_{10} par les congruences $\lambda_1 \equiv \lambda_2 \equiv \lambda_3 \equiv \lambda_4 \equiv \lambda_5 \bmod \mathfrak{a}$ et $\lambda_1 + \lambda_2 + \lambda_3 + \lambda_4 - \lambda_5 \equiv 0 \bmod \mathfrak{a}^2$ sur $\mathbb{Z}(\omega)^5$ muni de la forme hermitienne $\lambda_1\bar{\lambda}_1 + \lambda_2\bar{\lambda}_2 + \lambda_3\bar{\lambda}_3 + \lambda_4\bar{\lambda}_4 + 2\lambda_5\bar{\lambda}_5$. On voit que les 135 couples de vecteurs minimaux de K_{10} sont représentés par $3^4 = 81$ vecteurs de composantes de la forme ω^i et $6.9 = 54$ vecteurs obtenus par permutation des 4 premières composantes de vecteurs de la forme $(\omega^i(1 - \omega), -\omega^j(1 - \omega), 0, 0, 0)$. On obtient le réseau dual en remplaçant dans la forme hermitienne $2\lambda_5\bar{\lambda}_5$ par $\frac{1}{2}\lambda_5\bar{\lambda}_5$ et en divisant par $1 - \omega$, et les 120 couples de vecteurs minimaux proviennent de 81 vecteurs comme ci-dessus, de $4.9 = 36$ vecteurs obtenus par permutation des 4 premières composantes de vecteurs de la forme $(\omega^i(1 - \omega), 0, 0, 0, -\omega^j(1 - \omega))$ et des 3 vecteurs $(0, 0, 0, 0, 3\omega^i)$].

Grâce à des programmes de Batut, on vérifie qu'il existe dans les cas de K^*_{11} et de K'^{*}_9 une unique relation d'eutaxie. Pour une indexation convenable des directions de vecteurs minimaux, elles ont les formes respectives

$$\mathrm{Id} = \frac{3}{4}\sum_{i=2}^{12} p_i \quad \text{et} \quad \mathrm{Id} = p_1 + \frac{1}{4}\sum_{i=2}^{41} p_i \,.$$

On montre que la section de K_{11} (resp. de K_9) par l'hyperplan orthogonal à la première direction minimale définit le réseau K'_{10} (resp. K'_8), alors que les autres directions minimales sont asociées à K_{10} (resp. à des réseaux K''_8 isométriques au réseau P_8 de Barnes, noté $\mathbb{A}_8^{(2)}$ dans [C–S], ch. 8, § 6) [Ces propriétés d'eutaxie s'interprètent par l'existence de deux orbites de plans hexagonaux engendrés par des vecteurs minimaux dans $K'^*_{12} = K_{12}$ et dans K'^*_{10} ; signalons que K_{10} possède une section parfaite K''_9 de même déterminant (972) que K'_9, mais avec $s = 82$ au-lieu de $s = 81$; les réseaux K'_9 et K''_9 ont été trouvés par Barnes ([Bar], II, p. 221)].

8. — En plongeant K_{12} dans le réseau de Leech Λ_{24} et en utilisant la méthode du n° 2, on complète la suite K_n jusqu'à la dimension 24. Il est clair que l'on obtient une suite de sections de Λ_{24} qui sont de déterminant minimum parmi les réseaux contenant ou contenus dans K_{12}, et que l'on a la relation de symétrie $\det(K_n) = \det(K_{24-n})$.

On peut procéder de la même façon avec la série K'_n. On commence par munir Λ_{24} d'une $\mathbb{Z}[\omega]$–structure compatible avec le plongement $K_{12} \to \Lambda_{24}$; on indiquera dans la quatrième partie comment réaliser un tel plongement *sur un ordre maximal du corps de quaternion de centre* $\mathbb{Q}$ *ramifié en* $\{3, \infty\}$, ce qui est un résultat plus précis. La méthode du n° 2 permet de prolonger la suite K'_n jusqu'à la dimension 24, les réseaux obtenus étant des $\mathbb{Z}[\omega]$-réseaux pour n pair; on a les relations de symétrie $\det(K'_n) = \det(K'_{24-n})$ et les égalités $K'_{13} = K_{13}$ et $K'_n = \Lambda_n$ pour $n = 22, 23, 24$ (alors que la coïncidence de K_n et de Λ_n a lieu dès la dimension 18); on définit de même un réseau K''_{16} à partir de K''_8.

Pour étudier ces réseaux K'_n au-delà de la dimension 12, on utilise la détermination par Plesken et Pohst ([Pl-P]) des réseaux faiblement laminés pour la norme 4 au-dessus de K_{12}. Ces auteurs ont trouvé un réseau en dimension 13, qui est $K'_{13} = K_{13}$, deux en dimension 14 qui sont K_{14} et K'_{14}, puis, au-dessus de l'un d'eux, qui ne peut être que K'_{14}, une suite de réseaux de déterminants $\det(K'_n)$, uniques à isométrie près, sauf en dimension 16 où il y a deux réseaux, que l'on distingue par leurs invariants s, qui prennent les valeurs 1218 et 1224.

8.1. THÉORÈME. — *Le réseau* K'_{16} *est le réseau d'invariant* $s = 1224$.

Démonstration (1) (H. NAPIAS). On repère le réseau K'_{22} dans Λ_{24} à l'aide de matrices de Gram, on construit la suite descendante des K'_n jusqu'à la dimension 17, et l'on distingue les sections K'_{16} et K''_{16} par leurs orthogonaux.

[Elle a également montré qu'un seulement parmi les 37 vecteurs minimaux de K'^*_{17} a pour orthogonal K'_{16} dans K'_{17}, résultat analogue à ceux que l'on a observés pour K'^*_{11} et K'^*_9].

(2) Par adjonction à L_{2m}^m du vecteur $v_1 = \frac{1}{1-\omega}(1, 1, 1, 1, 1, 1, 0, 0, \ldots)$

pour $2m \geq 12$ puis du vecteur $v_2 = \frac{1}{1-\omega}(1,-1,0,1,-1,0,1,-1,0,0,\ldots)$ pour $2m \geq 16$, on obtient des réseaux de déterminant 3^m puis 3^{m-2}, obtenant K_{12} pour $m=6$, puis un réseau Λ de déterminant 3^6 pour $m=8$, qui contient visiblement K_{12} ainsi qu'une suite de sections en dimensions $n=15,14,13$ de déterminants $\det(K'_n)$. On a ainsi construit celui des deux réseaux de Plesken-Pohst qui est K'_{16}, et l'on vérifie facilement que l'on a $s(\Lambda)=1224$.
[On construit K'_{18} par adjonction à L_{18}^9 de v_1, v_2 et $v_3 = \frac{1}{1-\omega}(0,0,0,1,1,1,1,1,1)$, et l'on en déduit des constructions explicites de K'_{17} et de K''_{16}].

8.2. Remarque. — Les réseaux K_n et K'_n, $n \geq 12$ et K''_{16} sont parfaits. Cela se voit en contrôlant la perfection relative à partir de la dimension 12. Il est probable, mais non démontré, que les constructions par lamination *pour une norme donnée* donnent dans ce cas particulier les réseaux faiblement laminés au-dessus de K_{12}, un résultat qui entraînerait directement la perfection, comme dans le cas des réseaux Λ_n considérés par Conway et Sloane.

8.3. Remarque. — Prenons $A = \mathbb{Z}[\zeta_9]$ et $\mathfrak{a} = ((1-\zeta_9)^2)$, et soit R_{6m} le réseau défini par la congruence $\sum \lambda_i \equiv 0 \bmod \mathfrak{a}^4$ sur $((\mathfrak{a}^2)^m, \frac{1}{9}\mathrm{Tr})$. On a $R_6 \simeq \mathbb{E}_6$ ([Cra]), d'où $R_{12} \simeq K_{12}$ et $\langle R_{18}, \frac{1}{1-\zeta_9}(1,1,1)\rangle \simeq K'_{18}$; on trouvera dans [Bay–M] une construction de K_{12} comme module de rang 1 sur $\mathbb{Q}(\zeta_{21})$.

9. — Nous démontrons maintenant le th.7.1.

9.1. Lemme. — *Soit Λ un réseau pair de dimension n et de norme m. Supposons vérifiée l'inégalité $\det(\Lambda) < \frac{m+2}{m} m^n \gamma_n^{-n}$. Alors, Λ possède n vecteurs minimaux indépendants.*

Démonstration. L'inégalité de Minkowski sur les minima successifs d'un réseau montre qu'il existe des vecteurs $e_1, e_2, \ldots, e_n$ de Λ vérifiant l'inégalité $N(e_1)N(e_2)\ldots N(e_n) \leq \gamma_n^n \det(\Lambda)$, qui entraîne que l'on a $N(e_1)N(e_2)\ldots N(e_n) < \frac{m+2}{m} m^n$. On peut supposer ces vecteurs rangés par normes croissantes. On montre que ce sont des vecteurs minimaux en raisonnant par récurrence sur leur indice. On a en effet $N(e_1)\ldots N(e_{i-1})N(e_i)^{n-i+1} < \frac{m+2}{m} m^n$; si l'on suppose que $e_1, \ldots, e_{i-1}$ sont de norme m, on trouve pour la norme de e_i les inégalités $N(e_i) < m(\frac{m+2}{m})^{1/(n-i+1)} < m+2$, donc $N(e_i) \leq m$ puisque Λ est pair, d'où l'égalité $N(e_i) = m$. □

[Pour $n = 2, 3, \ldots, 8$, la borne du lemme est égale à 18, 48, 96, 192, 288, 324, 324].

9.2. Théorème. — *Soit Λ un $\mathbb{Z}[\omega]$-réseau entier de norme 4, de dimension 4 (resp.6), et de déterminant < 96 (resp. < 288). Alors, Λ est $\mathbb{Z}[\omega]$-semblable* *à $\mathbb{D}_4$ ou à L_4^2 (resp. à $\mathbb{E}_6$ ou à $\mathbb{E}_6^*$). En particulier, les $\mathbb{Z}[\omega]$-structures sur les réseaux $\mathbb{D}_4$, L_4^2 et E_6 sont uniques à $\mathbb{Z}[\omega]$-isométrie près.*

Démonstration. Le lemme montre que Λ contient $n = 4$ (resp. $n = 6$) vecteurs minimaux indépendants et donc, compte tenu de l'action de $\mathbb{Z}[\omega]$, qu'il contient un sous-réseau Λ' possédant une base de la forme $(x, \omega x, y, \omega y)$ (resp. $(x, \omega x, y, \omega y, z, \omega z)$). Le réseau Λ' est déterminé par la donnée des produits scalaires $a_1 = x.y, b_1 = x.\omega y$ (resp. $a_1 = x.y, b_1 = x.\omega y, a_2 = x.z, b_2 = x.\omega z, a_3 = y.z, b_3 = y.\omega z$) qui sont majorés par 2 en valeur absolue. On a $\det(\Lambda') \leq 2^n \det(\mathbb{A}_2)^{n/2} = 12^{n/2}$ et $\det(\Lambda') \geq 2^4 \det(\mathbb{D}_4)$ (resp. $\det(\Lambda') \geq 2^6 \det(\mathbb{E}_6)$, d'où la majoration $[\Lambda : \Lambda'] \leq 3$ avec égalité seulement pour $n = 6$, $\Lambda' \simeq \mathbb{A}_2 \perp \mathbb{A}_2 \perp \mathbb{A}_2$ et $[\Lambda : \Lambda'] = 3$ (car 2 est inerte dans $\mathbb{Q}[\omega]$), cas dans lequel Λ est de déterminant $2^6.3$, donc semblable à $\mathbb{E}_6$, et où l'existence d'autres vecteurs minimaux que ceux des orbites de x, y, z permet encore de supposer que l'on a $\Lambda' = \Lambda$.

On peut supposer que $x.y$ est ≥ 0 et minimum parmi les valeurs absolues des produits scalaires des vecteurs minimaux de Λ' appartenant à deux orbites distinctes, et utiliser l'automorphisme $\omega \mapsto \omega^2$ de $\mathbb{Z}[\omega]$ pour échanger $x.\omega y$ et $x.\omega^2 y$.

On voit tout de suite que, en dimension 4, $\Lambda = \Lambda'$ est obtenu en prenant pour (a_1, b_1) l'un des 4 couples $(0,0), (0,1), (0,2), (1,1)$, conduisant à des réseaux de déterminants respectifs $144, 121, 64, 81$, d'où le théorème dans ce cas.

Dans le cas de la dimension 6, nous avons d'abord prouvé "à la main" l'assertion d'unicité de la $\mathbb{Z}[\omega]$-structure de $\mathbb{E}_6^*$, qui entraîne le résultat analogue pour $\mathbb{E}_6$. On observe pour cela que le produit scalaire de deux vecteurs minimaux de E_6^* n'est jamais nul, ce qui permet de définir Λ' en prenant pour $(a_1, b_1, a_2, b_2, a_3, b_3)$ l'une des suites $(1,1,1,1,1,1)$ ou $(1,1,1,1,1,-2)$, et la seconde se ramène à la première en remplaçant y par $x + \omega^2 y$. On achève la démonstration en contrôlant sur ordinateur qu'il n'y a pas de déterminant dans l'intervalle $]192, 243[$, et que la valeur 243 du déterminant, lorsqu'elle ne provient pas d'une suite sans produit scalaire nul, correspond à un réseau de norme 2.

[Variante pour l'assertion d'unicité concernant $\mathbb{E}_6$: on considère un vecteur minimal x de $\mathbb{E}_6$; on vérifie que le réseau $\mathbb{E}_6 \cap (\mathbb{Z}[\omega]x)^\perp$, qui est de norme 2 et de déterminant 9, est isométrique à $\mathbb{A}_2 \perp \mathbb{A}_2$; on en déduit que $\mathbb{E}_6$ s'identifie à un réseau de la forme $\langle \mathbb{A}_2 \perp \mathbb{A}_2 \perp \mathbb{A}_2, \frac{1}{1-\omega}(x, y, z) \rangle$, et l'on observe que x, y, z doivent être des unités de $\mathbb{Z}[\omega]$ pour que le nombre de vecteurs minimaux soit supérieur à celui de $\mathbb{A}_2 \perp \mathbb{A}_2 \perp \mathbb{A}_2$. Par multiplications à droite par des unités, on se ramène au cas $x = y = z = 1$].

10. — Nous terminons cette première partie par quelques remarques sur la constante γ'_n. Sloane, dans une lettre à l'auteur ([Sl]), a donné pour certaines dimensions ≤ 24 des exemples de réseaux sur lesquels l'invariant γ'_n prend des valeurs relativement grandes. Pour $n \leq 9$, ce sont ceux de [Be–M1], 4.6 et 4.7. Pour $n = 10$, il indique la valeur 4 pour ${\gamma'_n}^2$, atteinte sur deux réseaux semblables à leur dual (dont $\mathbb{D}_{10}^+$), cf. [C-S3]; le couple $(K'_{10}, {K'_{10}}^*)$ fournit la même valeur. Le résultat proposé est le même pour γ'_{11}, atteint en particulier sur les réseaux A_{11}^3 et K_{11}, qui sont tous deux dual-extrêmes ([Be–M1], § 4, (a) pour le premier, n° 7 ci-dessus pour le second).

H. Napias a montré que l'on ${\gamma'_n}^2(K'_{18}) = 8$ et ${\gamma'_n}^2(K_{21}) = 9$. Nous avons rencontré pour la première fois le réseau K'_{18} dans un travail de Souvignier ([Sou]) consacré aux sous-groupes maximaux de $\mathrm{Gl}_n(\mathbb{Z})$, dont nous avons extrait le premier exemple d'un réseau L de dimension 21 avec $\gamma'_{21}(L) > \gamma'_{21}(\Lambda_{21})$ (L et son dual sont extrêmes, et l'on a $\gamma'_{21}(\Lambda_{21}) = 8 < \gamma'_{21}(L) = 8,4 < \gamma'_{21}(K'_{21}) = 9$). Le réseau K'_{20} donne le même résultat (${\gamma'_{20}}^2 = 8$) que Λ_{20} cité dans [Sl].

Tous ces réseaux sont dual–extrêmes.

TROISIÈME PARTIE : **autour du réseau de Barnes–Wall**

11. — Soit H_2 ou simplement H le corps de quaternions "usuels" de centre $\mathbb{Q}$ ramifié en 2 et à l'infini, muni de sa base $(1, i, j, k)$ vérifiant les relations $i^2 = j^2 = -1$ et $ij = -ji = k$, d'où l'on déduit les relations supplémentaires $k^2 = -1, jk = -kj = i, ki = -ik = j$, et soit $\mathfrak{M}_2$ ou simplement $\mathfrak{M}$ l'ordre maximal des quaternions de Hurwitz, de base $(1, i, j, \omega = \frac{-1+i+j+k}{2})$ sur $\mathbb{Z}$; c'est l'unique ordre de H contenant strictement l'ordre $\mathfrak{O}$ de base $(1, i, j, k)$. Notons $\mathfrak{a}$ l'idéal bilatère engendré (à gauche ou à droite) par $1 + i$; on a $\mathfrak{a} = \{x = a + bi + cj + dk \in \mathfrak{O} \mid a + b + c + d \equiv 0 \bmod 2\}$. Munis de la forme $\frac{1}{2}\mathrm{Trd}(x\overline{y})$, $\mathfrak{a}$ et $\mathfrak{M}$ s'identifient respectivement aux réseaux $\mathbb{D}_4$ et $\mathbb{D}_4^*$, et l'on obtient $\mathbb{E}_8$ en considérant sur $\mathfrak{M} \times \mathfrak{M}$ la congruence $\lambda \equiv \mu \bmod \mathfrak{a}$ (ou par adjonction à $\mathfrak{M} \times \mathfrak{M}$ muni de la forme $\mathrm{Trd}(x\overline{y})$ de l'élément $\frac{1}{1+i}(1, 1)$).

Soit $m \geq 0$ un entier et soit $n = 4m$. On on munit $\mathfrak{M}$ de la forme $\mathrm{Tr}(\lambda\overline{\mu})$ et l'on pose $J_n = \{(\lambda_1, \cdots, \lambda_m) \in \mathfrak{M}^m \mid \lambda_1 + \ldots \lambda_m \equiv 0 \bmod \mathfrak{a}\}$. On vérifie que J_n est un réseau de norme 4, primitif sauf pour $n = 4$ ou $n = 8$ où l'on trouve une renormalisation de $\mathbb{D}_4$ et de $\mathbb{E}_8$, dont le dual s'identifie à $\frac{1}{1+i}(\mathfrak{M})^m$. En identifiant $\mathbb{D}_4^*$ à la dernière composante de J_n^*, et en coupant par les orthogonaux des sections de $\mathbb{D}_4^*$ semblables à $\{0\}, \mathbb{A}_1, \mathbb{A}_2, \mathbb{A}_3, \mathbb{D}_4$, on obtient des réseaux $J_n, J_{n-1}, J_{n-2}, J_{n-3}$ et un réseau qui s'identifie à J_{n-4}, ce qui définit J_n pour tout n. On a ainsi construit les analogues pour $\mathfrak{M}$ des réseaux $L_n^{\lfloor n/2 \rfloor}$ construits par Barnes sur l'anneau des entiers d'Eisenstein.

11.1. PROPOSITION. — *Pour tout $n \geq 1$, J_n est un réseau entier de norme 4, qui est une section hyperplane de J_{n+1}. Il possède les invariants suivants :*

$n = 4h$	$\det(J_n) = 2^{2h+4}$	$s = 12h(4h-3)$
$n = 4h+1$	$\det(J_n) = 2^{2h+5}$	$s = 4h(12h-7)$
$n = 4h+2$	$\det(J_n) = 3.2^{(2h+4)}$	$s = 12h(4h-1)$
$n = 4h+3$	$\det(J_n) = 2^{2h+6}$	$s = 3(16h^2+4h+1)$

Il est parfait quelque soit n, extrême pour $n \equiv 0$ ou $1 \bmod 4$, et est dual-extrême lorsque n est divisible par 4. En outre, pour $n \leq 12$, J_n est un réseau laminé Λ_n, et l'on a plus précisément $J_{12} \simeq \Lambda_{12}^{\max}$ et $J_{11} \simeq \Lambda_{11}^{\max}$. Enfin, pour $n = 4h + \ell \geq 9$ et $\ell \in \{1,2,3\}$, J_n^* est de norme $\frac{\ell}{\ell+1}$ et la configuration $S(J_n^*)$ est semblable à $S(\mathbb{A}_\ell^*)$.

Démonstration. Le calcul du déterminant et du nombre de vecteurs minimaux ne présente pas de difficulté. On vérifie aussi facilement que J_n est relativement parfait par rapport à J_{n-1}, d'où l'assertion de perfection. On montre que J_n est extrême pour $n \equiv 0, 1 \bmod 4$ en montrant qu'il contient un réseau de norme 4 semblable à $\mathbb{D}_n$, ce qui assure qu'il est dual-extrême pour $n = 4m$ vu que $(\mathfrak{M})^m$ est eutactique. La suite des valeurs des déterminants pour $0 \leq n \leq 12$ montre tout de suite qu'il s'agit de réseaux laminés, que l'invariant s permet d'identifier en dimensions 11 et 12. Enfin, on détermine $S(J_n^*)$ par récurrence descendante en identifiant J_n^* à une projection de J_{n+1}^*.

12. — L'analogue pour l'ordre de Hurwitz des réseaux $\mathbb{D}_8^+ = \mathbb{E}_8$ et K_{12} est le réseau Λ_{16} de Barnes–Wall (la notation Λ_{16} des réseaux laminés est justifiée ci-dessous), que l'on peut définir au choix par le double système de congruences

$$\lambda_1 \equiv \lambda_2 \equiv \lambda_3 \equiv \lambda_4 \bmod \mathfrak{a} \quad \text{et} \quad \lambda_1 + \lambda_2 + \lambda_3 + \lambda_4 \equiv 0 \bmod \mathfrak{a}^2$$

sur $\mathfrak{M}^4$ muni de la forme $\frac{1}{2}\sum_{\ell=1}^4 \mathrm{Trd}(\lambda_\ell \overline{\mu_\ell})$ ou par adjonction à J_{16} de $\frac{1}{1+i}(1,1,1,1)$.

[Plus généralement, on définit de façon analogue un réseau J'_n pour tout $n \geq 16$ divisible par 8, ayant le même déterminant (4^m) que $\mathfrak{M}^m$].

On démontre comme dans les cas de $\mathbb{D}_n^+$ et de K_{12} le résultat suivant :

12.1. Proposition. — *L'application $q \mapsto q(1+i)^{-1}$ est une isométrie de Λ_{16} sur son dual; en particulier, Λ_{16}^* est de norme 2.*

Il résulte de cette proposition que, pour tout vecteur minimal x de Λ_{16}^*, $\mathfrak{M}x$ est un réseau de dimension 4 isométrique à $\mathbb{D}_4$. Les orthogonaux de ses sections minimales $\{0\}, \mathbb{A}_1, \mathbb{A}_2, \mathbb{A}_3, \mathbb{D}_4$ sont des réseaux relatifs de dimensions $16, 15, 14, 13, 12$ de déterminants respectifs $256, 512, 768, 1024, 1024$ dont le dernier est visiblement isométrique à $J_{12} \simeq \Lambda_{12}^{\max}$, ce qui prouve bien qu'il s'agit du réseau laminé de dimension 16, et que les sections considérées sont $\Lambda_{16}, \Lambda_{15}, \Lambda_{14}, \Lambda_{13}^{\max}, \Lambda_{12}^{\max}$.

De même qu'il y a 2 orbites de réseaux $\mathbb{A}_2$ dans K_{12}, il y a 3 orbites de réseaux $\mathbb{D}_4$ dans Λ_{16}, et l'on constate que les réseaux de dimension 12 que l'on obtient par orthogonalité sont $\Lambda_{12}^{\max}$, $\Lambda_{12}^{\text{mid}}$, $\Lambda_{12}^{\min}$. On construit $\Lambda_{13}^{\min}$ à l'aide des deux dernières orbites (et $\Lambda_{11}^{\min}$ à l'intérieur de $\Lambda_{12}^{\min}$ et de $\Lambda_{12}^{\text{mid}}$ comme il se doit), mais on vérifie qu'il n'est pas possible de plonger $\Lambda_{13}^{\text{mid}}$ dans Λ_{16}. Ce résultat se voit également en utilisant la construction des réseaux faiblement laminés jusqu'à la dimension 24 par Plesken et Pohst ([P-L]), qui ont en outre montré que le plongement de $\Lambda_{13}^{\text{mid}}$ est possible dans Λ_{17}.

13. — Considérons toujours la suite des réseaux laminés Λ_n, en nous limitant à ceux qui sont des $\mathfrak{M}$-modules, ce qui impose que n soit divisible par 4. Conway et Sloane ([C–S1]) ont construit la "série principale"

$$\Lambda_0 \subset \Lambda_4 \sim \mathbb{D}_4 \subset \Lambda_8 \sim \mathbb{E}_8 \subset \Lambda_{12}^{\max} \subset \Lambda_{16} \subset \Lambda_{20} \subset \Lambda_{24} \subset \ldots \subset \Lambda_{48}$$

dont le terme de dimension 24 est le réseau de Leech et dont les termes de dimension supérieure ne sont sans doute qu'une possibilité parmi d'autres. Il s'agit de réseaux qui sont laminés au sens fort en tant que $\mathbb{Z}$-réseaux et qui possèdent une $\mathfrak{M}$-structure, celle de Λ_{n-4} étant induite par celle de Λ_n. L'unicité de ces réseaux en tant que $\mathbb{Z}$-réseaux a été établie par Conway et Sloane jusqu'à la dimension 24 sauf bien sûr en dimension 12. La question de l'unicité en tant que $\mathfrak{M}$-réseaux se pose, ainsi que celle de l'existence pour les réseaux de dimension 12 autres que $\Lambda_{12}^{\max}$.

13.1. Proposition. — *Un $\mathfrak{M}$-réseau Λ de dimension 12, de déterminant 1024 et de norme 4 dont le dual est de norme 1 est semblable à $\Lambda_{12}^{\max}$ ou à $\Lambda_{12}^{\min}$, les configurations respectives des vecteurs minimaux étant respectivement celles de de $\mathbb{D}_4^* \perp \mathbb{D}_4^* \perp \mathbb{D}_4^*$ et de $\mathbb{D}_4^*$.*

Démonstration. Pour tout vecteur x minimal de Λ^*, $\mathfrak{M}x$ est un réseau isométrique à $\mathbb{D}_4^*$; les sections de Λ par les orthogonaux des réseaux semblables à $\{0\}, \mathbb{A}_1, \mathbb{A}_2, \mathbb{A}_3, \mathbb{D}_4$ qu'il contient constituent une suite décroissante de sections de Λ de normes au moins 4 et de déterminants $1024, 1024, 768, 512, 256$. Le dernier terme de la suite est d'invariant d'Hermite 2, et est donc de norme 4 et isométrique à $\Lambda_8 \sim \mathbb{E}_8$. Il s'en suit que ces sections prises pour les dimensions croissantes de 8 à 12 sont des réseaux laminés. Comme les $\mathfrak{M}$-réseaux ont un invariant s divisible par 12 (le nombre de couples $\pm u$ d'unités de $\mathfrak{M}$), on doit exclure $\Lambda_{12}^{\text{mid}}$. Enfin, la valeur des invariants s (respectivement 12 et 36) détermine les structures des ensembles de vecteurs minimaux des duals.

13.2. Proposition. — *Le réseau $\Lambda_{12}^{\min}$ possède une structure de $\mathfrak{M}$-réseau*

Démonstration. Sigrist([Si]) en utilisant une généralisation de l'algorithme de Voronoï pour les réseaux quaternioniens (cf. [Be–M–S]), puis Laïhem ([La]) au cours d'une recherche de réseaux quaternioniens entiers de norme 4 et de dimension 12, ont trouvé un réseau de dimension 12, de déterminant 1024, de dual de norme 1 avec $s = 312$. La proposition 13.1 entraîne que ce réseau est isométrique à $\Lambda_{12}^{\min}$, qui se trouve ainsi muni d'une structure de $\mathfrak{M}$-réseau.

En ce qui concerne l'unicité à isométrie hermitienne près des structures sur les réseaux laminés de petite dimension, elle est connue dans les cas suivants : Λ_4 (parce qu'il y a une seule classe dans $\mathfrak{M}$), Λ_8 et Λ_{24} (traités par Quebbemann dans [Q] à l'aide d'une formule de masse), Λ_{16} (Quebbemann, communication privée). Comme le sous-groupe unitaire de $\mathrm{Aut}(\Lambda_{16})$ est transitif sur $S(\Lambda_{16})$, les inclusions $\Lambda_{12} \subset \Lambda_{16}$ compatibles avec la $\mathfrak{M}$-structure Λ_{16} imposent que Λ_{12} soit en fait $\mathbb{Z}$-isométrique à $\Lambda_{12}^{\max}$. On a encore dans ce cas un résultat d'unicité :

13.3. Proposition. — *À $\mathfrak{M}$-isométrie près, le réseau $\Lambda_{12}^{\max}$ porte une unique $\mathfrak{M}$-structure.*

Démonstration. Son dual est de norme 1 et de déterminant $\frac{1}{1024}$, et ses vecteurs minimaux engendrent un réseau de déterminant 4^3 (prop. 13.1), donc d'indice 4, i.e. de "$\mathfrak{M}$-indice" $\mathfrak{a} = (1+i)$. Par homothétie, on obtient le réseau $\Lambda \simeq \mathfrak{M} \perp \mathfrak{M} \perp \mathfrak{M}$ tandis que $\Lambda_{12}^{\max *}$ devient un réseau Λ' engendré par adjonction à Λ d'un vecteur $\frac{1}{1+i}v$ où $v = (x, y, z)$, $x, y, z, \in \mathfrak{M}$. Comme on ne change pas Λ' en remplaçant x, y, z par des éléments qui leurs sont congrus modulo $\mathfrak{a}$, on peut les supposer dans $\{0, 1, \omega, \omega^2\}$, et l'égalité $S(\Lambda') = S(\Lambda)$ impose que x, y, z soient non nuls. Par multiplications à droite par l'une des unités $1, \omega, \omega^2$, on se ramène au cas où $x = y = z = 1$, ce qui montre immédiatement que Λ' est égal au dual de $\Lambda_{12}^{\max}$ muni de la $\mathfrak{M}$-structure qui a servi à le définir au n° 11.

[Un raisonnement analogue permet de traiter le cas de $\mathbb{E}_8$: on choisit un vecteur minimal x de $\mathbb{E}_8$; en considérant l'orthogonal de $\mathfrak{M}x$, on plonge $\mathfrak{M} \perp \mathfrak{M} \simeq \mathbb{D}_4 \perp \mathbb{D}_4$ dans $\mathbb{E}_8$, et l'on reconstruit $\mathbb{E}_8$ par adjonction à $\mathbb{D}_4 \perp \mathbb{D}_4$ d'un vecteur de la forme $\frac{1}{1+i}(x, y)$. On peut de même traiter le cas de Λ_{16} en l'identifiant au réseau défini par l'adjonction à $(1+i)(\mathfrak{M} \perp \mathfrak{M} \perp \mathfrak{M} \perp \mathfrak{M})$ des 4 vecteurs $(1,1,0,0), (0,1,1,0), (0,0,1,1), (1,\omega,\omega^2,1)$, qui engendrent un code de poids 2 sur $\mathbb{F}_4$, et aussi retrouver la prop. 13.3 en identifiant $\Lambda_{12}^{\max}$ à $\langle (1+i)(\mathfrak{M} \perp \mathfrak{M} \perp \mathfrak{M}), (1,1,0), (0,1,1) \rangle$].

Les résultats que nous venons de donner permettent de déterminer pour l'essentiel la suite des réseaux laminés munis de $\mathfrak{M}$-structures jusqu'à la dimension 24 : on a la série principale de [C-S1] décrite plus haut, une bifurcation en dimension 8 vers le réseau $\Lambda_{12}^{\min}$ qui est un cul-de-sac vu le résultat d'unicité pour la dimension 16 (et qui pourrait ne pas être unique en tant que $\mathfrak{M}$-réseau, mais c'est peu probable), et peut-être des bifurcations en dimension 16 vers des réseaux Λ_{20} munis de $\mathfrak{M}$-strucures exotiques, et qui seraient alors des culs-de-sac vu le résultat d'unicité pour la dimension 24; cette éventualité est elle aussi peu probable.

14. — On définit de façon naturelle le procédé de lamination (au sens faible comme au sens fort) au-dessus d'un $\mathfrak{M}$-réseau Λ_0 dans l'ensemble des $\mathfrak{M}$-réseaux : il s'agit de suites croissantes de $\mathfrak{M}$-réseaux de même norme que Λ_0, le plongement d'un réseau dans le suivant étant compatible à l'action de $\mathfrak{M}$, les déterminants vérifiant les conditions de minimalité forte ou faible. On s'intéresse ici aux laminations fortes dans le cas où Λ_0 est le réseau de dimension 0 auquel est attribuée la norme 4.

Supposons démontré pour une certaine dimension $n < 44$ que les laminations au sens ci-dessus conduisent à un réseau Λ_n laminé au sens usuel, et considérons le terme $\Lambda' = \Lambda'_{n+4}$ suivant. Posons $m = N(\Lambda'^*)$. Pour tout $x \in \Lambda^*$, le $\mathfrak{M}$-réseau $\mathfrak{M}x$ est semblable à $\mathbb{D}_4$, et possède donc des sections semblables à $\mathbb{D}_4, \mathbb{A}_3, \mathbb{A}_2$ et $\mathbb{A}_1$, qui sont les réseaux les plus denses dans les dimensions $4, 3, 2, 1$; leurs déterminants sont $\frac{1}{4}m^4, \frac{1}{2}m^3, \frac{3}{4}m^2$ et m. Il en résulte que les sections de déterminant minimum de Λ dans les dimensions $n-1, n-2, n-3, n-4$, que nous notons $\Lambda'_{n+3}, \Lambda'_{n+2}, \Lambda'_{n+1}, \Lambda'_n$, ont pour déterminants les déterminants des orthogonaux des réseaux contenus dans $\mathfrak{M}x$, c'est-à-dire $m\det(\Lambda), \frac{3}{4}m^2\det(\Lambda), \frac{1}{2}m^3\det(\Lambda)$ et $\frac{1}{4}m^4\det(\Lambda)$. Le caractère minimal de $\det(\Lambda'_n)$ montre que l'on a $\det(\Lambda'_n) \leq \det(\Lambda_n)$, donc en fait $\det(\Lambda'_n) = \det(\Lambda_n)$, ce qui entraîne l'inégalité $\det(\Lambda'_{n+1}) \geq \det(\Lambda_{n+1})$.

Or, Conway et Sloane ont calculé jusqu'à la dimension 48 les déterminants des Λ_n. Le déterminant d_n de Λ_n s'obtient à partir de sa valeur en dimensions ≤ 4 $(1, 4, 12, 32, 64)$ par les formules $d_n = 2^{16-n}d_{8-n}$ pour $0 \leq n \leq 8$, $d_n = 2^{16-n}d_{n-8}$ pour $8 \leq n \leq 16$, $d_n = d_{24-n}$ pour $12 \leq n \leq 24$ et $d_n = 2^{24-n}d_{n-24}$ pour $24 \leq n \leq 48$ (cf. [C–S], ch. 6). Il est alors facile de vérifier que la minoration de $\det(\Lambda'_{n+1})$ par $\det(\Lambda_{n+1})$ entraîne l'égalité de ces deux déterminants, et que la valeur de m qui s'en déduit entraîne l'égalité $\det(\Lambda'_{n+4}) = \det(\Lambda_{n+4})$, i.e. que les réseaux laminés sur l'ordre de Hurwitz en dimension $n+4$ sont des réseaux laminés au sens usuel.

Vu que $\Lambda_8 \sim \mathbb{E}_8$ est le plus dense des réseaux de dimension 8, on déduit du raisonnement ci-dessus que les $\mathfrak{M}$-réseaux laminés (en tant que $\mathfrak{M}$-réseaux) de dimension 12 sont $\Lambda_{12}^{\min}$ et $\Lambda_{12}^{\max}$. Le résultat analogue est probablement vrai dans les dimensions $16, 20, 24$, pour lesquelles il est généralement conjecturé que $\Lambda_{16}, \Lambda_{20}, \Lambda_{24}$ sont les réseaux les plus denses. Il semble possible de traiter le cas de la dimension 16 en vérifiant à la façon de [La] qu'il n'y a pas d'autres $\mathfrak{M}$-réseaux de norme 4 en dimension 12 que $\Lambda_{12}^{\min}$ et $\Lambda_{12}^{\max}$. Nous conjecturons plus, à savoir que ces deux réseaux réalisent le maximum de l'invariant γ_{12} sur les $\mathfrak{M}$-réseaux.

QUATRIÈME PARTIE : **au-delà de la dimension** 16

15. — Soit H un corps de quaternions *totalement défini*. Cela signifie que le centre F de H est un corps de nombres totalement réel et que la norme réduite $\mathrm{Nrd}_{H/F}$ est positive à toutes les places infinies de F. Étant donné $\alpha >> 0$ de F, la forme $\mathbb{Z}$-bilinéaire $(\lambda, \mu) \mapsto \mathrm{Tr}_{F/\mathbb{Q}}(\alpha \mathrm{Trd}_{H/F}(\lambda\overline{\mu}))$ est définie positive. Les ordres maximaux de H deviennent ainsi des réseaux de $\mathbb{R} \otimes H$, et l'on construit d'autres réseaux par congruences selon un procédé déjà utilisé, cf. n° 5 et n° 11. Nous examinons dans cette quatrième partie des exemples dans lesquels F est soit le corps $\mathbb{Q}$, le corps H n'ètant plus le corps des quaternions usuels, soit un corps quadratique réel, le corps H n'étant ramifié qu'aux deux places réelles de F, renvoyant à [Bay–M] pour d'autres exemples.

Soit $\mathfrak{M}$ un ordre maximal de H. Pour un idéal premier $\mathfrak{p}$ non nul de F, il y a seulement deux possibilités :

• CAS RAMIFIÉ. On a $\mathfrak{p}\mathfrak{M} = \mathfrak{P}^2$, $\mathfrak{M}/\mathfrak{P}$ est une extension quadratique de $\mathbb{Z}_F/\mathfrak{p}$, et le complété en $\mathfrak{P}$ de H est un corps gauche.

• CAS DÉCOMPOSÉ. On a $\mathfrak{M}/\mathfrak{p}\mathfrak{M} \simeq \mathcal{M}_2(\mathbb{Z}_F/\mathfrak{p})$ et le complété de H en $\mathfrak{p}\mathfrak{M}$ est isomorphe à $\mathcal{M}_2(F_\mathfrak{p})$.

On utilisera surtout la variante suivante des doubles congruences qui sont intervenues aux n° 5 et 11 : on choisit deux idéaux à gauche $\mathfrak{P}$ et $\mathfrak{P}'$ dans $\mathfrak{M}$ au-dessus d'un idéal maximal $\mathfrak{p}$ de F non ramifié dans H, en se limitant au cas où $\mathfrak{p}$ est l'idéal au-dessus de (2) supposé inerte dans $F/\mathbb{Q}$, et l'on considère pour un entier $m \geq 0$ le réseau de $\mathbb{R} \otimes H^m$ muni du produit scalaire $(\lambda, \mu) \mapsto \frac{1}{2}\mathrm{Tr}_{F/\mathbb{Q}}(\alpha \mathrm{Trd}_{H/F}(\lambda\overline{\mu}))$, de dimension $n = 4[F : \mathbb{Q}]m$, défini sur $\mathfrak{M}^m$ par

$$(*) \qquad \lambda_1 \equiv \lambda_2 \equiv \cdots \equiv \lambda_m \bmod \mathfrak{P} \quad \text{et} \quad \lambda_1 + \lambda_2 + \cdots + \lambda_m \equiv 0 \bmod \mathfrak{P}'.$$

[S'il y a de la décomposition au-dessus de 2 dans $F/\mathbb{Q}$, on peut choisir un couple $\mathfrak{P}, \mathfrak{P}'$ pour chaque idéal de F au-dessus de 2].

15.1. THÉORÈME. — *Sous les hypothèses ci-dessus, le réseau défini par la condition (∗) est entier et pair, et de norme ≥ 4 lorsque m est ≥ 3.*

Démonstration. La démonstration de l'intégralité se fait par complétion en $\mathfrak{p}$ (notée par le symbole $\widehat{\ }$), ce qui permet de ramener les calculs de norme réduite à des calculs de déterminants d'ordre 2. Par une identification convenable de l'algèbre locale à une algèbre de matrices, on peut faire en sorte que l'on ait

$$\widehat{\mathfrak{M}} = \begin{pmatrix} \widehat{\mathbb{Z}}_K & \widehat{\mathbb{Z}}_K \\ \widehat{\mathbb{Z}}_K & \widehat{\mathbb{Z}}_K \end{pmatrix}, \qquad \widehat{\mathfrak{P}} = \begin{pmatrix} \widehat{\mathbb{Z}}_K & \widehat{\mathfrak{p}} \\ \widehat{\mathbb{Z}}_K & \widehat{\mathfrak{p}} \end{pmatrix} \quad \text{et} \quad \widehat{\mathfrak{P}'} = \begin{pmatrix} \widehat{\mathfrak{p}} & \widehat{\mathbb{Z}}_K \\ \widehat{\mathfrak{p}} & \widehat{\mathbb{Z}}_K \end{pmatrix}.$$

Identifiant alors $\lambda_i \in \mathfrak{M}$ à une matrice de la forme $\begin{pmatrix} x_i & y_i \\ z_i & t_i \end{pmatrix}$, on constate que les congruences de la condition (*) deviennent

$$y_i \equiv y_1 \bmod \hat{\mathfrak{p}} \text{ et } t_i \equiv t_1 \bmod \hat{\mathfrak{p}} \ (1 \leq i \leq m), \quad \text{et} \ \sum_{i=1}^{m} x_i \equiv \sum_{i=1}^{m} z_i \equiv 0 \bmod \hat{\mathfrak{p}}.$$

Comme la norme réduite dans une algèbre de matrices n'est autre que le déterminant, on a

$$\sum_{i=1}^{m} \lambda_i \overline{\lambda_i} = \sum_{i=1}^{m} x_i t_i - y_i z_i \equiv (\sum_{i=1}^{m} x_i) t_1 - (\sum_{i=1}^{m} z_i) y_1 \equiv 0 \bmod \mathfrak{p},$$

ce qui prouve qu'il s'agit d'un réseau entier pair. Pour minorer la norme de $\lambda = (\lambda_1, \lambda_2, \ldots, \lambda_m)$ supposé non nul, on distingue trois cas :

• si les λ_i ne sont pas dans $\mathfrak{P}$, on a $N(\lambda) \geq m \min \mathrm{Nrd}(\lambda_i) \geq 3 \min \mathrm{Nrd}(\lambda_i)$, donc $N(\lambda) \geq 4$;

• Si les λ_i sont dans $\mathfrak{P}$ et si deux d'entre eux sont non nuls, on a $N(\lambda) \geq 4$ puisque les produits $\lambda_i \overline{\lambda_i}$ sont dans $\mathfrak{P} \cap \mathbb{Z}_F = 2\mathbb{Z}_F$;

• si un seul des λ_i est non nul, c'est un élément de $\mathfrak{P} \cap \mathfrak{P}' = \mathfrak{p}\mathfrak{M}$, d'où encore le résultat dans ce cas.

16. — Nous prenons maintenant pour H l'algèbre H_3 de centre $\mathbb{Q}$ ramifiée en 3 et à l'infini, munie de sa base $(1, i, j, k)$ vérifiant les relations $i^2 = -1, j^2 = -3, ij = -ji = k$, et donc les relations supplémentaires $k^2 = -3, jk = -kj = 3i, ki = -ik = j$, et pour ordre maximal l'ordre $\mathfrak{M}_3 = \mathfrak{M}$ de base $1, i, \omega, i\omega$ sur $\mathbb{Z}$ où $\omega = \frac{-1+j}{2}$ est une racine de l'unité d'ordre 3. (Le choix de $\mathfrak{M}$ importe peu, les ordres maximaux de H étant conjugués, comme dans le cas des quaternions de Hurwitz.) Les unités de $\mathfrak{M}$ sont $\{\pm 1, \pm\omega, \pm\omega^2, \pm i, \pm i\omega, \pm i\omega^2\}$; elles forment un groupe isomorphe au groupe quaternionien d'ordre 12.

Le théorème ci-dessous donne une construction explicite d'une structure de $\mathfrak{M}_3$-réseau sur K_{12}, dont l'existence a été prouvée il y a peu par Gross ([Gro]) :

16.1. THÉORÈME. — *Le réseau construit à l'aide de la condition (*) avec $m = 3$ sur l'ordre $\mathfrak{M}_3$ est isométrique au réseau de Coxeter–Todd.*

Démonstration. Le théorème 15.1 montre qu'il s'agit d'un réseau de norme au moins 4, dont on voit tout de suite qu'il est entier en tant que $\mathfrak{M}_3$-réseau et de déterminant 3^6. Il est donc unimodulaire en tant que $\mathfrak{M}_3$-réseau, et donc en particulier en tant que réseau sur l'anneau des

entiers d'Eisenstein. Le théorème de Feit ([Fe]) cité au n° 7 entraîne qu'il est isométrique à K_{12}.
[Le théorème de Feit montre qu'il s'agit même d'une isométrie en tant que $\mathbb{Z}[\omega]$-réseau; Ch. Bachoc vient de démontrer que K_{12} est même unique à $\mathfrak{M}_3$-isométrie près]

17. — Soit F un corps quadratique réel, de discriminant d impair. Nous considérons maintenant le corps de quaternions H ramifié exactement aux deux places infinies de F, et nous supposons que l'unité fondamentale ε de F est de norme -1, ce qui équivaut au fait que la différente de F possède un générateur totalement positif, en l'occurence $\alpha = \varepsilon\sqrt{d}$, si bien qu'un ordre maximal $\mathfrak{M}$ de H, muni de la forme $\mathrm{Tr}_{K/\mathbb{Q}}(\alpha^{-1}\mathrm{Trd}(\lambda\overline{\mu}))$, définit un réseau $\mathbb{Z}$-isométrique à $\mathbb{E}_8$.

Un corps de quaternions H_0 de centre $\mathbb{Q}$ peut être plongé (d'une infinité de façon) dans un corps gauche H du type ci-dessus : il suffit de choisir un corps F dans lequel les nombres premiers ramifiés dans H_0 sont inertes ou ramifiés dans $F/\mathbb{Q}$ et de prendre $H = F \otimes_{\mathbb{Q}} H_0$, les invariants locaux aux places finies de F étant alors tous nuls. Un ordre arbitraire $\mathfrak{O}$ de H_0 étant contenu dans l'ordre $\mathbb{Z}_F\mathfrak{O}$ de H, lequel est à son tour contenu dans un ordre maximal $\mathfrak{M}$ de H, on voit que $\mathbb{E}_8$ peut être muni d'une structure de $\mathfrak{O}$–réseau sur n'importe quel ordre de quaternions totalement défini sur $\mathbb{Z}$. On verra plus loin d'autres exemples du même type; signalons simplement ici qu'un résultat analogue s'applique à Λ_{16}.

En appliquant à l'ordre $\mathfrak{M}$ la construction par le double système de congruences (*), on obtient un réseau unimodulaire pair Λ de dimension $n = 8m$ que nous notons simplement U_n, sans mettre en évidence dans la notation sa dépendance *a priori* des choix de $H, \mathfrak{M}, \mathfrak{P}, \mathfrak{P}'$. Il est fort possible que la classe d'isométrie de U_n (en tant que $\mathbb{Z}$-réseau) ne dépende pas de ces choix. C'est ce qu'on constate en dimension 8 (resp. 24), puisque E_8 (resp. Λ_{24}) est alors l'unique réseau unimodulaire pair (resp. et de norme 4, théorème de Conway). Le cas de la dimension 32 a été résolu par Coulangeon ([Cou]), qui a caractérisé U_{32} comme le réseau unimodulaire pair d'invariant de Venkov maximum qui est associé au code de Reed-Muller. Quant à la dimension 16, on trouve $\mathbb{E}_8 \perp \mathbb{E}_8$, comme on le voit en considérant l'application $(\lambda, \mu) \mapsto (\lambda + \mu, \lambda - \mu)$.
[Pour $n \geq 40$, les répartitions des normes réduites dans les suites $(\lambda_1, \lambda_2, \ldots, \lambda_m)$ définissant des vecteurs minimaux sont des permutations de $(1, 1, 0, \ldots, 0)$; on en déduit l'égalité $s(U_n) = 15n(n-7)$ pour $n \geq 40$. Les résultats pour $n = 24$ et $n = 32$ (et aussi pour $n = 40$) découlent de la théorie des fonctions Θ; on a $s(U_{24}) = 98280$ et $s(U_{32}) = 73440$. Pour $n = 32$, on doit ajouter aux $15n(n-7) = 12000$ vecteurs ci-dessus 61440 vecteurs associés à la répartition $(1,1,1,1)$; pour $n = 24$, on ajoute à $15n(n-7) = 6120$ la contribution des

permutations de la répartition $(2, 1, 1)$, soit 92160 vecteurs.

On connaît[(*)] en dimension 40 quelques autres réseaux unimodulaires pairs de norme 4. Pour celui de McKay (cf. [C–S], ch. 8, § 5), d'après McKay, le groupe d'automorphismes ne serait pas transitif sur l'ensemble de ses vecteurs minimaux, ce qui entraînerait que U_{40} ne lui est pas isométrique. Nous ignorons si notre réseau U_{40} coïncide avec l'un des réseaux construits par Eva Bayer dans [Bay] ou par Ozeki dans [Oz]].

Revenons au double système de congruences $(*)$ défini par deux idéaux à gauche maximaux $\mathfrak{P}$ et $\mathfrak{P}'$ au-dessus de 2 d'un ordre maximal $\mathfrak{O}$ d'un corps de quaternions H_0 de centre $\mathbb{Q}$. Si on choisit un corps F dans lequel 2 et les nombres premiers ramifiés dans H_0 sont inertes, on plonge comme ci-dessus H_0 dans H et $\mathfrak{O}$ dans un ordre maximal $\mathfrak{M}$ de H, et ces plongements transforment le réseau Λ_0 associé au double système de congruences en un réseau défini de façon analogue sur $\mathfrak{M}$ à l'aide d'idéaux maximaux au-dessus de 2 contenant respectivement $\mathfrak{P}$ et $\mathfrak{P}'$.

Ceci s'applique en particulier au cas où $\mathfrak{M}_0$ est l'ordre noté $\mathfrak{M}_3$ au n° 16 en prenant $F = \mathbb{Q}(\sqrt{p})$ où p est n'importe quel nombre premier congru à 5 ou 11 modulo 24, par exemple $p = 5$. En prenant $m = 3$, on en déduit une construction du réseau de Leech Λ_{24} sur $\mathfrak{M}_3$, utilisée par Tits ([Ti]) dans le cas du corps $F = \mathbb{Q}(\sqrt{5})$, et un plongement de K_{12} dans Λ_{24} compatible avec la $\mathfrak{M}_3$-structure dont nous avons muni K_{12} au n° 16, qui justifie la construction de la série K'_n au-delà de la dimension 12 que nous avons faite au n° 8.

On peut faire une remarque analogue avec l'ordre $\mathfrak{M}_2$ de Hurwitz. On vérifie que la construction par double congruence des réseaux J'_{4m} faite au début du n° 12 conduit au réseau $\Lambda_{12}^{\max}$ lorsque l'on prend $m = 3$ (lorsque m est impair, le déterminant calculé au n° 12 doit être multiplié par 2^4), et l'on en déduit le plongement connu de $\Lambda_{12}^{\max}$ dans Λ_{24} en tant que réseaux sur l'ordre de Hurwitz.

Manuscrit reçu le 8 mars 1993

[(*)] Je remercie Eva Bayer pour les références concernant les réseaux de dimension 40

BIBLIOGRAPHIE

[Bar] E.S. BARNES. — *The construction of perfect and extreme forms I, II,* Acta Arith. **5** (1959), 57–79, 461–506.

[Bay] E. BAYER-FLUCKIGER. — *Definite unimodular lattices having an automorphism of given characteristic polynomial,* Comm. Math. Helvet. **59** (1984), 509–538.

[Bay–M] E. BAYER-FLUCKIGER et J. MARTINET. — *Formes quadratiques liées aux algèbres semi-simples,* J. reine angew. Math. (1994), à paraître.

[Be–M1] A.-M. BERGÉ et J. MARTINET. — *Sur un problème de dualité lié aux sphères en géométrie des nombres,* J. Number Theory **32** (1989), 14–42.

[Be–M2] A.-M. BERGÉ et J. MARTINET. — *Réseaux extrêmes pour un groupe d'automorphismes,* Astérisque **198–200** (1992), 41–66.

[Be–M–S] A.-M. BERGÉ, J. MARTINET et F. SIGRIST. — *Une généralisation de l'algorithme de Voronoï pour les formes quadratiques,* Astérisque **209** (1992), 137–158.

[C–S] J.H. CONWAY et N.J.A. SLOANE. — *Sphere Packings, Lattices and Groups,* Springer-Verlag, Grundlehren n° 290, Heidelberg, 1988.

[C–S1] J.H. CONWAY et N.J.A. SLOANE. — *Complex and integral laminated lattices,* Trans. Amer. Math. Soc. **280** (1983), 463–490.

[C–S2] J.H. CONWAY et N.J.A. SLOANE. — *Low-dimensional lattices. III. Perfect forms,* Proc. Royal Soc. London A, **418** (1988), 43–80.

[C–S3] J.H. CONWAY et N.J.A. SLOANE. — *On Lattices Equivalent to Their Duals,* à paraître.

[Cou] R. COULANGEON. — Exposé au Sém. Th. Nombres de Paris, (janvier 1993).

[Cox] H.S.M. COXETER. — *Extreme forms,* Canad. J. Math. **3** (1951), 391–441.

[Cox–T] H.S.M. COXETER and J.A. TODD. — *An extreme duodenary form,* Canad. J. Math. **5** (1953), 384–392.

[Cra] M. CRAIG. — *Extreme forms and cyclotomy,* Mathematika **25** (1967), 44–56.

[Fe] W. FEIT. — *Some Lattices over* $\mathbb{Q}(\sqrt{-3})$, J. Algebra **52** (1978), 248–263.

[Gro] B. Gross. — *Group representation and lattices*, J. Amer. Math. Soc. **3** (1990), 929–960.

[La] M. Laïhem. — Communication privée.

[Oz] M. Ozeki. — *Examples of even unimodular extremal lattices of rank* 40, J. Number Theory **28** (1989), 119–131.

[Pl–P2] W. Plesken and M. Pohst. — *Constructing Integral Lattices With Prescribed Minimum. II*, Math. Comp. **60** (1993), 817–825.

[Q] H.–G. Quebbemann. — *An application of Siegel's formula over quaternion orders*, Mathematatika **31** (1984), 12–16.

[Si] F. Sigrist. — *Lettre électronique du 11 septembre 1990 à l'auteur.*

[Sl] N.J.A. Sloane. — *Lettre à l'auteur du 11 mai 1992.*

[Sou] B. Souvignier. — *Diplomarbeit*, Aachen, 1991.

[Ti] J. Tits. — *Quaternions over* $\mathbb{Q}(\sqrt{5})$, *Leech's lattice and the sporadic group of Hall-Janko*, J. Algebra **63** (1980), 56–75.

Jacques Martinet
Mathématiques, Université Bordeaux I
351, cours de la Libération
33405 TALENCE cedex

Number Theory
Paris 1992–93

Construction of Elliptic Units in Function Fields

Hassan Oukhaba

1. — Introduction

Let k be a global function field and $\mathbb{F}_q$ be its field of constants. Fix a place ∞ of k. Let $\mathcal{O}_k$ be the Dedekind ring of elements of k regular outside of ∞, and k_∞ be the completion of k at ∞.

For each finite abelian extension F of k we let $\mathcal{O}_F$ be the integral closure of $\mathcal{O}_k$ in F. We know that $\mathcal{O}_F$ is a Dedekind ring with a finite ideal class number $h(\mathcal{O}_F)$. As usual we denote by $\mathcal{O}_F^\times$ the group of units of $\mathcal{O}_F$, and $\mu(F) \subset \mathcal{O}_F^\times$ the finite multiplicative group of non zero constants of F. We have $\mu(k) = \mathcal{O}_k^\times = \mathbb{F}_q^\times$, and in general the quotient group $\mathcal{O}_F^\times/\mu(F)$ is a free abelian group of rank $r_F - 1$, where r_F is the exact number of places of F sitting over ∞.
Now suppose that $F \subset k_\infty$, which means that the place ∞ splits completely in F/k. Suppose in addition that one of the following two conditions holds.

1) The extension F/k is unramified.

2) One, and only one, prime divisor of k ramifies in F/k and deg $(\infty) = 1$.

Then one knows that there exists a subgroup $\mathcal{E}_F$ of $\mathcal{O}_F^\times$ called the group of elliptic units of F. It is a Galois module generated by the torsion points of certain Drinfeld $\mathcal{O}_k$–modules. It's elements are also obtained as finite products of special values of elliptic functions. The group $\mathcal{E}_F$ has finite index in $\mathcal{O}_F^\times$, cf. [10]. Actually when only one prime does ramify in F/k we had succeed to construct subgroups of finite index in $\mathcal{O}_F^\times$ even if deg $(\infty) > 1$, cf. [9]. Unfortunately, the index formula obtained then contains a factor depending on deg (∞) and which is hard to control. When deg $(\infty) = 1$ this factor is equal to 1 also and the index formula is just what one can expect. But in general this factor increases proportionaly to deg (∞). This means that the subgroups so constructed are not sufficiently large when deg $(\infty) > 1$. Hence, one could suppose that there is possibility to obtain larger subgroups of $\mathcal{O}_F^\times$, in other words to obtain more units of $\mathcal{O}_F$, using new techniques of constructions. This is what we propose to do in the present paper. Our aim here is to define $\mathcal{E}_F$, the group of elliptic units of F. We shall expose some of its interesting properties, precise the nature of its elements and calculate its index in $\mathcal{O}_F^\times$. As we shall see the description of $\mathcal{E}_F$ is rather easy and almost canonical. Moreover, the "exponential function", which we are going to redefine in the next section, is the only basic material of its construction.

Finally we would like to draw the attention to the work of D. Kersey, cf. [7] chap. 12 and 13, which was one of our source of inspiration.

Some supplementary notations

Let $F \subset k_\infty$ be a finite abelian extension of k such that the place ∞ splits completely in F/k. let $\mathfrak{b} \subset \mathcal{O}_k$ be an ideal of $\mathcal{O}_k$ prime to the conductor of F/k. Then we will write $(\mathfrak{b}, F/k)$ for the automorphism of F/k associated to $\mathfrak{b}$ by the Artin map. Moreover if $\mathfrak{q}$ is a prime ideal of $\mathcal{O}_k$ then $\mathfrak{q}_F$ will denote the product of the prime ideals of $\mathcal{O}_F$ sitting over $\mathfrak{q}$. Finally, if $\mathfrak{m} \subset \mathcal{O}_k$ is an ideal of $\mathcal{O}_k$ then we know that there exists a maximal finite abelian extension of k whose conductor divides $\mathfrak{m}$ and which is contained in k_∞. It will be denoted by $H_\mathfrak{m}$.

2. — Some preliminaries

In this section we recall some definitions and results, necessary in the sequel. The reader is invited to consult [1], [4], [9] or [11], where are proved all the results stated here. Let Ω be the completion at ∞ of the algebraic closure of k_∞. Then we call a lattice of Ω every finitely generated projective $\mathcal{O}_k$-module, contained into Ω. To such a lattice $\Gamma \subset \Omega$ one can associate its exponential function defined on Ω by :

$$e_\Gamma(z) \overset{dfn}{=} z \prod_{\substack{\gamma\in\Gamma \\ \gamma\neq 0}} \left(1 - \frac{z}{\gamma}\right).$$

We know that e_Γ is defined everywhere and is entire and $\mathbb{F}_q$–linear. It is also an epimorphism and we have $e_\Gamma(z) = 0$ if, and only if, $z \in \Gamma$. Moreover the equation $e_{x\Gamma}(xz) = x\, e_\Gamma(z)$ holds for every $x \in \Omega^\times$ and $z \in \Omega$.

When Γ is contained into a lattice $\overline{\Gamma}$ of Ω such that Γ and $\overline{\Gamma}$ have the same rank as $\mathcal{O}_k$–modules then e_Γ and $e_{\overline{\Gamma}}$ are related by the formula :

$$e_{\overline{\Gamma}}(z) = P\left(\overline{\Gamma}/\Gamma\,;\, e_\Gamma(z)\right), \tag{1}$$

where $P(\overline{\Gamma}/\Gamma\,;\, t)$ is a linear polynomial whose roots are all simples and constitute the finite group $e_\Gamma(\overline{\Gamma})$. Its leading coefficient is

$$\delta(\Gamma,\overline{\Gamma}) \overset{\text{dfn}}{=} \Bigg(\prod_{\substack{\rho\in\overline{\Gamma}/\Gamma\\ \rho\neq 0}} e_\Gamma(\rho)\Bigg)^{-1}$$

where ρ describe a complete system of non zero representatives of $\overline{\Gamma}$ modulo Γ.

Let $K(\infty)$ be the constant field of k_∞. It is a finite extension of $\mathbb{F}_q$. We have $[K(\infty) : \mathbb{F}_q] = \deg(\infty)$. Let us choose s (once of all) a sign-function of k_∞, i.e., a co-section of the inclusion map $K(\infty)^\times \hookrightarrow k_\infty^\times$ such that $s(z) = 1$ if $|z-1|_\infty < 1$. Then one can associate to each lattice Γ of Ω *of rank* 1 its *s-discriminant* $\Delta_s(\Gamma) \in \Omega^\times$ and σ_Γ an $\mathbb{F}_q$-automorphism of $K(\infty)$ such that

$$x\delta(\Gamma, x^{-1}\Gamma) = \Delta_s(\Gamma)^{\frac{Nx-1}{w_\infty}} s(x)^{\sigma_\Gamma}$$

for all $x \in \mathcal{O}_k\backslash\{0\}$.
In the above formula, w_∞ is just the number of non zero elements of $K(\infty)$, i.e., $w_\infty = \sharp K(\infty)^\times$. On the other hand Nx is by definition the exact number of congruence classes of $\mathcal{O}_k$ modulo the ideal $x\mathcal{O}_k$. One can show that $\Delta_s(z\Gamma) = z^{-w_\infty}\Delta_s(\Gamma)$ for all $z \in \Omega^\times$. Moreover we have the equation

$$\delta(\Gamma,\overline{\Gamma})^{w_\infty} = \Delta_s(\Gamma)^{[\overline{\Gamma}:\Gamma]}/\Delta_s(\overline{\Gamma})$$

whenever $\Gamma \subset \overline{\Gamma}$ are lattices of Ω which have rank 1 as $\mathcal{O}_k$-modules. The invariants $\Delta_s(\mathfrak{a})$ associated to fractional ideals $\mathfrak{a}$ of $\mathcal{O}_k$ are used to express the special values of L-functions at $s = 0$ and hence are related to analytic class number formulas, cf. [3], [5] and [6]. In other respects J. Yu has shown that they are transcendental over k, cf. [13]. However, as in the classical theory, it is possible to construct elements of Ω which are algebraic over k using the above invariants. Indeed let $k^{ab} \subset k_\infty$ be the abelian closure of k in k_∞. Let $H_{(1)}$ be the maximal subextension of k^{ab} such that $H_{(1)}/k$ is unramified. Then one knows that the quotient $\Delta_s(\mathfrak{a}_1)/\Delta_s(\mathfrak{a}_2) \in H_{(1)}$ for all fractional ideals $\mathfrak{a}_1$ and $\mathfrak{a}_2$ of $\mathcal{O}_k$. In fact this last quotient generates in $\mathcal{O}_{H_{(1)}}$ the ideal $(\mathfrak{a}_2\mathfrak{a}_1^{-1}\mathcal{O}_{H_{(1)}})^{w_\infty}$.

Now suppose that Γ is a lattice of Ω of rank 1. Then the lattice $t\Gamma$ is well defined for all idele t of k. And if ρ is an element of the k-vector space $k\Gamma$ generated by Γ then one can check, using the strong approximation theorem, that there exists $u \in \Omega$ such that

$$u \equiv t_v\rho \mod. t_v\Gamma_v$$

for all the places $v \neq \infty$ of k, where t_v is the component of t at v and Γ_v is the completion of Γ at v. The elements u that verify the property above define the same class modulo $t\Gamma$. We shall write $t\rho$ for any representative of this class. Let $\sigma \in \mathrm{Gal}(\Omega/k)$ and t be a idele of k such that the automorphism $[t,k]$ of k^{ab}/k, associated to t by the Artin map coincide with the restriction of σ to k^{ab}, cf. [11], then there exists a non zero element $\Lambda_\sigma(t,\Gamma) \in \Omega^\times$ which verifies the formula

$$e_\Gamma(\rho)^\sigma = \Lambda_\sigma(t,\Gamma)e_{t^{-1}\Gamma}(t^{-1}\rho) \tag{2}$$

for all $\rho \in k\Gamma$.
It is possible to describe the behavior of $\Lambda_\sigma(t,\Gamma)$ as σ, t or Γ varies, cf. [11]. In fact if we suppose that $\Delta_s(\Gamma) = 1$ and if the component t_v of t at each place $v \neq \infty$ is integral, then we have

$$\Lambda_\sigma(t,\Gamma) = \delta(\Gamma, t^{-1}\Gamma)^{-1},$$

provided that $s(t_\infty) = 1$.
Let us observe that the automorphism $[s,k]$ of k^{ab}/k is equal to the identity map if, and only if, $t \in k^\times \times k_\infty^\times$. Therefore the formula (2) implies that the quotient $e_\Gamma(\rho_1)/e_\Gamma(\rho_2) \in k^{ab}$, for all ρ_1 and ρ_2 in $k\Gamma$. Moreover we have :

$$\left(\frac{e_\Gamma(\rho_1)}{e_\Gamma(\rho_2)}\right)^{[t,k]} = \frac{e_{t^{-1}\Gamma}(t^{-1}\rho_1)}{e_{t^{-1}\Gamma}(t^{-1}\rho_2)} \tag{3}$$

for all idele t of k.

3. — The ramified part of the group of elliptic units

Let $F \subset k_\infty$ be a finite abelian extension of k. Suppose that the conductor of F/k is $\mathfrak{q}^n$, where $\mathfrak{q}$ is a prime ideal of $\mathcal{O}_k$ and n is a positive integer. Then it is possible to construct elements of $F^\times$ using the values $e_{\mathfrak{a}^{-1}\mathfrak{q}^n}(1)$, where $\mathfrak{a}$ is any ideal of $\mathcal{O}_k$ prime to $\mathfrak{q}$. These elements will constitute the ramified part of the group of elliptic units of F.

PROPOSITION 1. — *Let B be a finite set of ideals of $\mathcal{O}_k$ all prime to $\mathfrak{q}$. Then the product*

$$\prod_{\mathfrak{b}\in B} e_{\mathfrak{b}^{-1}\mathfrak{q}^n}(1)^{n_\mathfrak{b}} \tag{4}$$

belongs to $H_{\mathfrak{q}^n}$ provided that the rational integers $n_\mathfrak{b}, \mathfrak{b} \in B$ verify the condition

$$\sum_{\mathfrak{b}\in B} n_\mathfrak{b} = 0. \tag{5}$$

This proposition is equivalent to the following one.

PROPOSITION 2. — *The quotient*

$$e_{\mathfrak{a}^{-1}\mathfrak{q}^n}(1)/e_{\mathfrak{b}^{-1}\mathfrak{q}^n}(1) \in H_{\mathfrak{q}^n}$$

for all ideals $\mathfrak{a}$ *and* $\mathfrak{b}$ *of* $\mathcal{O}_k$ *which are prime to* $\mathfrak{q}$.

Proof : let $\sigma \in \mathrm{Gal}\,(\Omega/k)$ and t a idele of k, choosen so that σ is equal to the automorphism $[t,k]$ of k^{ab}/k. Then Corollary 3.3 of [11] (see also the previous section) implies that

$$e_{\mathfrak{a}^{-1}\mathfrak{q}^n}(1)^\sigma = \Lambda_\sigma(t,\mathfrak{a}^{-1}\mathfrak{q}^n)e_{t^{-1}\mathfrak{a}^{-1}\mathfrak{q}^n}(t^{-1}.1)$$

$$e_{\mathfrak{b}^{-1}\mathfrak{q}^n}(1)^\sigma = \Lambda_\sigma(t,\mathfrak{b}^{-1}\mathfrak{q}^n)e_{t^{-1}\mathfrak{b}^{-1}\mathfrak{q}^n}(t^{-1}.1)\,,$$

where $\Lambda_\sigma(t,\mathfrak{a}^{-1}\mathfrak{q}^n)$ and $\Lambda_\sigma(t,\mathfrak{b}^{-1}\mathfrak{q}^n)$ are elements of $\Omega^\times$; which are equal in this case, cf. [11]. So that we get the formula

$$\left(\frac{e_{\mathfrak{a}^{-1}\mathfrak{q}^n}(1)}{e_{\mathfrak{b}^{-1}\mathfrak{q}^n}(1)}\right)^\sigma = \frac{e_{t^{-1}\mathfrak{a}^{-1}\mathfrak{q}^n}(t^{-1}.1)}{e_{t^{-1}\mathfrak{b}^{-1}\mathfrak{q}^n}(t^{-1}.1)}\,. \tag{6}$$

Now suppose that σ is the identity map on k^{ab}, then we know that $t \in k^\times \times k_\infty^\times$. In this case the quotient on the right of the above formula is equal to $e_{\mathfrak{a}^{-1}\mathfrak{q}^n}(1)/e_{\mathfrak{b}^{-1}\mathfrak{q}^n}(1)$. This means in particular that this last quotient belongs to k^{ab}. In fact using class field theory we see also that it is in $H_{\mathfrak{q}^n}$. ◇

PROPOSITION 3. — *Let* $\mathfrak{a}, \mathfrak{b}$ *and* $\mathfrak{d}$ *be integral ideals of* $\mathcal{O}_k$, *all prime to* $\mathfrak{q}$. *Then we have*

$$\left(\frac{e_{\mathfrak{a}^{-1}\mathfrak{q}^n}(1)}{e_{\mathfrak{b}^{-1}\mathfrak{q}^n}(1)}\right)^{(\mathfrak{d},H\mathfrak{q}^n/k)} = \frac{e_{\mathfrak{d}^{-1}\mathfrak{a}^{-1}\mathfrak{q}^n}(1)}{e_{\mathfrak{d}^{-1}\mathfrak{b}^{-1}\mathfrak{q}^n}(1)}\,. \tag{7}$$

Proof : proposition 3 is easily derived from above formula (6) and the well known properties of the Artin map. ◇

PROPOSITION 4. — *Let* B *be a finite set of integral ideals of* $\mathcal{O}_k$, *all prime to* $\mathfrak{q}$. *Let* $n_\mathfrak{b}, \mathfrak{b} \in B$, *be rational integers which verify (4). Then the product*

$$\prod_{\mathfrak{b}\in B} e_{\mathfrak{b}^{-1}\mathfrak{q}^n}(1)^{n_\mathfrak{b}}$$

generates in the integral closure $\mathcal{O}_{H_{\mathfrak{q}^n}}$ *of* $\mathcal{O}_k$ *in* $H_{\mathfrak{q}^n}$ *the ideal*

$$\left(\prod_{\mathfrak{b}\in B} \mathfrak{b}^{n_\mathfrak{b}}\right)^{-1}.$$

Proof : all we have to prove is that the quotient $e_{\mathfrak{a}^{-1}\mathfrak{q}^n}(1)/e_{\mathfrak{q}^n}(1)$ generates in $\mathcal{O}_{H_{\mathfrak{q}^n}}$ the ideal $(\mathfrak{a}\,\mathcal{O}_{H_{\mathfrak{q}^n}})^{-1}$. So, put $\sigma \overset{dfn}{=} (\mathfrak{a}, H_{\mathfrak{q}^n}/k)$ and consider the element $\varphi(\sigma)$ of $\Omega^\times$ defined as follows

$$\varphi(\sigma) \overset{dfn}{=} \Delta_s(\mathfrak{a}^{-1}\mathfrak{q}^n)\left[e_{\mathfrak{a}^{-1}\mathfrak{q}^n}(1)\right]^{w_\infty}. \tag{8}$$

It is obvious that $\varphi(\sigma)$ depends only on σ (and not on the ideal $\mathfrak{a}$). The behavior of this invariant is well known, cf. [9] chap. IV. In particular $\varphi(\sigma) \in \mathcal{O}_{H_{\mathfrak{q}^n}}$ and we have $\varphi(\sigma)^\tau = \varphi(\sigma\tau)$, for all $\sigma, \tau \in$ Gal $(H_{\mathfrak{q}^n}/k)$. Moreover it verifies the following norm formula

$$N_{H_{\mathfrak{q}^n}/H_{(1)}}\Big(\varphi(\sigma)\Big)^{w_k} = \frac{\Delta_s(\mathfrak{a}^{-1})}{\Delta_s(\mathfrak{a}^{-1}\mathfrak{q})},$$

which implies that $\varphi(\sigma)$ generates in $\mathcal{O}_{H_{\mathfrak{q}^n}}$ the ideal $\mathfrak{q}_{H_{\mathfrak{q}^n}}^{\frac{w_\infty}{w_k}}$, where $w_k \overset{\mathrm{dfn}}{=} \sharp F_q^\times$. Indeed the quotient $\Delta_s(\mathfrak{a}^{-1})/\Delta_s(\mathfrak{a}^{-1}\mathfrak{q})$ generates in $\mathcal{O}_{H_{(1)}}$ the ideal $\mathfrak{q}_{H_{(1)}}$. On the other hand the extension $H_{\mathfrak{q}^n}/H_{(1)}$ is totally ramified at each prime factor of $\mathfrak{q}_{H_{(1)}}$. Now since we have the relation

$$\left(\frac{e_{\mathfrak{a}^{-1}\mathfrak{q}^n}(1)}{e_{\mathfrak{q}^n}(1)}\right)^{w_\infty} = \frac{\varphi(\sigma)}{\varphi(1)}\,\frac{\Delta_s(\mathfrak{q}^n)}{\Delta_s(\mathfrak{a}^{-1}\mathfrak{q}^n)}$$

and since $\Delta_s(\mathfrak{q}^n)/\Delta_s(\mathfrak{a}^{-1}\mathfrak{q}^n)$ generates in $\mathcal{O}_{H_{(1)}}$ the ideal $(\mathfrak{a}^{-1}\mathcal{O}_{H_{(1)}})^{w_\infty}$, cf. [11] Proposition 3.7, we can deduce that the quotient $e_{\mathfrak{a}^{-1}\mathfrak{q}^n}(1)/e_{\mathfrak{q}^n}(1)$ generates in $\mathcal{O}_{H_{\mathfrak{q}^n}}$ the ideal $(\mathfrak{a}^{-1}\,\mathcal{O}_{H_{\mathfrak{q}^n}})$. ◇

DEFINITION 1. — *Let $F \subset k_\infty$ be a finite abelian extension of k. Suppose that the conductor of F/k is equal to $\mathfrak{q}^n$. Then we set S_F to be the sub-group of $F^\times$ generated by all the norms*

$$N_{H\mathfrak{q}^n/F}\left(\frac{e_{\mathfrak{a}^{-1}\mathfrak{q}^n}(1)}{e_{\mathfrak{q}^n}(1)}\right)$$

where $\mathfrak{a}$ describes the set of integral ideals of $\mathcal{O}_k$ which are prime to $\mathfrak{q}$.

4. — The unramified part of the group of elliptic units

We describe below the method of construction of non zero elements of those unramified abelian extensions of k which are included into k_∞. The connexion with the torsion points of certain Drinfeld $\mathcal{O}_k$-modules is explained in [10], proof of Theorem 2.2.

Let $\Gamma \subset \overline{\Gamma}$ be lattices of Ω such that the index $[\overline{\Gamma} : \Gamma]$ of Γ in $\overline{\Gamma}$ is finite. Then one can define the function

$$\Psi(z\,;\,\Gamma,\overline{\Gamma}) \overset{\text{dfn}}{=} \delta(\Gamma,\overline{\Gamma})\frac{e_\Gamma(z)^{[\overline{\Gamma}:\Gamma]}}{e_{\overline{\Gamma}}(z)}, \quad z \in \Omega,$$

which is well defined on the complement of $\overline{\Gamma}\backslash\Gamma$ in Ω. It vanishes on Γ and, in fact, is elliptic (i.e. periodic) with respect to Γ. Moreover, as a rational function of $e_\Gamma(z)$ its divisor is

$$[\overline{\Gamma} \;:\; \Gamma](0)_\Gamma - \sum_{\rho\in\overline{\Gamma}/\Gamma} (\rho)_\Gamma.$$

On the other hand, we have the homogeneity formula

$$\Psi(\lambda z\,;\,\lambda\Gamma,\lambda\overline{\Gamma}) = \Psi(z\,;\,\Gamma,\overline{\Gamma}), \text{ for all } \lambda \in \Omega^\times.$$

Also if $\Gamma_1 \subset \Gamma_2 \subset \Gamma_3$ are lattices of Ω such that the index $[\Gamma_3 \;:\; \Gamma_1]$ of Γ_1 into Γ_3 is finite, then we have

$$\Psi(z\,;\,\Gamma_1,\Gamma_3) = \Psi(z\,;\,\Gamma_1,\Gamma_2)^{[\Gamma_3:\Gamma_2]}\Psi(z\,;\,\Gamma_2,\Gamma_3)\,. \tag{9}$$

PROPOSITION 5. — *Let $M, \overline{M}$ and Γ be lattices of Ω such that $M \subset \overline{M}$, $M \subset \Gamma$ and $\Gamma \cap \overline{M} = M$. Consider the lattice $\overline{\Gamma} \overset{dfn}{=} \Gamma + \overline{M}$ and choose S a complete system of representatives of Γ modulo M. Then we have the distributivity formulas*

$$\Psi(z\,;\,\Gamma,\overline{\Gamma}) = \prod_{\rho\in S} \Psi(z+\rho\,;\,M,\overline{M}), \tag{10}$$

$$\frac{\delta(\Gamma,\overline{\Gamma})}{\delta(M,\overline{M})} = \prod_{\substack{\rho\in S\\ \rho\neq 0}} \Psi(\rho\,;\,M,\overline{M}). \tag{11}$$

Proof : this is a simple consequence of the formula (1). ◇

Now suppose that $\Gamma \subset \overline{\Gamma}$ are lattices of Ω of rank 1. Then the value $\Psi(\rho\,;\,\Gamma,\overline{\Gamma}) \in k^{ab}$ for all $\rho \in k\Gamma$. Indeed, $\Psi(\rho\,;\,\Gamma,\overline{\Gamma})$ is a product of quotients of the form $e_\Gamma(\rho_1)/e_\Gamma(\rho_2)$, $\rho_1, \rho_2 \in k\Gamma$, which are elements of k^{ab} by Theorem 3.2 of [11] (see also §2 above). Moreover, the formula (3) implies the following property

$$\Psi(\rho\,;\,\Gamma,\overline{\Gamma})^{[t,k]} = \Psi(t^{-1}\rho\,;\,t^{-1}\Gamma,t^{-1}\overline{\Gamma}) \tag{12}$$

for all idele t of k.

PROPOSITION 6. — *Let $\mathfrak{m} \neq 0$ be a proper ideal of $\mathcal{O}_k$. Let $\mathfrak{a}$ be a integral ideal of $\mathcal{O}_k$ prime to $\mathfrak{m}$. Then we have*

$$\Psi(1\,;\,\mathfrak{m},\mathfrak{a}^{-1}\mathfrak{m}) \in H_{\mathfrak{m}}\,.$$

Moreover if $\mathfrak{b}$ is a ideal of $\mathcal{O}_k$ prime to $\mathfrak{m}$ then the automorphism $(\mathfrak{b}, H_{\mathfrak{m}}/k)$ of $H_{\mathfrak{m}}/k$ applied to $\Psi(1\,;\,\mathfrak{m},\mathfrak{a}^{-1}\mathfrak{m})$ gives

$$\Psi(1\,;\,\mathfrak{m},\mathfrak{a}^{-1}\mathfrak{m})^{(\mathfrak{b},H_{\mathfrak{m}}/k)} = \Psi(1\,;\,\mathfrak{b}^{-1}\mathfrak{m},\mathfrak{a}^{-1}\mathfrak{b}^{-1}\mathfrak{m})\,.$$

Proof. this Proposition is a direct consequence of (12). See also the alternative proof given in [10]. ◇

The above Proposition 5 and Proposition 6 have the following remarkable consequence.

PROPOSITION 7. — *Let $\mathfrak{p}$ be a prime ideal of $\mathcal{O}_k$ and $n > 0$ be a positive integer. Let $\mathfrak{a}$ be a integral ideal of $\mathcal{O}_k$ prime to $\mathfrak{p}$. Then we have*

$$N_{H_{\mathfrak{p}^{n+1}}/H_{\mathfrak{p}^n}}\Big(\Psi(1\,;\,\mathfrak{p}^{n+1},\mathfrak{a}^{-1}\mathfrak{p}^{n+1})\Big) = \Psi(1\,;\,\mathfrak{p}^{n},\mathfrak{a}^{-1}\mathfrak{p}^{n}), \tag{13}$$

$$N_{H_{\mathfrak{p}}/H_{(1)}}(\Psi(1\,;\,\mathfrak{p},\mathfrak{a}^{-1}\mathfrak{p}))^{w_k} = \frac{\delta(\mathcal{O}_k,\mathfrak{a}^{-1})}{\delta(\mathfrak{p},\mathfrak{a}^{-1}\mathfrak{p})}. \tag{14}$$

Moreover the ideal of $\mathcal{O}_{H_{\mathfrak{p}^n}}$ generated by the value $\Psi(1\,;\,\mathfrak{p}^{n},\mathfrak{a}^{-1}\mathfrak{p}^{n})$ is $\mathfrak{p}_{H_{\mathfrak{p}^n}}^{\frac{N\mathfrak{a}-1}{w_k}}$, where $w_k = \sharp\mathbb{F}_q^{\times}$.

Proof : let X be a complete system of representatives in $\mathcal{O}_k$ of $(\mathcal{O}_k/\mathfrak{p})^{\times}$ modulo the group $\mathbb{F}_q^{\times}$. Then the elements of Gal $(H_{\mathfrak{p}}/H_{(1)})$ are precisely the automorphisms $(x, H_{\mathfrak{p}}/k), x \in X$. On the other hand if we put $M = \mathfrak{p}$, $\overline{M} = \mathfrak{a}^{-1}\mathfrak{p}$ and $\Gamma = \mathcal{O}_k$ then we have $\overline{\Gamma} \overset{\text{dfn}}{=} \Gamma + \overline{M} = \mathfrak{a}^{-1}$; moreover the set $\{\xi x, \xi \in \mathbb{F}_q^{\times} \text{ and } x \in X\}$ is a complete system of non zero representatives of Γ modulo M. Therefore the formula (14) is just the identity (11) applied to this precise case. The formula (13) is obtained from (10) proceeding as above. Now let us observe that we have the equation

$$\Psi(1\,;\,\mathfrak{p}^{n},\mathfrak{a}^{-1}\mathfrak{p}^{n})^{w_\infty} = \varphi(1)^{(N\mathfrak{a}-(\mathfrak{a},H_{\mathfrak{p}^n}/k))}$$

which implies in particular that $\Psi(1\,;\,\mathfrak{p}^{n},\mathfrak{a}^{-1}\mathfrak{p}^{n})$ generates in $\mathcal{O}_{H_{\mathfrak{p}^n}}$ the ideal $\mathfrak{p}_{H_{\mathfrak{p}^n}}^{\frac{N\mathfrak{a}-1}{w_k}}$ since the ideal generated by $\varphi(1)$ is $\mathfrak{p}_{H_{\mathfrak{p}^n}}^{\frac{w_\infty}{w_k}}$, cf. § 3. ◇

Now let $L \subset k_\infty$ be an unramified abelian extension of k. Then, for a given prime ideal $\mathfrak{p}$ of $\mathcal{O}_k$ we shall write $\mathcal{R}_{\mathfrak{p},L}$ for the subgroup of $L^\times$ generated by $\mu(L)$ and by all the norms

$$N_{H_{\mathfrak{p}^n}/L}\Big(\Psi(1\,;\,\mathfrak{p}^n, \mathfrak{a}^{-1}\mathfrak{p}^n)\Big)$$

where n is any positive integer and $\mathfrak{a}$ is any integral ideal of $\mathcal{O}_k$ prime to $\mathfrak{p}$. In fact n can be fixed according to the formula (13). On the other hand if $\mathfrak{p}'$ is a prime ideal of $\mathcal{O}_k$ such that the automorphism $(\mathfrak{p}', L/k)$ is equal to $(\mathfrak{p}, L/k)$ then we have, cf. [10] Theorem 3.2,

$$\mathcal{R}_{\mathfrak{p},L} \cap \mathcal{O}_L^\times = \mathcal{R}_{\mathfrak{p}',L} \cap \mathcal{O}_L^\times.$$

The group $\mathcal{R}_{\mathfrak{p},L} \cap \mathcal{O}_L^\times$ will be noted $\mathcal{E}_{L,\sigma}$ if $\sigma = (\mathfrak{p}, L/k)$.

DEFINITION 2. — *We define $\mathcal{R}_L$ to be the subgroup of $L^\times$ generated by all $\mathcal{R}_{\mathfrak{p},L}$, where $\mathfrak{p}$ is any prime ideal of $\mathcal{O}_k$. Hence we have*

$$\mathcal{R}_L \overset{dfn}{=} \prod_{\mathfrak{p}} \mathcal{R}_{\mathfrak{p},L}.$$

Remark 1 : the group $\mathcal{E}_L$ of the elements of $\mathcal{R}_L$ which are units of $\mathcal{O}_L$ is called the group of elliptic units of L. We have

$$\mathcal{E}_L \overset{dfn}{=} \mathcal{R}_L \cap \mathcal{O}_L^\times = \prod_{\sigma \in \mathrm{Gal}\,(L/k)} \mathcal{E}_{L,\sigma}.$$

The quotient group $\mathcal{O}_L^\times/\mathcal{E}_L$ is finite, cf. [10]. We have the index formula

$$[\mathcal{O}_L^\times \,:\, \mathcal{E}_L] = \frac{h(\mathcal{O}_L)}{[H_{(1)} \,:\, L]}. \tag{15}$$

Remark 2 : the formula

$$\Psi(1\,;\,\mathfrak{b}^{-1}\mathfrak{p}^n, \mathfrak{a}^{-1}\mathfrak{b}^{-1}\mathfrak{p}^n) = \Psi(1\,;\,\mathfrak{p}^n, \mathfrak{a}^{-1}\mathfrak{b}^{-1}\mathfrak{p}^n)/\Psi(1\,;\,\mathfrak{p}^n, \mathfrak{b}^{-1}\mathfrak{p}^n)^{N\mathfrak{a}},$$

verified for all prime ideal $\mathfrak{p}$ of $\mathcal{O}_k$ and all integral ideals $\mathfrak{a}$ and $\mathfrak{b}$ of $\mathcal{O}_k$ not divisible by $\mathfrak{p}$, shows clearly that the group $\mathcal{R}_{\mathfrak{p},L}$ is stable under the action of the Galois group of L/k. Thus the groups $\mathcal{R}_L$ and $\mathcal{E}_L$ are also stable under the action of Gal (L/k). Therefore, using formula (10), one can prove that $\mu(L)\mathcal{R}_L^{w_k}$ is generated by $\mu(L)$ and by all the quotients

$$N_{H_{(1)}/L}\left(\frac{\delta(\mathcal{O}_k, \mathfrak{b}^{-1})}{\delta(\mathfrak{a}, \mathfrak{b}^{-1}\mathfrak{a})}\right)$$

where $\mathfrak{a}$ and $\mathfrak{b}$ are any integral ideals of $\mathcal{O}_k$ which are coprime. Hence, the group $\mathcal{R}_L^{w_k w_\infty}$ is generated by all the elements of $L^\times$ of the form

$$N_{H_{(1)}/L}\left(\left(\frac{\Delta_s(\mathcal{O}_k)}{\Delta_s(\mathfrak{a})}\right)^{N\mathfrak{b}} \frac{\Delta_s(\mathfrak{b}^{-1}\mathfrak{a})}{\Delta_s(\mathfrak{b}^{-1})}\right)$$

where $\mathfrak{a}$ and $\mathfrak{b}$ are as above.

Remark 3 : using the description of $\mu(L)\mathcal{R}_L^{w_k}$ just given above and the fact that the order w_k of $\mathbb{F}_q^\times$ is the g.c.d. of the integers $N\mathfrak{a} - 1$, $\mathfrak{a}$ ideal of $\mathcal{O}_k$, one can prove that we have

$$N_{H_{(1)}/L}\Big(\prod_{\mathfrak{b}\in B} \delta(\mathcal{O}_k, \mathfrak{b}^{-1})^{n_\mathfrak{b}}\Big) \in \mathcal{R}_L\,,$$

where $\mathcal{B}$ is a finite set of integral ideals of $\mathcal{O}_k$ and $n_\mathfrak{b}$, $\mathfrak{b} \in \mathcal{B}$ are rational integers such that

$$\sum_{\mathfrak{b}\in\mathcal{B}} n_\mathfrak{b}(N\mathfrak{b} - 1) = 0.$$

DEFINITION 3. — *Let $\tau \in$ Gal (L/k).*

i) *If $L = H_{(1)}$, then we put $h = h(\mathcal{O}_k)$ and*

$$\partial_{H_{(1)}}(\tau) \overset{dfn}{=} \alpha^{w_\infty} \Delta_s(\mathfrak{b})^h$$

where $\mathfrak{b}$ is any fractionnal ideal of $\mathcal{O}_k$ such that $(\mathfrak{b}, H_{(1)}/k) = \tau^{-1}$ and $\mathfrak{b}^h = \alpha\mathcal{O}_k$, with $\alpha \in k$.

ii) *In general, we put :*

$$\partial_L(\tau) \overset{\mathrm{dfn}}{=} \prod_{\substack{\tilde{\tau} \in \mathrm{Gal}\,(H_{(1)}/k) \\ \tilde{\tau}|_L = \tau}} \partial_{H_{(1)}}(\tilde{\tau}).$$

One can show, cf. [11] Lemma 3.5, that the quotient $\partial_L(\tau_1)/\partial_L(\tau_2)$ is a unit of $\mathcal{O}_L$. Moreover if $\tau \in \mathrm{Gal}\,(L/k)$ then we have the property

$$\left(\frac{\partial_L(\tau_1)}{\partial_L(\tau_2)}\right)^{\tau} = \frac{\partial_L(\tau_1\tau)}{\partial_L(\tau_1\tau)},$$

which implies in particular that

$$N_{H_{(1)}/L}\left(\frac{\partial_{H_{(1)}}(\tilde{\tau}_1)}{\partial_{H_{(1)}}(\tilde{\tau}_2)}\right) = \frac{\partial_L(\tau_1)}{\partial_L(\tau_2)}$$

where $\tilde{\tau}_1$ and $\tilde{\tau}_2$ are automorphisms of $H_{(1)}/k$ such that $\tilde{\tau}_{i_{|L}} = \tau_i, i = 1, 2$.

Finally we see that the group $\mathcal{R}_L^{w_k w_\infty h}$ is generated by all the elements of $L^\times$ which have the form

$$\left(\frac{\partial_L(1)}{\partial_L(\tau)}\right)^{N\mathfrak{a}_{\tau'}} \left(\frac{\partial_L(\tau\tau')}{\partial_L(\tau')}\right) x^{w_\infty(1-N\mathfrak{a}_{\tau'})[H_{(1)}:L]}$$

where $\tau, \tau' \in \mathrm{Gal}\,(L/k)$ and $\mathfrak{a}_{\tau'}$ is any integral ideal of $\mathcal{O}_k$ such that $(\mathfrak{a}_{\tau'}, L/k) = \tau'$. The element x of $k^\times$ is such that there exists $\mathfrak{b}$ a ideal of $\mathcal{O}_k$ which verifies $(\mathfrak{b}, L/k) = \tau$ and $\mathfrak{b}^h = x\mathcal{O}_k$.
In particular, for any $\lambda \in k^\times$ take $\mathfrak{b} = \lambda\mathcal{O}_k$ and $x = \lambda^h$ so that $\tau \stackrel{\mathrm{dfn}}{=} (\mathfrak{b}, L/k) = 1$ and

$$\lambda^{hw_\infty(N\mathfrak{a}-1)[H_{(1)}\,:\,L]}, \quad \mathfrak{a} \text{ any ideal of } \mathcal{O}_k,$$

belongs to $\mathcal{R}_L^{w_k w_\infty h}$. As $w_k = \sum_{\mathfrak{a}\in\mathcal{U}} n_\mathfrak{a}(N\mathfrak{a} - 1)$ for a well suited finite set $\mathcal{U}$ of integral ideals of $\mathcal{O}_k$, and convenient integers $n_\mathfrak{a}, \mathfrak{a} \in \mathcal{U}$, we get

COROLLARY. — *The group* $k^{\times w_k w_\infty h[H_{(1)}\,:\,L]}$ *is contained into* $\mathcal{R}_L^{w_k w_\infty h}$.

4. — The group of elliptic units

Let $F \subset k_\infty$ be a finite abelian extension of k such that the conductor of F/k is equal to $\mathfrak{q}^n$, where $\mathfrak{q}$ is a prime ideal of $\mathcal{O}_k$ and n is a positive integer. We put $F_{(1)} \stackrel{\mathrm{dfn}}{=} F \cap H_{(1)}$ and we define a subgroup $\mathcal{R}_F$ of $F^\times$ by setting

$$\mathcal{R}_F \stackrel{dfn}{=} \mathcal{R}_{F_{(1)}}.S_F \tag{16}$$

DEFINITION 4. — *Let* $\mathcal{E}_F$ *be the intersection of* $\mathcal{R}_F$ *with the group of units of* $\mathcal{O}_F$, *i.e.*,

$$\mathcal{E}_F \stackrel{\mathrm{dfn}}{=} \mathcal{R}_F \cap \mathcal{O}_F^\times. \tag{17}$$

We call $\mathcal{E}_F$ the group of elliptic units of F.

Our goal in the present section is to describe the group $\mathcal{E}_F^{w_k w_\infty h}$ in a manner which will allow us to calculate its index in $\mathcal{O}_F^\times$. Therefore we have first to introduce some "new" elements of $\mathcal{O}_F$ defined as the norm from $H_{\mathfrak{q}^n}$ down to F of the invariants $\varphi(\sigma), \sigma \in \mathrm{Gal}\,(H_{\mathfrak{q}^n}/k)$, defined by the formula (8).

DEFINITION 5. — *Let $\tau \in \mathrm{Gal}\,(F/k)$. Then we put*

$$\varphi_F(\tau) \stackrel{dfn}{=} N_{H_{\mathfrak{q}^n}/F}(\varphi(\tilde{\tau})),$$

where $\tilde{\tau} \in \mathrm{Gal}\,(H_{\mathfrak{q}^n}/k)$ is such that $\tilde{\tau}_{|F} = \tau$.

Remark 5 : let $M \subset k_\infty$ be a finite abelian extension of k. Let $\mathfrak{a}$ be a integral ideal of $\mathcal{O}_k$ prime to the conductor of M/k. Then we know that

$$\xi^{(\mathfrak{a},M/k)} = \xi^{N\mathfrak{a}}, \quad \text{for all} \quad \xi \in \mu(M)\,.$$

In particular if $(\mathfrak{a}', M/k) = (\mathfrak{a}, M/k)$ then $N\mathfrak{a}' \equiv N\mathfrak{a} \,\mathrm{mod}\,.\, w_M$ where $w_M \stackrel{\mathrm{dfn}}{=} \sharp\mu(M)$. This means that we have a well defined Dirichlet character

$$\psi_M \;:\; \mathrm{Gal}\,(M/k) \to (\mathbb{Z}/w_M\mathbb{Z})^\times$$

$$\tau \mapsto \psi_M(\tau)$$

given by the condition

$$\psi_M(\tau) \equiv N\mathfrak{b}\ (\mathrm{mod}.\ w_M),$$

where $\mathfrak{b}$ is any integral ideal of $\mathcal{O}_k$ such that $\tau = (\mathfrak{b}, M/k)$.
We have $\Psi_M(\tau) \equiv 1(\mathrm{mod}.\ w_k)$ for any $\tau \in \mathrm{Gal}\ (M/k)$ so that we can make the following construction :

For I_M the augmentation ideal of the group ring $\mathbb{Z}[\mathrm{Gal}\ (M/k)]$, and each integer $\ell \geq 1$, we define

$$\Psi_M^{(\ell)} \;:\; I_M^\ell \longrightarrow \mathbb{Z}/m\mathbb{Z}, \text{ with } m = w_M/w_k,$$

to be the following surjective morphism

$$\Psi_M^{(\ell)}\left[\sum_{\tau_1,\tau_2,\ldots,\tau_\ell} n_{\tau_1,\tau_2,\ldots,\tau_\ell}\Big((\tau_1-1)(\tau_2-1)\cdots(\tau_\ell-1)\Big)\right]$$

$$\stackrel{\mathrm{dfn}}{=} \sum_{\tau_1,\tau_2,\ldots,\tau_\ell} n_{\tau_1,\tau_2,\ldots,\tau_\ell}\left(\frac{\Psi_M(\tau_1)-1}{w_k}\right)\left(\frac{\Psi_M(\tau_2)-1}{w_k}\right)\cdots\left(\frac{\Psi_M(\tau_\ell)-1}{w_k}\right).$$

These operators are well defined, as will be made clear in some further work; here we just need to have $\Psi_M^{(1)}$ and $\psi_M^{(2)}$ to our disposition, and the following lemma relating them :

LEMMA 1. — *For any element A of I_M^2 we have*

$$\Psi_M^{(2)}(w_k A) \equiv \Psi_M^{(1)}(A) \quad \left(\bmod . \frac{w_M}{w_k}\right).$$

Proof : obvious.

PROPOSITION 8. — *Let $F_{(1)} \overset{\mathrm{dfn}}{=} F \cap H_{(1)}$ and, for $\mathfrak{q}^n$ the conductor of F/k; put $d_1 \overset{\mathrm{dfn}}{=} [H_{\mathfrak{q}^n} : FH_{(1)}]$ and $d_2 \overset{\mathrm{dfn}}{=} [FH_{(1)} : F] = [H_{(1)} : F_{(1)}]$ so that $d_1 d_2 = [H_{\mathfrak{q}^n} : F]$. Let $\Psi_{F_{(1)}}^{(1)} : I_{F_{(1)}} \to \mathbb{Z}/m\mathbb{Z}$ and $\Psi_{F_{(1)}}^{(2)} : I_{F_{(1)}}^2 \to \mathbb{Z}/m\mathbb{Z}$, with $m = w_{F_{(1)}}/w_k$, be the surjective morphisms defined as above. Also put $W = w_k w_\infty h$.*

Then the group $\mathcal{E}_F^W$ is formed of all the products

$$\left(\prod_{\sigma \in \mathrm{Gal}\,(F/k)} \varphi_F(\sigma)^{m_\sigma}\right)^{w_k h} \prod_{\tau \in \mathrm{Gal}\,(F_{(1)}/k)} \partial_{F_{(1)}}(\tau)^{n_\tau} \tag{18}$$

when the elements $\Sigma m_\sigma(\sigma)$ of $\mathbb{Z}[\mathrm{Gal}\,(F/k)]$ and $\Sigma m_\tau(\tau)$ of $\mathbb{Z}[\mathrm{Gal}\,(F_{(1)}/k)]$ are such that

1) $\Sigma m_\sigma(\sigma) \in I_F$;

2) $\Sigma n_\tau(\tau) \in I_{F_{(1)}}^2$, *i.e.*, $\Sigma n_\tau(\tau) \in I_{F_{(1)}}$ *and* $\prod \tau^{n_\tau} = 1$;

3) *consider the other element $\Sigma M_\tau(\tau)$ of $I_{F_{(1)}}$ defined by $\tau_{\mathfrak{q}^n} \overset{\mathrm{dfn}}{=} (\mathfrak{q}^n, F_{(1)}/k)$ and $M_\tau \overset{\mathrm{dfn}}{=} \sum_{\substack{\sigma \in \mathrm{Gal}\,(F/k) \\ \sigma|_{F_{(1)}} = \tau\tau_{\mathfrak{q}^n}}} m_\sigma$, $\tau \in \mathrm{Gal}\,(F_{(1)}/k)$; then request*

$$d_1 \Psi_{F_{(1)}}^{(1)}\left(\sum_\tau M_\tau(\tau)\right) + \Psi_{F_{(1)}}^{(2)}\left(\sum_\tau n_\tau(\tau)\right) \equiv 0 \left(\bmod . \frac{w_{F_{(1)}}}{w_k}\right).$$

Proof : let us see that any element α of $F^\times$ satisfying the above conditions 1) to 3) belongs to $\mathcal{E}_F^W$: so we fix elements $\Sigma m_\sigma(\sigma)$ of $\mathbb{Z}[\mathrm{Gal}\,(F/k)]$ and $\Sigma n_\tau(\tau)$ of $\mathbb{Z}[\mathrm{Gal}\,(F_{(1)}/k)]$ satisfying 1) to 3); in particular we define M_τ for $\tau \in \mathrm{Gal}\,(F_{(1)}/k)$ by

$$M_\tau \overset{\mathrm{dfn}}{=} \sum_{\substack{\sigma \in \mathrm{Gal}\,(F/k) \\ \sigma|_{F_{(1)}} = \tau\tau_{\mathfrak{q}^n}}} m_\sigma$$

a) First by conditions 1) and 2) we have that $\Sigma m_\sigma(\sigma)$ and $\Sigma n_\tau(\tau)$ are elements of the augmentation ideals I_F and $I_{F_{(1)}}$ respectively, so that we have :

$$\alpha \in \mathcal{O}_F^\times$$

and the only thing we have to prove is :

$$\alpha \in \mathcal{R}_{F_{(1)}}^W \cdot S_F^W$$

b) Arbitrarily, choose a finite set Z of integral ideals of $\mathcal{O}_k$, all prime to $\mathfrak{q}$, such that the Artin map

$$\mathfrak{a} \mapsto (\mathfrak{a}, F/k)$$

define a bijection from Z to Gal (F/k); thereafter put

$$m_{\mathfrak{b}} \stackrel{\text{dfn}}{=} m_{(\mathfrak{b},\, F/k)}.$$

Also, for each $\tau \in \mathrm{Gal}\,(F_{(1)}/k)$, choose an integral ideal $\mathfrak{a}_\tau$ of $\mathcal{O}_k$ and an element x_τ of $k^\times$ such that $(\mathfrak{a}_\tau, F_{(1)}/k) = \tau$ and $\mathfrak{a}_\tau^h = x_\tau \mathcal{O}_k$.

c) Then α can be written as the product ABC where A, B and C are as follows

$$A \stackrel{\text{dfn}}{=} N_{H\mathfrak{q}^n/F}\left(\prod_{\mathfrak{b} \in Z} e_{\mathfrak{b}^{-1}\mathfrak{q}^n}(1)^{m_{\mathfrak{b}}}\right)^W,$$

$$B \stackrel{\text{dfn}}{=} \left(\prod_{\sigma \in \mathrm{Gal}\,(F/k)} \partial_{F_{(1)}}(\tau)^{d_1 m_\sigma}\right)^{w_k} \left(\prod_{\sigma \in \mathrm{Gal}\,(F/k)} x_\sigma^{d_1 m_\sigma}\right)^{d_2 w_k w_\infty},$$

where σ and τ are related by $\sigma_{|F_{(1)}} = \tau\tau_{\mathfrak{q}^n}$ so that $\tau = (\mathfrak{b}\mathfrak{q}^{-n}, F_{(1)}/k)$ and $x_\sigma \in k^\times$ is defined by $(\mathfrak{b}\mathfrak{q}^{-n})^h = x_\sigma \mathcal{O}_k$;

$$C \stackrel{\text{dfn}}{=} \prod_{\tau \in \mathrm{Gal}\,(F_{(1)}/k)} \partial_{F_{(1)}}(\tau)^{n_\tau}.$$

By condition 1) we have $A \in S_F^W$.

d) But we know that $w_k = \sum_{\mathfrak{a} \in \mathcal{U}} n_{\mathfrak{a}}(N\mathfrak{a} - 1)$ where $\mathcal{U}$ is a finite set of integral ideals of $\mathcal{O}_k$ and $n_{\mathfrak{a}}$, $\mathfrak{a} \in \mathcal{U}$, are rational integers. So, just put $\tau_{\mathfrak{a}} = (\mathfrak{a}, F_{(1)}/k), \mathfrak{a} \in \mathcal{U}$, and observe that the product BC is also equal to the product $B'C'$ with

$$B' \stackrel{\text{dfn}}{=} \prod_{\sigma \in \mathrm{Gal}\,(F/k)} B_\sigma'^{d_1 m_\sigma} \text{ and } B'_\sigma \stackrel{\text{dfn}}{=} \prod_{\mathfrak{a} \in \mathcal{U}} \left[\left(\frac{\partial_{F_{(1)}}(\tau)}{\partial_{F_{(1)}}(1)}\right)^{N\mathfrak{a}} \left(\frac{\partial_{F_{(1)}}(\tau_{\mathfrak{a}})}{\partial_{F_{(1)}}(\tau\tau_{\mathfrak{a}})}\right) x_\sigma^{(N\mathfrak{a}-1)w_\infty d_2}\right]^{n_{\mathfrak{a}}}$$

where as above $\tau = \tau_{\mathfrak{q}^n}^{-1}\sigma_{|F_{(1)}} = (\mathfrak{b}\mathfrak{q}^{-n}, F_{(1)}/k)$ and $x_\sigma \mathcal{O}_k = (\mathfrak{b}\mathfrak{q}^{-n})^h$; for defining C' regroup all the m_σ with $\sigma_{|F_{(1)}} = \tau\tau_{\mathfrak{q}^n}$ so that

$$C' \stackrel{\text{dfn}}{=} \prod_{\tau \in \mathrm{Gal}\,(F_{(1)}/k)} \left\{ \prod_{\mathfrak{a}\in\mathcal{U}} \left(\frac{\partial_{F_{(1)}}(1)}{\partial_{F_{(1)}}(\tau)} \, \frac{\partial_{F_{(1)}}(\tau_\mathfrak{a}\tau)}{\partial_{F_{(1)}}(\tau_\mathfrak{a})} \right)^{n_\mathfrak{a}} \right\}^{d_1 M_\tau} \prod_{\tau \in \mathrm{Gal}\,(F_{(1)}/k)} \partial_{F_{(1)}}(\tau)^{n_\tau}.$$

Nota Bene. We have that AB and AB' are units in $\mathcal{O}_F$ so that C and C' are units in $\mathcal{O}_{F_{(1)}}$, as was obvious by their very definition for the quotients $\partial_{F_{(1)}}(\tau_1)/\partial_{F_{(1)}}(\tau_2)$ are units.

Observe now that by the end of § 4 we have $B'_\sigma \in \mathcal{R}^W_{F_{(1)}}$. Moreover, as we have

$$\Psi^{(2)}_{F_{(1)}}\Big[\sum_{\mathfrak{a}\in\mathcal{U}} n_\mathfrak{a}(\tau-1)(\tau_\mathfrak{a}-1)\Big] \equiv \Big(\frac{\Psi(\tau)-1}{w_k}\Big).\sum_{\mathfrak{a}\in\mathcal{U}} n_\mathfrak{a}\Big(\frac{N\mathfrak{a}-1}{w_k}\Big)$$
$$\equiv \Psi^{(1)}_{F_{(1)}}(\tau-1)(\mathrm{mod.}\ w_{F_{(1)}}/w_k),\ \tau\in\mathrm{Gal}\,(F_{(1)}/k),$$

we deduce from the condition 3) the fact that

$$\Psi^{(2)}_{F_{(1)}}\Bigg(\sum_\tau d_1 M_\tau\Big(\sum_{\mathfrak{a}\in\mathcal{U}} n_\mathfrak{a}(\tau-1)(\tau_\mathfrak{a}-1)\Big)+\sum_\tau n_\tau(\tau)\Bigg)\equiv 0\big(\mathrm{mod.}\ w_{F_{(1)}}/w_k\big).$$

Yet, see [10], the condition $\Psi^{(2)}_{F_{(1)}}\big(\sum_t n_t(t)\big)\equiv 0\big(\mathrm{mod.}\ w_{F_{(1)}}/w_k\big)$ is a necessary and sufficient condition for the product

$$\prod_{t\in\mathrm{Gal}\,(F_{(1)}/k)} \partial_{F_{(1)}}(t)^{n_t},\ \ \sum_t n_t(t)\in I^2_{F_{(1)}},$$

to belong to $\mathcal{E}^W_{F_{(1)}} = \mathcal{R}^W_{F_{(1)}} \cap \mathcal{O}^W_{F_{(1)}}$; whence $C' \in \mathcal{R}^W_{F_{(1)}}$ and $\alpha \in \mathcal{R}^W_{F_{(1)}}.S^W_F$.

On the other hand, let us prove that any

$$\alpha \in \left(\mathcal{R}^W_{F_{(1)}}.S^W_F\right)\cap\mathcal{O}^\times_F$$

satisfy the conditions 1) to 3) : by Definition 1 and the observation at the end of § 4, the unit α may be written as a product AB with

$$A \stackrel{\text{dfn}}{=} \prod_{(\mathfrak{a},\mathfrak{a}')\in Y\times Y} \left\{ \left(\frac{\partial_{F_{(1)}}(1)}{\partial_{F_{(1)}}(\tau_\mathfrak{a})}\right)^{N\mathfrak{a}'} \frac{\partial_{F_{(1)}}(\tau_\mathfrak{a}\tau_{\mathfrak{a}'})}{\partial_{F_{(1)}}(\tau_{\mathfrak{a}'})} [\mathfrak{a}^{hw_\infty}]^{(1-N\mathfrak{a}')d_2} \right\}^{n_{\mathfrak{a},\mathfrak{a}'}},$$

$$B \stackrel{\text{dfn}}{=} \left[\prod_{\mathfrak{b}\in Z} N_{H\mathfrak{q}^n/F}\left(\frac{e_{\mathfrak{b}^{-1}\mathfrak{q}^n}(1)}{e_{\mathfrak{q}^n}(1)}\right)^{m_\mathfrak{b}}\right]^W,$$

where we have made the following conventions. Y is a finite set of integral ideals of $\mathcal{O}_k$. Z is a finite set of integral ideals of $\mathcal{O}_k$, all prime to $\mathfrak{q}$; $n_{\mathfrak{a},\mathfrak{a}'}$, $(\mathfrak{a},\mathfrak{a}') \in Y \times Y$, and $m_{\mathfrak{b}}$, $\mathfrak{b} \in Z$, are rational integers. $\tau_{\mathfrak{a}} = (\mathfrak{a}, F_{(1)}/k), \sigma_{\mathfrak{b}} = (\mathfrak{b}, F/k)$ or its restriction to $F_{(1)}$. Finally $[\mathfrak{a}^{w_\infty h}]$ is the w_∞–power of any element $x \in k^\times$ such that $\mathfrak{a}^h = x\mathcal{O}_k$.
Now using formula (8) and Definition 3, we can write

$$N_{H\mathfrak{q}^n/F}\left(\frac{e_{\mathfrak{b}^{-1}\mathfrak{q}^n}(1)}{e_{\mathfrak{q}^n}(1)}\right)^W = \left(\frac{\varphi_F(\sigma_{\mathfrak{b}})}{\varphi_F(1)}\right)^{w_k h} \times N_{H_{(1)}/F_{(1)}}\left(\frac{\Delta_s(\mathfrak{q}^n)}{\Delta_s(\mathfrak{b}^{-1}\mathfrak{q}^n)}\right)^{w_k h d_1}$$

$$= \left(\frac{\varphi_F(\sigma_{\mathfrak{b}})}{\varphi_F(1)}\right)^{w_k h} \times \left(\frac{\partial_{F_{(1)}}(\tau_{\mathfrak{q}^n}^{-1})}{\partial_{F_{(1)}}(\sigma_{\mathfrak{b}}\tau_{\mathfrak{q}^n}^{-1})}\right)^{w_k d_1} [\mathfrak{b}^{-h w_\infty}]^{w_k d_1 d_2}.$$

In particular we get

$$\prod_{\mathfrak{a}\in Y} [\mathfrak{a}^{h w_\infty}]^{\nu_{\mathfrak{a}} d_2} \prod_{\mathfrak{b}\in Z} [\mathfrak{b}^{h w_\infty}]^{m_{\mathfrak{b}} w_k d_1 d_2} = 1,$$

where $\nu_{\mathfrak{a}} = \sum_{\mathfrak{a}'\in Y} n_{\mathfrak{a},\mathfrak{a}'}(N\mathfrak{a}' - 1)$; whence in terms of automorphisms

$$\prod_{\mathfrak{a}\in Y} \tau_{\mathfrak{a}}^{\frac{\nu_{\mathfrak{a}}}{w_k}} \prod_{\mathfrak{b}\in Z} \tau_{\mathfrak{b}}^{m_{\mathfrak{b}} d_1} = 1.$$

This equality is a necessary and sufficient condition for the sum

$$(*) \qquad \sum_{\mathfrak{a}\in Y} \frac{\nu_{\mathfrak{a}}}{w_k}(1 - \tau_{\mathfrak{a}}) + d_1 \sum_{\mathfrak{b}\in Z} m_{\mathfrak{b}} \tau_{\mathfrak{q}^n}^{-1}(1 - \tau_{\mathfrak{b}})$$

to be an element of $I^2_{F_{(1)}}$; yet this sum itself (which belongs to $w_k I_{F_{(1)}}$) is congruent modulo $I^2_{F_{(1)}}$ to our $\sum_\tau n_\tau(\tau)$ which here is

$$\sum_{\mathfrak{a},\mathfrak{a}'\in Y} n_{\mathfrak{a},\mathfrak{a}'}(N\mathfrak{a}' - \tau_{\mathfrak{a}'})(1 - \tau_{\mathfrak{a}}) + w_k d_1 \sum_{\mathfrak{b}\in Z} m_{\mathfrak{b}} \tau_{\mathfrak{q}^n}^{-1}(1 - \tau_{\mathfrak{b}});$$

hence condition 2) is proved. Condition 1) is trivial.
On the other hand, by the Lemma 1 applied to the sum (*) we have

$$\psi^{(2)}_{F_{(1)}}\left(\sum_\tau n_\tau(\tau)\right) \equiv \sum_{\mathfrak{a},\mathfrak{a}'\in Y} n_{\mathfrak{a},\mathfrak{a}'} \left(\frac{\psi(\tau_{\mathfrak{a}'}) - N\mathfrak{a}'}{w_k}\right)\left(\frac{\psi(\tau_{\mathfrak{a}}) - 1}{w_k}\right)$$

$$+ d_1 \sum_{\mathfrak{b}\in Z} m_{\mathfrak{b}}\, \psi(\tau_{\mathfrak{q}^n}^{-1}) \left(\frac{1 - \psi(\tau_{\mathfrak{b}})}{w_k}\right) \quad \left(\text{mod. } \frac{w_{F_{(1)}}}{w_k}\right)$$

$$\equiv d_1 \sum_{\mathfrak{b}\in Z} m_{\mathfrak{b}}\, \psi(\tau_{\mathfrak{q}^n}^{-1}) \left(\frac{1 - \psi(\tau_{\mathfrak{b}})}{w_k}\right) \quad \left(\text{mod. } \frac{w_{F_{(1)}}}{w_k}\right)$$

$$\equiv -d_1\, \psi^{(1)}_{F_{(1)}}\left(\sum_\tau M_\tau(\tau)\right) \quad \left(\text{mod. } \frac{w_{F_{(1)}}}{w_k}\right)$$

where as above we have put

$$M_\tau \overset{dfn}{=} \sum_{\substack{\sigma_{\mathfrak{b}} \in \mathrm{Gal}\,(F/k) \\ \sigma_{\mathfrak{b}|F_{(1)}} = \tau\tau_{\mathfrak{q}^n}}} m_{\mathfrak{b}} \ ;$$

hence condition 3) is also satisfied.
This concludes the proof of Proposition 8.

6. — The index formula

Take $F \subset k_\infty$ to be, as in section 5, a finite abelian extension of k such that the conductor of F/k is equal to $\mathfrak{q}^n$, where $\mathfrak{q}$ is a prime ideal of $\mathcal{O}_k$. We want to calculate the index of the group $\mathcal{E}_F$ in $\mathcal{O}_F^\times$. The technique we will use is well known, cf [9] or [12]. Let $\sigma \in \mathrm{Gal}\,(F/k)$. Then for each rational integer $a > 0$ we put

$$t_{a,F}(\sigma) \overset{\mathrm{dfn}}{=} \partial_{F_{(1)}}(\tilde{\sigma})^a \varphi_F(\sigma)$$

where $\tilde{\sigma} \in \mathrm{Gal}\,(F_{(1)}/k)$ is such that $\tilde{\sigma} = \sigma_{|F_{(1)}}$. It is obvious that $t_{a,F}(\sigma_1)/t_{a,F}(\sigma_2) \in \mathcal{O}_F^\times$, for all $\sigma_1, \sigma_2 \in \mathrm{Gal}\,(F/k)$. Moreover we have the action

$$\left(\frac{t_{a,F}(\sigma_1)}{t_{a,F}(\sigma_2)}\right)^\sigma = \frac{t_{a,F}(\sigma_1\sigma)}{t_{a,F}(\sigma_2\sigma)}, \quad \text{for all } \sigma \in \mathrm{Gal}\,(F/k).$$

Let us denote $T_{a,F}$ the subgroup of $\mathcal{O}_F^\times$ generated by the quotients $t_{a,F}(\sigma)/t_{a,F}(\sigma')$, σ, $\sigma' \in \mathrm{Gal}\,(F/k)$. We know that the group $T_{a,F}$ has finite index in $\mathcal{O}_F^\times$, cf. [9] or [12]. We have

$$[\mathcal{O}_F^\times : T_{a,F}] = w_k e_a(F)\, \frac{h(\mathcal{O}_F)}{h}\, (w_\infty)^{[F:k]-1} \tag{19}$$

where $e_a(F)$ is a **positive** integer, equal to 1 if $F \cap H_{(1)} = k$; otherwise we have

$$e_a(F) \overset{\mathrm{dfn}}{=} \prod_{\chi \neq 1} \Big(1 - \chi((\mathfrak{q}, F \cap H_{(1)}/k)) + a w_k h [F : F \cap H_{(1)}]\Big) \tag{20}$$

where χ runs through the set of all <u>non trivial</u> characters of $\mathrm{Gal}\,(F \cap H_{(1)}/k)$.

The fact that $e_a(F) \neq 0$ implies that the quotients $t_{a,F}(\sigma)/t_{a,F}(1)$, $\sigma \in \mathrm{Gal}\,(F/k)$ and $\sigma \neq 1$, constitute a maximal system of independant elements of $\mathcal{O}_F^\times$. In particular we have

$$N_{F/F_{(1)}}\Big(T_{a,F}\Big) = T_{a,F} \cap H_{(1)}.$$

In other words the group $T_{a,F} \cap H_{(1)}$ is generated by the quotients $t'_{a,F}(\tau)/t'_{a,F}(1), \tau \in \mathrm{Gal}\,(F_{(1)}/k)$ where we have put $F_{(1)} = F \cap H_{(1)}$ and

$$t'_{a,F}(\tau) \stackrel{dfn}{=} \prod_{\substack{\tilde{\tau} \in \mathrm{Gal}\,(F/K) \\ \tilde{\tau}|_{F_{(1)}} = \tau}} t_{a,F}(\tilde{\tau}), \quad \text{for all } \tau \in \mathrm{Gal}\,(F_{(1)}/k).$$

This leads to the identity

$$\text{(21)} \qquad \Big(T_{a,F} \cap H_{(1)}\Big)^n = T^n_{a,F} \cap H_{(1)}, \text{ for all } n > 0.$$

Moreover the group $T_{a,F} \cap H_{(1)}$ has finite index in $\mathcal{O}^\times_{F_{(1)}}$ given by the formula

$$\text{(22)} \qquad \Big[\mathcal{O}^\times_{F_{(1)}} : T_{a,F} \cap H_{(1)}\Big] = w_k e_a(F)\, \frac{h(\mathcal{O}_{F_{(1)}})}{h}\, (w_\infty)^{[F_{(1)}:k]-1}.$$

Nota Bene. Let us recall also that the subgroup $\Delta_{F_{(1)}}$ of $\mathcal{O}^\times_{F_{(1)}}$ formed of all the products

$$\prod_{\tau \in \mathrm{Gal}\,(F_{(1)}/k)} \partial_{F_{(1)}}(\tau)^{n_\tau}$$

such that $\sum_{\tau \in \mathrm{Gal}\,(F_{(1)}/k)} n_\tau(\tau) \in I^2_{F_{(1)}}$ has a finite index in $\mathcal{O}^\times_{F_{(1)}}$, cf. [9] or [12], given by the following formula

$$\text{(23)} \qquad \big[\mathcal{O}^\times_{F_{(1)}} : \Delta_{F_{(1)}}\big] = w_k (w_k w_\infty h)^{[F_{(1)}:k]-1}\, \frac{h(\mathcal{O}_{F_{(1)}})}{[H_{(1)} : F_{(1)}]}.$$

PROPOSITION 9. — *Let a be a positive integer. Then the group*

$$Z_{a,F} \stackrel{\mathrm{dfn}}{=} T^{w_k h}_{a,F} \Delta_{F_{(1)}}$$

has finite index in $\mathcal{O}^\times_F$, given by the formula

$$\text{(24)} \qquad \big[\mathcal{O}^\times_F : Z_{a,F}\big] = w_k (w_k w_\infty h)^{[F:k]-1}\, \frac{h(\mathcal{O}_F)}{[H_{(1)} : F_{(1)}]}.$$

Proof : on one hand we have the isomorphism $Z_{a,F}/T^{w_k h}_{a,F} \simeq \Delta_{F(1)}/T^{w_k h}_{a,F} \cap \Delta_{F(1)}$. On the other hand one can check that $T^{w_k h}_{a,F} \cap \Delta_{F_{(1)}} = T^{w_k h}_{a,F} \cap H_{(1)}$. This leads to the following identity

$$\big[\mathcal{O}^\times_F : Z_{a,F}\big]\, \big[\mathcal{O}^\times_{F_{(1)}} : (T_{a,F} \cap H_{(1)})^{w_k h}\big] = \big[\mathcal{O}^\times_F : T^{w_k h}_{a,F}\big]\, \big[\mathcal{O}^\times_{F_{(1)}} : \Delta_{F_{(1)}}\big]$$

which allows us to conclude the proof using Formulas (19), (21) and (23).◇

One can notice the inclusion $\mathcal{E}_F^{w_k w_\infty h} \subset Z_{a,F}$, for all $a > 0$. In fact when $w_\infty \mid a$ then the group $\mathcal{E}_F^{w_k w_\infty h}$ can be characterized as follows. It is the group of the products

$$(25) \qquad \Big(\prod_{\sigma \in \mathrm{Gal}\,(F/k)} t_{a,F}(\sigma)^{m_\sigma} \Big)^{w_k h} \prod_{\tau \in \mathrm{Gal}\,(F(1)/k)} \partial_{F_{(1)}}(\tau)^{n_\tau}$$

such that

1') $\sum_{\sigma \in \mathrm{Gal}\,(F/k)} m_\sigma \in I_F\,;$

2') $\sum_{\tau \in \mathrm{Gal}\,(F_{(1)}/k)} n_\tau \in I^2_{F_{(1)}}\,;$

3') We have the congruence

$$d_1 \Psi^{(1)}_{F_{(1)}} \Big(\sum_{\tau \in \mathrm{Gal}\,(F_{(1)}/k)} M_\tau(\tau) \Big) + \Psi^{(2)}_{F_{(1)}} \Big(\sum_{\tau \in \mathrm{Gal}\,(F_{(1)}/k)} n_\tau(\tau) \Big) \equiv 0 \Big(\mathrm{mod.}\ \frac{w_{F_{(1)}}}{w_k} \Big),$$

where $\sum_\tau M_\tau(\tau)$ is the element of $I_{F_{(1)}}$ such that

$$M_\tau = \sum_{\substack{\sigma \in \mathrm{Gal}\,(F/k) \\ \sigma|_{F_{(1)}} = \tau \tau_{q^n}}} m_\sigma, \quad \tau \in \mathrm{Gal}\,(F_{(1)}/k).$$

LEMMA 2. — *We have a well defined morphism*

$$Z_{a,F} \longrightarrow (\mathbb{Z}/m\mathbb{Z}),\, m = w_{F_{(1)}}/w_k,$$

which associate to the element (25) of $Z_{a,F}$ the lefthand side of the above congruence 3'). This morphism is onto and its kernel is just the group $\mathcal{E}_F^{w_k w_\infty h}$, so that we have

$$\Big[Z_{a,F} \ :\ \mathcal{E}_F^{w_k w_\infty h} \Big] = \frac{w_{F_{(1)}}}{w_k}.$$

Proof : all we have to prove is that the congruence

$$d_1\, \Psi^{(1)}_{F_{(1)}} \Big(\sum_{\tau \in \mathrm{Gal}\,(F_{(1)}/k)} M_\tau(\tau) \Big) + \Psi^{(2)}_{F_{(1)}} \Big(\sum_{\tau \in \mathrm{Gal}\,(F_{(1)}/k)} n_\tau(\tau) \Big) \equiv 0 \Big(\mathrm{mod}\,.\ \frac{w_{F_{(1)}}}{w_k} \Big)$$

occurs whenever the element

$$\Big(\prod_{\sigma \in \mathrm{Gal}\,(F/k)} t_{a,F}(\sigma)^{m_\sigma} \Big)^{w_k h} \prod_{\tau \in \mathrm{Gal}\,(F_{(1)}/k)} \partial_{F_{(1)}}(\tau)^{n_\tau}$$

of $Z_{a,F}$ is equal to 1.
But in this case one can see easily that the product

$$\prod_{\sigma \in \mathrm{Gal}\,(F/k)} t_{a,F}(\sigma)^{m_\sigma} \in T_{a,F} \cap H_{(1)},$$

which means that $m_\sigma = m_{\sigma'}$, if $\sigma_{|F_{(1)}} = \sigma'_{|F_{(1)}}$; and then one can write using the definition of $t_{a,F}(\sigma)$ for $\sigma \in \mathrm{Gal}\,(F/k)$

$$\prod_{\sigma \in \mathrm{Gal}\,(F/k)} t_{a,F}(\sigma)^{m_\sigma} = \Big(\prod_{\tau \in \mathrm{Gal}\,(F_{(1)}/k)} \partial_{F_{(1)}}(\tau)^{m'_\tau} \Big)^{a[F\,:\,F_{(1)}]} \times \prod_{\tau \in \mathrm{Gal}\,(F_{(1)}/k)} \Big(\prod_{\substack{\sigma \in \mathrm{Gal}\,(F/k) \\ \sigma_{|F_{(1)}} = \tau}} \varphi_F(\sigma) \Big)^{m'_\tau},$$

where we have put $m'_\tau = m_\sigma$ if $\sigma \in \mathrm{Gal}\,(F/k)$ and $\tau \in \mathrm{Gal}\,(F_{(1)}/k)$ are such that $\sigma_{|F_{(1)}} = \tau$. Now the formula

$$\Big(\prod_{\substack{\sigma \in \mathrm{Gal}\,(F/k) \\ \sigma_{|F_{(1)}} = \tau}} \varphi_F(\sigma)^{m_\sigma} \Big)^{w_k h} = \frac{\partial_{F_{(1)}}(\tau)}{\partial_{F_{(1)}}(\tau \tau_{\mathfrak{q}}^{-1})}, \ \tau \in \mathrm{Gal}\,(F_{(1)}/k),$$

already proved in [9], chap. IV, leads to the equality :

$$\prod_{\tau \in \mathrm{Gal}\,(F_{(1)}/k)} \partial_{F_{(1)}}(\tau)^{[n_\tau + m'_\tau + m'_\tau a w_k h [F\,:\,F_{(1)}] - m'_{\tau\tau_{\mathfrak{q}}}]} = 1$$

which is equivalent to the condition

$$n_\tau + m'_\tau + m'_\tau a w_k h [F\,:\,F_{(1)}] - m'_{\tau\tau_{\mathfrak{q}}} = 0 \tag{26}$$

for all $\tau \in \mathrm{Gal}\,(F_{(1)}/k)$.
Now we have

$$\begin{aligned} d_1 \psi^{(1)}_{F_{(1)}} \Big(\sum_{\tau \in \mathrm{Gal}\,(F_{(1)}/k)} M_\tau(\tau) \Big) &\equiv d_1 \psi^{(1)}_{F_{(1)}} \Big(\sum_{\tau \in \mathrm{Gal}\,(F_{(1)}/k)} [F\,:\,F_{(1)}]\, m'_{\tau\tau_{\mathfrak{q}^n}}(\tau) \Big) \\ &\equiv [H\mathfrak{q}^n : H_{(1)}]\, \psi^{(1)}_{F_{(1)}} \Big(\sum_{\tau \in \mathrm{Gal}\,(F_{(1)}/k)} m'_\tau (\tau\, \tau_{\mathfrak{q}^n}^{-1}) \Big) \\ &\equiv N\mathfrak{q}^{n-1} \Big(\frac{N\mathfrak{q} - 1}{w_k} \Big) \sum_{\tau \in \mathrm{Gal}\,(F_{(1)}/k)} m'_\tau \Big(\frac{\psi(\tau \tau_{\mathfrak{q}^n}^{-1}) - 1}{w_k} \Big) \\ &\equiv -\Psi^{(2)}_{F_{(1)}} \Big(\sum_{\tau \in \mathrm{Gal}\,(F_{(1)}/k)} m'_\tau (1 - \tau)(1 - \tau_{\mathfrak{q}}^{-1}) \Big). \end{aligned}$$

On the other hand the above condition (26) allows us to write

$$\begin{aligned}\Psi^{(2)}_{F_{(1)}}\Big(\sum_{\tau\in \mathrm{Gal}\,(F_{(1)}/k)} n_\tau(\tau)\Big) &\equiv \Psi^{(2)}_{F_{(1)}}\Big(\sum_{\tau\in \mathrm{Gal}\,(F_{(1)}/k)} (m'_{\tau\tau_{\mathfrak{q}}} - m'_\tau)(\tau)\Big)\\ &\equiv \Psi^{(2)}_{F_{(1)}}\Big(\sum_{\tau\in \mathrm{Gal}\,(F_{(1)}/k)} m'_\tau(1-\tau)(1-\tau_{\mathfrak{q}}^{-1})\Big).\end{aligned}$$

The Lemma 2 is proved.

PROPOSITION 10. — *The quotient group $\mathcal{O}_F^\times/\mathcal{E}_F$ is finite. The exact number of its elements is given by the index formula*

$$[\mathcal{O}_F^\times : \mathcal{E}_F] = \frac{h(\mathcal{O}_F)}{[H_{(1)} : F_{(1)}]}.$$

Manuscrit reçu le 4 décembre 1993

References

[1] V.G. Drinfeld. — *Elliptic modules*, Math. USSR-Sbornik, **23** (1974), 561-592.

[2] S. Galovitch, M. Rosen. — *The class number of cyclotomic function fields*, Journal of number theory, **13** (1981), 363-375.

[3] S. Galovitch, M. Rosen. — *Units and class groups in cyclotomic function fields*, Journal of number theory, **14** (1982), 156-184.

[4] D.R. Hayes. — *Explicit class field theory for global function fields*, Studies in algebra and number theory (Rota G.C (ed)), New-York, Academic Press, (1979), 173-217.

[5] D.R. Hayes. — *Elliptic units in function fields, In proc. of a conference on modern developments related to Fermat's last theorem*, D. Goldfeld ed. Birkhaüser, Boston, 1982.

[6] D.R. Hayes. — *Stickelberger elements in function fields*, Compositio. Math, **55** (1985), 209-239.

[7] D. Kubert, S. Lang. — *Modular Units*, Grundleh der Math. Wiss., **244** (1981), ed. Springer.

[8] H. Oukhaba, G. Robert. — *Etude d'un idéal particulier associé à un caractère de Dirichlet d'un groupe fini*, Séminaire de Théorie des Nombres de Bordeaux **3** (1991), 117-127.

[9] H. Oukhaba. — *Fonctions discriminant, formules pour le nombre de classes et unités elliptiques; le cas des corps de fonctions (associés à des courbes sur des corps finis)*, Thèse (Grenoble, Institut Fourier, Juin 1991).

[10] H. Oukhaba. — *Groups of elliptic Units in Global function fields*, in Proceedings of the Workshop at the Ohio State University, June 17-26, 1991.

[11] H. Oukhaba. — *On discriminant functions associated to Drinfeld Modules of rank* 1, Journal of number theory, **47** (1994).

[12] G. Robert. — *Unités elliptiques*, Bulletin Soc. Math. France, Supplément **36**, Décembre 1973.

[13] J. Yu. — *Transcendence and Drinfeld modules*, Invent. Math. **83** (1986), 507-517.

Hassan OUKHABA
Equipe de Mathématiques
URA CNRS 741
16, Route de Gray
France – 25030 Besançon Cedex

Number Theory
Paris 1992–93

Arbres, ordres maximaux et formes quadratiques entières

Isabelle Pays

On sait depuis Lagrange que tout entier naturel est une somme de quatre carrés, et, d'après Jacobi (1828), que le nombre de représentations d'un entier en somme de quatre carrés est

$$r_4(m) = 8 \sum_{\substack{d \mid m \\ 4 \nmid d}} d \quad (m \geq 1),$$

les représentations obtenues en permutant l'ordre ou en changeant le signe des composantes étant comptées séparément. Les preuves connues de cette formule sont de nature analytique (analyse complexe, formes modulaires, fonctions elliptiques, ...). Parmi les nombreuses références, citons E. Landau [8, pp. 146–150] qui détermine le nombre de représentations d'un entier en somme de quatre carrés en utilisant les formules sur le nombre de décompositions d'entiers en somme de deux carrés (qu'il a établies auparavant de manière tout a fait élémentaire); G.H. Hardy et E.M. Wright [7, p. 314], J.V. Uspensky et M.A. Heaslet [15, pp. 450-458], ainsi que E. Grosswald [6, pp. 30-36] donnent des preuves basées sur des identités qui peuvent être dérivées des propriétés des fonctions elliptiques ou simplement vérifiées "à la main"; dans [9, p. 333], W. Scharlau exploite le fait que la somme de quatre carrés est une forme quadratique avec un seul élément dans son genre pour déduire la formule de Jacobi; A. Robert [11] et B. Gordon [4] établissent la formule de Jacobi à partir de résultats sur les formes modulaires (ce qui nécessite un peu d'analyse complexe). Toutes ces preuves utilisent ou bien des identités un peu "mystérieuses", ou alors du matériel assez sophistiqué. Pour des références concernant l'origine et les développements historiques, nous renvoyons le lecteur au recueil de L.E. Dickson [3, Chap. VIII, p. 285]. Signalons aussi un article de G. Rousseau [12], où l'auteur donne un moyen pour construire des représentations d'un entier en somme de quatre carrés à partir de fractions continues.

Nous proposons ici une nouvelle preuve à caractère purement algébrique et géométrique de la formule de Jacobi. Cette preuve est tout à fait élémentaire : les prérequis sont à peine un peu plus qu'un cours de premier cycle en algèbre. La preuve que nous présentons découle de résultats plus généraux sur le nombre de représentations d'une puissance quelconque d'un nombre premier par certaines formes quadratiques à quatre variables, obtenus au moyen d'actions de "groupes de quaternions" sur "l'arbre de $\mathrm{SL}_2(\mathbb{Q}_p)$". L'article se présente comme suit : Au §1, on rappelle la définition d'une algèbre de quaternions. Les ordres maximaux dans une algèbre de quaternions rationnelle permettent de définir les formes quadratiques entières que l'on examine plus loin. Au §2 on décrit la construction de l'arbre à partir des ordres maximaux de $M_2(\mathbb{Q}_p)$. C'est au §3 que l'on explique la relation entre l'action d'un certain groupe sur l'arbre et les représentations d'une puissance d'un nombre premier par les formes quadratiques associées (au §1) aux ordres maximaux. On montre au §4 que, lorsque l'ordre maximal est principal, on peut obtenir le nombre de représentations d'un entier quelconque (et non plus uniquement d'une puissance d'un nombre premier). Cela conduit à une nouvelle preuve de la formule de Jacobi.

1. — Algèbres de quaternions et ordres

Nous renvoyons aux ouvrages [1], [10] et [16] pour les preuves détaillées des résultats mentionnés dans ce paragraphe.

Soit K un corps de caractéristique différente de 2 et soient a et b deux éléments non nuls de K. L'*algèbre de quaternions* $(a,b)_K$ est l'algèbre admettant une base de quatre éléments sur K, notés $1, i, j, k$, avec la multiplication définie par les relations $i^2 = a$, $j^2 = b$, $k = i.j = -j.i$. Le *conjugué* du quaternion $q = x_1 + x_2 i + x_3 j + x_4 k$, noté $\overline{q}$, est défini par $\overline{q} = x_1 - x_2 i - x_3 j - x_4 k$. La *norme réduite* du quaternion q, notée $n(q)$, est définie par $n(q) = q.\overline{q} = x_1^2 - a x_2^2 - b x_3^2 + ab x_4^2$. La *trace réduite* du quaternion q, notée $t(q)$, est définie par $t(q) = q + \overline{q} = 2x_1$.

Il est bien connu qu'une algèbre de quaternions est soit à division, soit isomorphe à l'algèbre de matrices $M_2(K)$.

Les corps considérés ici sont soit le corps (global) $\mathbb{Q}$ des nombres rationnels, soit un des corps (locaux) $\mathbb{Q}_p$ des nombres p-adiques ou $\mathbb{R}$ le corps des nombres réels.

Sur un corps local (ici $\mathbb{R}$ ou $\mathbb{Q}_p$), il y a une unique algèbre de quaternions à division, à isomorphisme près. Sur $\mathbb{R}$, il s'agit de l'algèbre des quaternions de Hamilton, $\mathbb{H} = (-1,-1)_{\mathbb{R}}$.

Soit $H = (a,b)_{\mathbb{Q}}$. Quitte à multiplier a et b par des carrés convenables, on peut supposer que a et b sont dans $\mathbb{Z}$. Pour reconnaître si $H_p = (a,b)_{\mathbb{Q}_p}$ est à division, on utilise le symbole de Hilbert $(a,b)_p$. L'algèbre $(a,b)_{\mathbb{Q}_p}$ est à division si et seulement si $(a,b)_p = -1$. Nous renvoyons le lecteur à [14, p.39] pour le calcul de ce symbole. Notons toutefois que $(a,b)_p = 1$ pour presque tout p (c'est-à-dire pour tout p sauf un nombre fini d'entre eux).

Le *discriminant* de H est le produit des nombres premiers p pour lesquels l'algèbre de quaternions $H \otimes \mathbb{Q}_p$ est à division

$$\mathrm{disc}(H) = \prod_{\substack{p \text{ premier} \\ (a,b)_p = -1}} p.$$

Soit R un anneau principal de caractéristique différente de 2, K son corps de fractions et H une algèbre de quaternions sur K (nous envisageons en particulier le cas où $R = \mathbb{Z}$ ou $\mathbb{Z}[\frac{1}{p}] = \{ap^n | a, n \in \mathbb{Z}\}$ avec $K = \mathbb{Q}$ ou alors $R = \mathbb{Z}_p$, l'anneau des entiers p-adiques, avec $K = \mathbb{Q}_p$). Nous désignons par $R^\times$ le groupe multiplicatif des éléments inversibles de R.

Un *ordre de H sur R* est un sous-R-module de H de rang 4 qui est aussi un anneau. Les éléments d'un ordre ont la propriété d'être entiers sur R, c'est-à-dire que leur trace et leur norme appartiennent à R. Un *ordre maximal* est un ordre qui n'est contenu proprement dans aucun autre ordre.

Voici deux exemples qui nous seront utiles.

Exemple 1. *Dans $H = (-1,-1)_\mathbb{Q}$ le $\mathbb{Z}$-module $\mathcal{O}'$ de base $(1, i, j, k)$ est un ordre de H sur $\mathbb{Z}$. De même, le $\mathbb{Z}$-module $\mathcal{O}$ engendré par 1, i, j, $a = (1+i+j+k)/2$ est un ordre de H. On note que $t(a) = 1$ et $n(a) = 1$. L'ordre $\mathcal{O}'$ n'est pas maximal car il est contenu dans l'ordre $\mathcal{O}$.*

Exemple 2. *Soit R un anneau principal et K son corps de fractions. Alors $M_2(R)$ est un ordre de $M_2(K)$.*

Les formes quadratiques que nous allons examiner sont les formes normes d'ordres sur $\mathbb{Z}$ d'une algèbre de quaternions H sur $\mathbb{Q}$. Soit $\mathcal{O}$ un tel ordre et soit (e_1, e_2, e_3, e_4) une base de $\mathcal{O}$ sur $\mathbb{Z}$. La *forme norme de $\mathcal{O}$ par rapport à la base* e est

$$n(x) = n(\sum_{i=1}^{4} X_i e_i) = \sum_i X_i^2 n(e_i) + \sum_{i<j} X_i X_j t(e_i \overline{e_j})$$

où $x \in \mathcal{O}$. (Nous la notons parfois $q_\mathcal{O}$.) Le fait que $\mathcal{O}$ est un ordre assure que la forme quadratique obtenue est à coefficients entiers. Le choix d'une autre base de $\mathcal{O}$ conduit à une forme $\mathbb{Z}$-équivalente. Par abus de langage nous appelons *forme norme de $\mathcal{O}$* un représentant quelconque de la classe d'équivalence.

On vérifie que deux ordres $\mathcal{O}$ et $\mathcal{O}'$ conjugués par un automorphisme intérieur de H (c'est-à-dire $\mathcal{O}' = h\mathcal{O}h^{-1}$ pour un certain élément inversible h de H) donnent lieu à des formes quadratiques $\mathbb{Z}$-équivalentes.

Dans l'exemple 1 ci-dessus, la forme quadratique associée à l'ordre $\mathcal{O}$, exprimée par rapport à la base $(1, i, j, a)$, est

$$q_{\mathcal{O}}(X_1, X_2, X_3, X_4) = X_1^2 + X_2^2 + X_3^2 + X_4^2 + X_1X_4 + X_2X_4 + X_3X_4,$$

tandis que celle associée à l'ordre $\mathcal{O}'$, exprimée par rapport à la base $(1, i, j, k)$, est la somme de quatre carrés :

$$q_{\mathcal{O}'}(X_1, X_2, X_3, X_4) = X_1^2 + X_2^2 + X_3^2 + X_4^2.$$

L'étude de ces formes quadratiques nécessite une connaissance plus approfondie des ordres maximaux sur $\mathbb{Z}$ d'une algèbre de quaternions rationnelle.

Pour commencer, nous allons expliquer comment on peut voir facilement si un ordre est maximal. Rappelons d'abord que l'on peut munir H naturellement de la forme bilinéaire b_t induite par la trace en posant $b_t(x, y) = t(xy)$. On a aussi besoin des définitions suivantes. Un *R-réseau* d'une algèbre de quaternions H sur K est un sous-R-module de H de rang 4. Le *discriminant* d'un R-réseau M de H, noté $\text{disc}(M)$, est le déterminant de la matrice de l'application bilinéaire b_t dans une base de M. On voit, en examinant la formule de changement de base, que cet élément de $K^\times$ est défini à un carré de $R^\times$ près. De plus, si L est un R-réseau contenu dans M, alors $\text{disc}(L) = r^2\text{disc}(M)$ pour un certain $r \in R$ et $L = M$ si et seulement si $r \in R^\times$. Notons aussi que le discriminant d'un ordre est un élément non nul de $R/R^{\times^2}$.
Le réseau "standard" M de base $(1, i, j, k)$ a pour discriminant $-(4ab)^2R^{\times^2}$. Dès lors, à nouveau par changement de bases, on voit que le discriminant d'un réseau de H est toujours l'opposé d'un carré de $K^\times$ (modulo $R^{\times^2}$). Notons que si $R = \mathbb{Z}$, $R^{\times^2}$ est réduit à l'unité.

COROLLAIRE. — *Soit R un anneau principal et K son corps de fractions. Alors :*
1. *$M_2(R)$ est un ordre maximal de $M_2(K)$,*
2. *$M_2(R)$ est un anneau principal,*
3. *tous les ordres maximaux de $M_2(K)$ sont conjugués à $M_2(R)$.*

Démonstration : 1) Il est clair que $M_2(R)$ est un ordre. Comme $\text{disc}(M_2(R)) = 1 \text{mod} R^{\times^2}$, on déduit du comportement du discriminant par rapport à l'inclusion des réseaux que $M_2(R)$ est maximal.
2) Soit I un idéal à gauche de $M_2(R)$ et soit $((x_1, x_2), (y_1, y_2))$ une base du R-réseau de R^2 engendré par les lignes des matrices de I. On vérifie aisément que la matrice $\mathcal{A}$ ayant pour première ligne (x_1, x_2) et pour seconde ligne (y_1, y_2) est dans I et que $I = M_2(R) \cdot \mathcal{A}$.

Si I est un idéal à droite de $M_2(R)$, on procède de manière semblable en considérant cette fois le réseau engendré par les colonnes des matrices de I.
3) Notons $\mathcal{O} = M_2(R)$ et soit $\mathcal{O}'$ un autre ordre maximal de $M_2(R)$. Alors, comme $\mathcal{O}$ est maximal, $\mathcal{O}'\mathcal{O}$ est un idéal à droite de $\mathcal{O}$ et est donc principal. On peut donc écrire $\mathcal{O}'\mathcal{O} = x\mathcal{O}$ pour un certain x dans $GL_2(K)$. Par ailleurs, comme $\mathcal{O}'$ est maximal, l'ordre de stabilisateurs à gauche de $\mathcal{O}'\mathcal{O}$ est égal à $\mathcal{O}'$ (l'ordre de stabilisateurs à gauche, $\mathcal{O}_g(I)$, d'un idéal I est $\mathcal{O}_g(I) = \{x \in H | xI \subseteq I\}$). Comme $\mathcal{O}_g(\mathcal{O}'\mathcal{O}) = \mathcal{O}_g(x\mathcal{O}) = x\mathcal{O}x^{-1}$, on obtient $\mathcal{O}' = x\mathcal{O}x^{-1}$. ■

Voici le critère qui permet de reconnaître les ordres maximaux :

Critère. *Un ordre $\mathcal{O}$ sur $\mathbb{Z}$ d'une algèbre de quaternions H sur $\mathbb{Q}$ est maximal si et seulement si son discriminant est égal à l'opposé du carré du discriminant de H, c'est-à-dire si disc($\mathcal{O}$) $= -(\mathrm{disc} H)^2$.*

2. — L'arbre

Les arbres qui nous seront utiles pour étudier les nombres de représentations sont les arbres associés aux groupes SL_2 sur les corps locaux, qui sont des cas particuliers des immeubles de Bruhat–Tits. Nous les réalisons ici à l'aide des ordres maximaux dans une algèbre de quaternions déployée sur un corps local $\mathbb{Q}_p$ (c'est-à-dire une algèbre isomorphe à $M_2(\mathbb{Q}_p)$).

Soit H une telle algèbre et $\mathcal{O}$ un ordre maximal de H sur l'anneau $\mathbb{Z}_p$ des entiers p-adiques. D'après le corollaire ci–dessus, on sait que tous les ordres maximaux de H sont conjugués à $\mathcal{O}$, ce qui va permettre de définir une distance entre ces ordres maximaux. On introduit pour cela la valuation p–adique $v : \mathbb{Q}_p \to \mathbb{Z}_p \cup \infty$ normalisée par la condition $v(p) = 1$, et la fonction $\mu : H \to \mathbb{Z} \cup \infty$ définie par :

$$\mu(x) = \max\{n \in \mathbb{Z} \mid x \in p^n\mathcal{O}\} \qquad \text{pour } x \neq 0$$

et

$$\mu(0) = \infty.$$

Cette fonction satisfait les propriétés suivantes :

PROPRIÉTÉS. — *Pour $x, y \in H$ et $\alpha \in \mathbb{Q}_p$, on a :*
1. $\mu(x) = \infty$ *si et seulement si* $x = 0$.
2. $\mu(x + y) \geq \min\{\mu(x), \mu(y)\}$.
3. $\mu(xy) \geq \mu(x) + \mu(y)$.
4. $\mu(x\alpha) = \mu(x) + v(\alpha)$.
5. $\mu(\overline{x}) = \mu(x)$.

De plus, en désignant respectivement par $\mathcal{O}^\times$ et par $H^\times$ les groupes multiplicatifs des éléments inversibles de $\mathcal{O}$ et de H,

a) *Pour $x \in H^\times$, on a $x \in \mathcal{O}^\times$ si et seulement si $\mu(x) = \mu(x^{-1}) = 0$.*

b) *Pour $x \in H^\times$, les conditions suivantes sont équivalentes :*

i) $x \in p^\alpha \mathcal{O}^\times$ *pour un certain* $\alpha \in \mathbb{Z}$.

ii) $\mu(x) + \mu(x^{-1}) = 0$.

iii) $x\mathcal{O}x^{-1} = \mathcal{O}$.

c) *Pour $x, y \in H$, si x satisfait les conditions équivalentes de la propriété précédente, on a : $\mu(xy) = \mu(x) + \mu(y) = \mu(yx)$ et $\mu(xyx^{-1}) = \mu(y)$.*

d) *Pour $x \in H^\times$, $\mu(x^{-1}) = \mu(x) - v(n(x))$ (où n désigne la norme de H).*

Démonstration : les propriétés 1 à 4 sont toutes évidentes. La propriété 5 découle immédiatement du fait que tout ordre d'une algèbre de quaternions est stable par la conjugaison quaternionienne (car $\overline{x} = t(x).1 - x$). Si $x \in \mathcal{O}^\times$, alors $\mu(x) = 0$ car les éléments de $p\mathcal{O}$ ne sont pas inversibles dans $\mathcal{O}$; on a alors de même $\mu(x^{-1}) = 0$. Réciproquement, si $\mu(x) = \mu(x^{-1}) = 0$, alors x et x^{-1} sont tous deux dans $\mathcal{O}$, donc $x \in \mathcal{O}^\times$. Cela prouve la propriété **(a)**. Si $x \in p^\alpha\mathcal{O}^\times$ pour un certain $\alpha \in \mathbb{Z}$, alors $\mu(x) = \alpha$ et $\mu(x^{-1}) = -\alpha$, donc $\mu(x) + \mu(x^{-1}) = 0$. Inversement, si $x \in H^\times$ est tel que $\mu(x) + \mu(x^{-1}) = 0$, soit $x = p^\alpha y$ pour $\alpha = \mu(x)$ et pour un certain $y \in \mathcal{O} \setminus p\mathcal{O}$. On a alors $\mu(y) = 0$ et $\mu(x^{-1}) = -\alpha + \mu(y^{-1})$. La relation $\mu(x) + \mu(x^{-1}) = 0$ entraîne alors : $\mu(y^{-1}) = 0$, donc $y \in \mathcal{O}^\times$ par la propriété **(a)**. Cela démontre l'équivalence des conditions (i) et (ii) de **(b)**. Par ailleurs, la condition (i) entraîne évidemment (iii). Réciproquement, si x satisfait la condition (iii), on écrit encore $x = p^\alpha y$ pour $\alpha = \mu(x)$ et pour un certain $y \in \mathcal{O} \setminus p\mathcal{O}$; comme $\mathcal{O} \simeq M_2(\mathbb{Z}_p)$, il n'est pas difficile de vérifier qu'alors $\mathcal{O}y\mathcal{O} = \mathcal{O}$. Or, de la relation $x\mathcal{O}x^{-1} = \mathcal{O}$, on déduit que $y\mathcal{O} = \mathcal{O}y$; on a donc

$$y\mathcal{O} = \mathcal{O}y = \mathcal{O}y\mathcal{O} = \mathcal{O},$$

ce qui montre que $y \in \mathcal{O}^\times$ et $x \in p^\alpha\mathcal{O}^\times$ et achève la démonstration de la propriété **(b)**. Pour établir la propriété **(c)**, on observe que, d'après la propriété 3,

$$\mu(xy) \geq \mu(x) + \mu(y)$$
$$\mu(y) = \mu(x^{-1}xy) \geq \mu(x^{-1}) + \mu(xy).$$

Lorsque $\mu(x^{-1}) = -\mu(x)$, on en déduit immédiatement que $\mu(xy) = \mu(x) + \mu(y)$. La relation $\mu(yx) = \mu(x) + \mu(y)$ se démontre de manière analogue, et la relation $\mu(xyx^{-1}) = \mu(y)$ se déduit des deux précédentes. Enfin, la propriété **(d)** résulte des propriétés 4 et 5, car $x^{-1} = \overline{x}.n(x)^{-1}$. ∎

Soient maintenant $\mathcal{O}_1$ et $\mathcal{O}_2$ des ordres maximaux de H, et soient x_1 et $x_2 \in H^\times$ tels que :

$$\mathcal{O}_1 = x_1\mathcal{O}x_1^{-1} \qquad \text{et} \qquad \mathcal{O}_2 = x_2\mathcal{O}x_2^{-1}.$$

On pose

$$d(\mathcal{O}_1, \mathcal{O}_2) = -\mu(x_1^{-1}x_2) - \mu(x_2^{-1}x_1) \in \mathbb{Z}.$$

Pour voir que la fonction d est bien définie, il faut vérifier que le second membre ne dépend pas du choix de x_1 et x_2. Si $x'_i \in H^\times$ est tel que $\mathcal{O}_i = x'_i\mathcal{O}x'^{-1}_i$ pour $i = 1, 2$, alors $x'^{-1}_i x_i \mathcal{O} x_i^{-1} x'_i = \mathcal{O}$ pour $i = 1, 2$, donc, par la propriété **(c)** ci-dessus, on a

$$\mu(x'^{-1}_1 x'_2) = \mu(x'^{-1}_1 x_1.x_1^{-1}x_2.x_2^{-1}x'_2) = \mu(x'^{-1}_1 x_1) + \mu(x_1^{-1}x_2) + \mu(x_2^{-1}x'_2)$$

et, de même, $\mu(x'^{-1}_2 x'_1) = \mu(x'^{-1}_2 x_2) + \mu(x_2^{-1}x_1) + \mu(x_1^{-1}x'_1)$. Dès lors, $\mu(x'^{-1}_1 x'_2) + \mu(x'^{-1}_2 x'_1) = \mu(x_1^{-1}x_2) + \mu(x_2^{-1}x_1)$, ce qui prouve que d est bien définie. On a en fait $d(\mathcal{O}_1, \mathcal{O}_2) \geq 0$, car, d'après la propriété 3,

$$\mu(x_1^{-1}x_2) + \mu(x_2^{-1}x_1) \leq \mu(x_1^{-1}x_2.x_2^{-1}x_1) = 0.$$

Proposition 2.1. — *La fonction d est une distance sur l'ensemble des ordres maximaux de H. Cette distance est invariante par conjugaison, c'est-à-dire que pour $x \in H^\times$ et pour $\mathcal{O}_1$, $\mathcal{O}_2$ des ordres maximaux de H,*

$$d(x\mathcal{O}_1x^{-1}, x\mathcal{O}_2x^{-1}) = d(\mathcal{O}_1, \mathcal{O}_2).$$

Démonstration : il est clair par définition que la fonction d est symétrique. De la propriété **(b)**, on déduit que $d(\mathcal{O}_1, \mathcal{O}_2) = 0$ si et seulement si $\mathcal{O}_1 = \mathcal{O}_2$. L'inégalité triangulaire découle de la propriété 3 et l'invariance par conjugaison est évidente puisque $(xx_1)^{-1}(xx_2) = x_1^{-1}x_2$. ■

On obtient alors l'arbre des ordres maximaux de H.

Théorème 2.2. — *Le graphe X dont les sommets sont les ordres maximaux de H et dont les arêtes sont les couples $(\mathcal{O}_1, \mathcal{O}_2)$ d'ordres maximaux tels que $d(\mathcal{O}_1, \mathcal{O}_2) = 1$ est un arbre, c'est-à-dire un graphe connexe et sans circuit. De plus, cet arbre est $(p+1)$-régulier, c'est-à-dire que chaque sommet est l'origine de $p+1$ arêtes.*

Démonstration : montrons d'abord que le graphe X est connexe. Il suffit de montrer que tout ordre maximal $\mathcal{O}'$ est lié à $\mathcal{O}$ par un chemin du graphe. On raisonne par induction sur la distance de $\mathcal{O}$ à $\mathcal{O}'$. L'énoncé est évident si cette distance est 1, puisqu'alors $\mathcal{O}$ et $\mathcal{O}'$ sont liés par une arête. Il suffit donc de prouver que si la distance de $\mathcal{O}$ à $\mathcal{O}'$ est $n > 1$, alors il existe un ordre $\mathcal{O}''$ à distance $n-1$ de $\mathcal{O}$ et à distance 1 de $\mathcal{O}'$.

Soit $\mathcal{O}' = x\mathcal{O}x^{-1}$. Quitte à multiplier x par une puissance convenable de p, on peut supposer $\mu(x) = 0$, c'est-à-dire que $x \in \mathcal{O} \smallsetminus p\mathcal{O}$. Comme $\mathcal{O}/p\mathcal{O} \simeq M_2(\mathbb{F}_p)$, la trace induit une forme bilinéaire non dégénérée sur

$\mathcal{O}/p\mathcal{O}$; on peut donc trouver $u \in \mathcal{O} \smallsetminus p\mathcal{O}$ tel que $t(xu) \notin p\mathbb{Z}_p$, ce qui entraîne bien sûr que $xu \in \mathcal{O} \smallsetminus p\mathcal{O}$. Soit alors $y = xu + p^{-1}n(x)$. Comme $d(\mathcal{O}, \mathcal{O}') = n > 1$, on a, par la propriété **(d)**,

$$-\mu(x^{-1}) = v(n(x)) = n > 1,$$

d'où $y \in \mathcal{O} \smallsetminus p\mathcal{O}$, c'est-à-dire, $\mu(y) = 0$. Par ailleurs,

$$x^{-1}y = u + p^{-1}\overline{x},$$

donc $\mu(x^{-1}y) = -1$, et de la relation

$$n(x^{-1}y) = n(u) + p^{-1}t(xu) + p^{-2}n(x)$$

on tire : $v(n(x^{-1}y)) = -1$. D'après la propriété **(d)**, on en déduit

$$-\mu(x^{-1}y) - \mu(y^{-1}x) = 1.$$

Par ailleurs, comme $v(n(x)) = n$, on doit avoir $v(n(y)) = n - 1$, donc

$$-\mu(y) - \mu(y^{-1}) = n - 1.$$

Dès lors, l'ordre $\mathcal{O}'' = y\mathcal{O}y^{-1}$ possède les propriétés requises.

Montrons ensuite que le graphe X ne contient pas de circuit. Soit $\mathcal{O}_1, \ldots, \mathcal{O}_n$ un chemin sans aller–retour, c'est-à-dire,

$$\text{(1)} \qquad \begin{cases} d(\mathcal{O}_i, \mathcal{O}_{i+1}) = 1 \text{ pour } i = 1, \ldots, n-1 \\ d(\mathcal{O}_i, \mathcal{O}_{i+2}) > 0 \text{ pour } i = 1, \ldots, n-2\,. \end{cases}$$

Pour prouver que ce chemin n'est pas un circuit, il suffit de montrer que $d(\mathcal{O}_1, \mathcal{O}_n) = n - 1$.

Ecrivons $\mathcal{O}_i = x_i\mathcal{O}x_i^{-1}$ pour $i = 1, \ldots, n$ et, pour $i = 1, \ldots, n-1$: $x_i^{-1}x_{i+1} = p^{\alpha_i}y_i$ pour un certain $y_i \in \mathcal{O} \smallsetminus p\mathcal{O}$ et $\alpha_i = \mu(x_i^{-1}x_{i+1}) \in \mathbb{Z}$. D'après la propriété **(d)**, on a, pour $i = 1, \ldots, n-1$,

$$\mu(x_{i+1}^{-1}x_i) = \alpha_i - v(n(p^{\alpha_i}y_i)) = -\alpha_i - v(n(y_i)).$$

Dès lors, $d(\mathcal{O}_i, \mathcal{O}_{i+1}) = -\mu(x_i^{-1}x_{i+1}) - \mu(x_{i+1}^{-1}x_i) = v(n(y_i))$, et les conditions (1) ci-dessus s'écrivent :

$$\begin{cases} v(n(y_i)) = 1 \text{ pour } i = 1, \ldots, n-1 \\ v(n(y_iy_{i+1})) - 2\mu(y_iy_{i+1}) > 0 \text{ pour } i = 1, \ldots, n-2. \end{cases}$$

Vu la multiplicativité de la norme, la première condition entraîne : $v(n(y_i y_{i+1})) = 2$ pour $i = 1, \ldots, n-2$; par ailleurs, comme $y_i \in \mathcal{O} \setminus p\mathcal{O}$ pour $i = 1, \ldots, n-1$, on a $y_i y_{i+1} \in \mathcal{O}$ pour $i = 1, \ldots, n-2$, donc $\mu(y_i y_{i+1}) \geq 0$. Ces observations conduisent à réécrire les conditions ci-dessus sous la forme :

$$(2) \qquad \begin{cases} v(n(y_i)) = 1 \text{ pour } i = 1, \ldots, n-1 \\ \mu(y_i y_{i+1}) = 0 \text{ pour } i = 1, \ldots, n-2. \end{cases}$$

Montrons alors, par induction sur m, que $\mu(y_1 \ldots y_m) = 0$. C'est clair pour $m = 2$. Supposons donc $\mu(y_1 \ldots y_{m-1}) = 0$ et $\mu(y_1 \ldots y_m) > 0$, c'est-à-dire,

$$y_1 \ldots y_m \in p\mathcal{O}.$$

On a alors

$$(3) \qquad (y_1 \ldots y_{m-1} + p\mathcal{O}).(y_m + p\mathcal{O}) = 0 \qquad \text{dans } \mathcal{O}/p\mathcal{O}.$$

Comme $\mathcal{O}/p\mathcal{O}$ est isomorphe à une algèbre de matrices carrées d'ordre 2 sur $\mathbb{F}_p$, on peut considérer $y_1 + p\mathcal{O}, \ldots, y_m + p\mathcal{O}$ comme des opérateurs linéaires sur un espace vectoriel de dimension 2 sur $\mathbb{F}_p$. Ces opérateurs sont non nuls puisque $y_i \notin p\mathcal{O}$, et non inversibles puisque $v(n(y_i)) = 1$; ils sont donc tous de rang 1. De même, $y_1 \ldots y_{m-1} + p\mathcal{O}$ est de rang 1 puisque $\mu(y_1 \ldots y_{m-1}) = 0$. L'équation (3) indique alors que :

$$\operatorname{Im}(y_m + p\mathcal{O}) = \operatorname{Ker}(y_1 \ldots y_{m-1} + p\mathcal{O}).$$

Par ailleurs, on a aussi

$$\operatorname{Ker}(y_1 \ldots y_{m-1} + p\mathcal{O}) = \operatorname{Ker}(y_{m-1} + p\mathcal{O}),$$

donc

$$(y_{m-1} + p\mathcal{O})(y_m + p\mathcal{O}) = 0$$

et par conséquent $\mu(y_{m-1} y_m) > 0$, contrairement à l'hypothèse. On a donc bien $\mu(y_1 \ldots y_m) = 0$ pour tout $m = 1, \ldots, n-1$. Un calcul direct donne alors

$$d(\mathcal{O}_1, \mathcal{O}_n) = v(n(y_1 \ldots y_{n-1})) - 2\mu(y_1 \ldots y_{n-1}) = n - 1,$$

ce qui achève de démontrer que le graphe X ne contient pas de circuit.

Pour prouver que X est $(p+1)$-régulier, comme la conjugaison par tout élément de $H^\times$ induit un automorphisme de X et que tout ordre maximal est conjugué à $\mathcal{O}$, il suffit de montrer qu'il y a $p + 1$ ordres à distance 1 de $\mathcal{O}$. Or, il y a une correspondance bijective entre l'ensemble des ordres à distance 1 de $\mathcal{O}$ et l'ensemble des idéaux à droite I de $\mathcal{O}$ tels que $\mathcal{O} \supsetneq I \supsetneq p\mathcal{O}$, qui associe à tout idéal I son ordre de stabilisateurs à gauche :

$$\mathcal{O}_g(I) = \{x \in H \mid xI \subseteq I\}$$

et à tout ordre $\mathcal{O}'$ à distance 1 de $\mathcal{O}$ l'idéal $p\mathcal{O}'\mathcal{O}$ (pour établir la bijectivité de cette correspondance, il est utile de remarquer que si $x \in \mathcal{O} \setminus p\mathcal{O}$ et $v(n(x)) = 1$, alors $\mathcal{O}\overline{x}\mathcal{O}$ est un idéal bilatère de $\mathcal{O}$ contenant proprement $p\mathcal{O}$, donc $\mathcal{O}\overline{x}\mathcal{O} = \mathcal{O}$ puisque $\mathcal{O}/p\mathcal{O} \simeq M_2(\mathbb{F}_p)$ est simple. Si à présent $\mathcal{O}'$ est un ordre à distance 1 de $\mathcal{O}$, on peut écrire :

$$\mathcal{O}' = x\mathcal{O}x^{-1}$$

pour un certain x comme ci-dessus, d'où

$$p\mathcal{O}'\mathcal{O} = x\mathcal{O}\overline{x}\mathcal{O} = x\mathcal{O}.$$

Réciproquement, $\mathcal{O}_g(x\mathcal{O}) = x\mathcal{O}x^{-1}$).

Comme par ailleurs les idéaux à droite I tels que $\mathcal{O} \supsetneq I \supsetneq p\mathcal{O}$ sont en bijection avec les idéaux à droite non triviaux de $\mathcal{O}/p\mathcal{O} \simeq M_2(\mathbb{F}_p)$, qui sont au nombre de $p+1$, il y a bien $p+1$ ordres à distance 1 de $\mathcal{O}$. ■

3. — Actions de sous-groupes et représentations d'entiers

Soit H une algèbre de quaternions sur $\mathbb{Q}$ et $\mathcal{O}$ un ordre maximal de H sur $\mathbb{Z}$. La forme quadratique quaternaire à coefficients entiers que nous allons étudier est la forme norme sur $\mathcal{O}$, c'est-à-dire (voir aussi §1) que si $e = (e_1, e_2, e_3, e_4)$ désigne une base de $\mathcal{O}$, la forme s'écrit

$$n(\sum_{i=1}^{4} X_i e_i) = \sum_i X_i^2 n(e_i) + \sum_{i<j} X_i X_j t(e_i \overline{e_j}).$$

Etant donné un nombre premier p, on se propose dans cette section d'étudier les représentations des puissances de p par cette forme, c'est-à-dire les solutions $(x_1, x_2, x_3, x_4) \in \mathbb{Z}^4$ de

$$n(\sum_{i=1}^{4} x_i e_i) = \pm p^n$$

ou, ce qui revient au même, les éléments x de $\mathcal{O}$ tels que $n(x) = \pm p^n$. L'ensemble de ces éléments est noté $R(p^n)$:

$$R(p^n) = \{x \in \mathcal{O} \mid n(x) = \pm p^n\}.$$

On note aussi $Rp(p^n)$ l'ensemble des solutions *primitives*, c'est-à-dire,

$$Rp(p^n) = \{x \in \mathcal{O} \setminus p\mathcal{O} \mid n(x) = \pm p^n\}.$$

Les résultats sont très différents suivant que p divise le discriminant de H ou non. Lorsque p ne divise pas le discriminant, l'algèbre $H_p = H \otimes \mathbb{Q}_p$ est isomorphe à $M_2(\mathbb{Q}_p)$, et l'ordre $\mathcal{O}_p = \mathcal{O} \otimes \mathbb{Z}_p$ en est un ordre maximal. Le groupe des inversibles de $\mathcal{O}[\frac{1}{p}]$, que l'on note $\mathcal{O}[\frac{1}{p}]^\times$, s'identifie à un sous-groupe de $H_p^\times$.

Le théorème suivant met en relation l'action de ce sous-groupe sur l'arbre X_p des ordres maximaux de H_p (par conjugaison) et l'ensemble des représentations primitives de $\pm p^n$ par la forme norme de $\mathcal{O}$. Soit Δ_n l'ensemble des sommets de l'arbre X_p à distance n de $\mathcal{O}_p$ qui sont dans la même orbite que $\mathcal{O}_p$ par l'action de $\mathcal{O}[\frac{1}{p}]^\times$. Soit $\sim$ la relation d'équivalence sur $\mathcal{O} \setminus \{0\}$ définie par $x \sim y$ si x et y sont associés (à droite) dans $\mathcal{O}$ c'est-à-dire si $x^{-1}y$ est dans $\mathcal{O}^\times$. Il est clair que si $x \in Rp(p^n)$, alors tout élément y tel que $y \sim x$ est aussi contenu dans $Rp(p^n)$. On peut donc considérer l'ensemble quotient $Rp(p^n)/\sim$. On a le :

THÉORÈME 3.1. — *L'application qui à un élément x de $\mathcal{O}$ fait correspondre l'ordre $x\mathcal{O}_p x^{-1}$ de H_p définit une bijection entre les ensembles $Rp(p^n)/\sim$ et Δ_n.*

Démonstration : si $x \in Rp(p^n)$, alors $\mu(x) = 0$ et $v(n(x)) = n$, donc, par la propriété **(d)** de la section précédente, $\mu(x^{-1}) = -n$. Dès lors, $d(\mathcal{O}_p, x\mathcal{O}_p x^{-1}) = n$.

Montrons que l'application définie dans l'énoncé est surjective : soit $\mathcal{O}'$ un élément de Δ_n ; il existe alors $\gamma \in \mathcal{O}[\frac{1}{p}]^\times$ tel que $\mathcal{O}' = \gamma\mathcal{O}_p\gamma^{-1}$ et $d(\mathcal{O}_p, \gamma\mathcal{O}_p\gamma^{-1}) = n$. Comme les scalaires agissent trivialement, on peut supposer, quitte à multiplier γ par une puissance adéquate de p, que $\mu(\gamma) = 0$, c'est-à-dire que $\gamma \in \mathcal{O} \setminus p\mathcal{O}$. La condition $d(\mathcal{O}_p, \gamma\mathcal{O}_p\gamma^{-1}) = n$ se traduit alors par $v(n(\gamma)) = n$. Par ailleurs, comme $\gamma \in \mathcal{O}[\frac{1}{p}]^\times$, on doit avoir

$$n(\gamma) \in \mathbb{Z}[\tfrac{1}{p}]^\times = \{\pm p^k \mid k \in \mathbb{Z}\}.$$

Les conditions précédentes entraînent : $n(\gamma) = \pm p^n$, donc l'ordre $\mathcal{O}'$ est l'image de γ par l'application décrite dans l'énoncé.

Montrons pour terminer que l'application est aussi injective. Deux éléments x, y de $Rp(p^n)$ définissent le même sommet si et seulement si

$$x\mathcal{O}_p x^{-1} = y\mathcal{O}_p y^{-1}.$$

D'après la propriété (b) de la section précédente, cette condition est équivalente à

$$x^{-1}y \in p^{\alpha}\mathcal{O}_p^{\times}$$

pour un certain $\alpha \in \mathbb{Z}$. Comme $n(x^{-1}y) = \pm 1$, on en déduit que $x^{-1}y \in H^{\times} \cap \mathcal{O}_p^{\times}$. Par ailleurs, $x^{-1}y \in \mathcal{O}[\frac{1}{p}]$ puisque $x^{-1} = \pm\overline{x}/p^n$ et que $\overline{x} \in \mathcal{O}$. Donc,

$$x^{-1}y \in \mathcal{O}[\tfrac{1}{p}]^{\times} \cap (H^{\times} \cap \mathcal{O}_p^{\times}) = \mathcal{O}^{\times}.$$

■

Voici un cas où le nombre d'éléments de Δ_n est particulièrement facile à calculer.

THÉORÈME 3.2. — *Si l'ordre maximal $\mathcal{O}$ est principal, alors $\mathcal{O}[\frac{1}{p}]^{\times}$ agit transitivement sur les sommets de l'arbre X_p, et par conséquent*

$$|\Delta_n| = p^{n-1}(p+1)$$

pour tout $n \geq 1$.

Démonstration : soit $\mathcal{O}'$ un ordre maximal de H_p. On sait que tous les ordres maximaux de H_p sont conjugués, donc $\mathcal{O}' = x\mathcal{O}_p x^{-1}$ pour un certain $x \in H_p^{\times}$. Quitte à multiplier x par une puissance convenable de p, on peut choisir $x \in \mathcal{O}_p$. Considérons alors l'idéal à droite I de $\mathcal{O}$ défini par

$$I = \left[\bigcap_{q \neq p} (\mathcal{O}_q \cap H)\right] \bigcap (x\mathcal{O}_p \cap H),$$

c'est-à-dire l'idéal dont les localisés sont $I_q = \mathcal{O}_q$ pour $q \neq p$ et $I_p = x\mathcal{O}_p$. D'après l'hypothèse, cet idéal est principal, donc $I = y\mathcal{O}$ pour un certain $y \in \mathcal{O}$. Comme $I_q = \mathcal{O}_q$ pour $q \neq p$, on a $y \in \mathcal{O}_q^{\times}$ pour $q \neq p$, donc $y \in \mathcal{O}[\frac{1}{p}]^{\times}$. Par ailleurs, de la relation

$$I_p = x\mathcal{O}_p = y\mathcal{O}_p\,,$$

on déduit que l'ordre des stabilisateurs à gauche de I_p est :

$$\mathcal{O}_g(I_p) = x\mathcal{O}_p x^{-1} = y\mathcal{O}_p y^{-1},$$

c'est-à-dire que $\mathcal{O}' = y\mathcal{O}_p y^{-1}$. Cela prouve que $\mathcal{O}[\frac{1}{p}]^{\times}$ agit transitivement sur les sommets de l'arbre X_p. Il en résulte en particulier que Δ_n est l'ensemble de tous les sommets à distance n de $\mathcal{O}_p$. Cet ensemble contient

$p^{n-1}(p+1)$ éléments, puisque l'arbre X_p est $(p+1)$-régulier, d'après le théorème 2.2. ∎

Remarque : plus généralement, Vignéras [16, p.147, Prop.3.3] a montré que le nombre d'orbites de sommets de X_p sous l'action de $\mathcal{O}[\frac{1}{p}]$ est égal au nombre de classes de $\mathcal{O}$. La démonstration n'est pas aussi élémentaire que celle du théorème précédent, car elle utilise un théorème puissant d'Eichler.

On a choisi au début de ce paragraphe un nombre premier p qui ne divisait pas le discriminant de H. Voici maintenant ce qui se produit lorsqu'au contraire p divise le discriminant de H, c'est-à-dire lorsque $H_p = H \otimes \mathbb{Q}_p$ est une algèbre à division. Comme précédemment, on dit que deux éléments $x, y \in \mathcal{O}$ sont associés (à droite) s'il existe un élément inversible $u \in \mathcal{O}^\times$ tel que $x = yu$; on note alors $x \sim y$.

THÉORÈME 3.3. — *Lorsque p divise le discriminant de H, alors les éléments de $R(p^n)$ sont tous associés, pour tout $n \geq 0$, de sorte que le quotient $R(p^n)/\sim$ contient un seul élément, si $R(p^n)$ n'est pas vide. Si l'ordre $\mathcal{O}$ est principal, alors $R(p^n)$ est non vide, pour tout $n \geq 1$.*

Démonstration : supposons que $R(p^n)$ est non vide. Soient $x, y \in R(p^n)$; alors x et y sont inversibles dans $\mathcal{O}_q = \mathcal{O} \otimes \mathbb{Z}_q$ pour $q \neq p$, et donc,

$$x\mathcal{O}_q = y\mathcal{O}_q = \mathcal{O}_q.$$

En p, l'algèbre de quaternions $H_p = H \otimes \mathbb{Q}_p$ est isomorphe à l'unique algèbre de quaternions à division sur $\mathbb{Q}_p$, et $\mathcal{O}_p$ est l'anneau de valuation de H_p ([10, §12]). De plus, tout idéal à droite de $\mathcal{O}_p$ est bilatère et principal et est donc de la forme $\pi_p^k \mathcal{O}_p$, où π_p est une uniformisante de $\mathcal{O}_p$. Comme la valuation p-adique de $n(\pi_p)$ est 1, on a en particulier

$$x\mathcal{O}_p = \pi^n \mathcal{O}_p = y\mathcal{O}_p.$$

Ainsi, $x\mathcal{O}_p = y\mathcal{O}_p$ pour tout p, donc $x\mathcal{O} = y\mathcal{O}$ et $x \sim y$. Cela prouve que tous les éléments de $R(p^n)$ sont associés.

Si $\mathcal{O}$ est principal, alors l'idéal I dont les localisés sont $\mathcal{O}_q$ pour $q \neq p$ et $\pi_p\mathcal{O}_p$ en p est principal; soit $I = \pi\mathcal{O}$ pour un certain $\pi \in \mathcal{O}$. On a alors

$$\mathcal{O} \supsetneq \pi\mathcal{O} \supsetneq p\mathcal{O},$$

donc $p = \pi\pi'$ pour un certain $\pi' \in \mathcal{O}$, et

$$n(\pi)n(\pi') = p^2.$$

Si $n(\pi) = \pm 1$, alors $\pi\mathcal{O} = \mathcal{O}$; si $n(\pi) = \pm p^2$, alors $\pi\mathcal{O} = p\mathcal{O}$. Comme ces deux égalités sont exclues, on doit avoir

$$n(\pi) = \pm p,$$

donc $R(p)$ est non vide. De plus, pour tout $n \geq 1$, on a $n(\pi^n) = \pm p^n$, donc $R(p^n)$ est non vide pour tout $n \geq 1$. ■

Dans le cas particulier où la forme norme est définie positive, ce qui revient à dire que l'algèbre de quaternions H est telle que $H \otimes \mathbb{R}$ est isomorphe à l'algèbre $(-1,-1)_\mathbb{R}$ des quaternions d'Hamilton, il est clair que les équations $n(x) = -p^n$ n'ont pas de solution et que les équations $n(x) = p^n$ n'en ont qu'un nombre fini. Les résultats précédents permettent de dénombrer ces solutions. L'ordre $\mathcal{O}$ de l'algèbre de quaternions rationnelle H étant fixé, notons $r(p^n)$ (resp. $rp(p^n)$) le nombre de solutions (resp. de solutions primitives) $x \in \mathcal{O}$ de l'équation $n(x) = p^n$, c'est à dire le nombre d'éléments de $R(p^n)$ (resp. $Rp(p^n)$).

COROLLAIRE 3.4. — *Supposons que la forme norme soit définie positive. Si p est un nombre premier qui ne divise pas le discriminant de l'algèbre H, alors pour tout $n \geq 1$,*

$$rp(p^n) = |\mathcal{O}^\times| \cdot |\Delta_n|$$

où Δ_n est l'ensemble des sommets de l'arbre X_p qui sont dans la même orbite que $\mathcal{O}_p$ sous l'action de $\mathcal{O}[\frac{1}{p}]^\times$ et à distance n de $\mathcal{O}_p$, et

$$r(p^n) = |\mathcal{O}^\times| \cdot \left(\sum_{k=0}^{[n/2]} |\Delta_{n-2k}| \right),$$

où $[n/2]$ est le plus grand entier inférieur ou égal à $n/2$.

Si p est un nombre premier qui divise le discriminant de H, alors pour tout $n \geq 1$,

$$r(p^n) = 0 \text{ ou } |\mathcal{O}^\times|.$$

Si de plus l'ordre $\mathcal{O}$ est principal, alors pour tout nombre premier p qui ne divise pas le discriminant de H,

$$rp(p^n) = |\mathcal{O}^\times| \cdot p^{n-1}(p+1) \quad \text{pour tout } n \geq 1$$

et

$$r(p^n) = |\mathcal{O}^\times| \cdot \frac{p^{n+1}-1}{p-1} \qquad \text{pour tout } n \geq 1,$$

et si p divise le discriminant de H,

$$r(p^n) = |\mathcal{O}^\times| \qquad \text{pour tout } n \geq 1.$$

Démonstration : si p ne divise pas le discriminant de H, les formules pour $rp(p^n)$ résultent directement des théorèmes 3.1 et 3.2 ci-dessus, puisque

$$rp(p^n) = |\mathcal{O}^\times| \cdot |Rp(p^n)/\sim|.$$

Les formules pour $r(p^n)$ s'en déduisent, car les solutions non primitives de $n(x) = p^n$ sont de la forme $x = p^k x_k$ où x_k est solution primitive de $n(x) = p^{n-2k}$. De même, si p divise le discriminant de H, les formules pour $r(p^n)$ découlent du théorème 3.3. ■

Pour compléter l'information donnée dans ce corollaire, remarquons que la structure du groupe $\mathcal{O}^\times$ est connue pour les ordres maximaux des algèbres de quaternions définies positives :

PROPOSITION 3.5. — *Si $\mathcal{O}$ est un ordre maximal d'une algèbre de quaternions H sur $\mathbb{Q}$ telle que $H \otimes \mathbb{R}$ est isomorphe à $(-1,-1)_\mathbb{R}$, alors $\mathcal{O}^\times/\{\pm 1\}$ est cyclique d'ordre 1, 2 ou 3 sauf dans deux cas :*

si $H = (-1,-1)_\mathbb{Q}$, alors $\mathcal{O}^\times/\{\pm 1\}$ est isomorphe au groupe alterné A_4 ;
si $H = (-1,-3)_\mathbb{Q}$, alors $\mathcal{O}^\times/\{\pm 1\}$ est isomorphe au groupe symétrique S_3.

Démonstration : voir [17, th. 5, p. 269]. ■

Exemple : revenons à l'exemple 1 avec l'ordre $\mathcal{O}$ de base $(1, i, j, a)$ dans l'algèbre $H = (-1,-1)_\mathbb{Q}$. Le discriminant de $\mathcal{O}$ vaut -4 et le discriminant de H est égal à 2. Dès lors, $\mathcal{O}$ est maximal. Pour voir que c'est un anneau principal, on montre que les éléments de $\mathcal{O}$ satisfont un algorithme de division euclidienne ([13, p. 98, lemme 3]). L'ordre du groupe $\mathcal{O}^\times$ des éléments inversibles de $\mathcal{O}$ est 24, d'après la proposition 3.5. Dès lors, pour tout $p \neq 2$, le nombre de représentations primitives de p^n par la forme :

$$q_\mathcal{O}(X_1, X_2, X_3, X_4) = X_1^2 + X_2^2 + X_3^2 + X_4^2 + X_1X_4 + X_2X_4 + X_3X_4.$$

est égal à

$$rp(p^n) = 24(p+1)p^{n-1} \qquad (p \neq 2),$$

et le nombre de représentations de 2 est donné par :

$$r(2^n) = 24 \qquad n \geq 0.$$

4. — Ordres principaux

Dans cette section, $\mathcal{O}$ désigne un ordre maximal principal dans une algèbre de quaternions rationnelle H dont la forme norme est définie positive. On se propose de montrer que, grâce au fait que $\mathcal{O}$ est un anneau principal, il est possible de donner le nombre de représentations d'un entier positif quelconque (et non pas seulement des puissances d'un nombre premier) par la forme norme de $\mathcal{O}$.

Lemme 4.1. — *Soient a et b des entiers positifs premiers entre eux. Si $x \in \mathcal{O}$ est tel que $n(x) = ab$, alors il existe $y, z \in \mathcal{O}$ tels que $x = yz$ et $n(y) = a, n(z) = b$.*
Si y et z sont des éléments de $\mathcal{O}$ tels que $n(y) = a$ et $n(z) = b$, alors $yz\mathcal{O} + a\mathcal{O} = y\mathcal{O}$ et $\mathcal{O}yz + \mathcal{O}b = \mathcal{O}z$.

Démonstration : soit $x \in \mathcal{O}$ tel que $n(x) = ab$. On considère l'idéal $x\mathcal{O} + a\mathcal{O}$ de $\mathcal{O}$. Comme $\mathcal{O}$ est principal, on a :

$$x\mathcal{O} + a\mathcal{O} = y\mathcal{O}$$

pour un certain y de $\mathcal{O}$. Soient $x = yz$ et $a = yy'$. Alors $n(x) = ab = n(y)n(z)$ et $a^2 = n(y)n(y')$, donc $n(y)$ est un commun diviseur de ab et de a^2. Comme a et b sont premiers entre eux, $n(y)$ divise a. Par ailleurs, de la relation $x\mathcal{O} + a\mathcal{O} = y\mathcal{O}$, on tire aussi

$$xx' + aa' = y$$

pour certains x', a' de $\mathcal{O}$. En prenant la norme des deux côtés, on déduit :

$$\begin{aligned} n(y) &= n(xx') + n(aa') + t(xx'\overline{aa'}) \\ &= n(x)\,n(x') + a^2 n(a') + a\,t(xx'\overline{a'}) \\ &= a(b\,n(x') + a\,n(a') + t(xx'\overline{a'})). \end{aligned}$$

Comme x, x' et $\overline{a'}$ sont dans $\mathcal{O}$, on a $t(xx'\overline{a'}) \in \mathbb{Z}$, donc le facteur de a dans le membre de droite est un entier. Il en résulte que a divise $n(y)$. Comme a et $n(y)$ sont tous les deux positifs, on a que $n(y) = a$ et donc aussi $n(z) = b$. Par ailleurs, si y et z sont des éléments de $\mathcal{O}$ tels que $n(y) = a$ et $n(z) = b$ alors, comme $\mathcal{O}$ est principal, on a :

$$yz\mathcal{O} + a\mathcal{O} = d\mathcal{O}$$

pour un certain d dans $\mathcal{O}$. Les arguments du début montrent que $n(d) = a$. Par ailleurs, de $y\overline{y} = a$, on déduit que $a\mathcal{O} \subset y\mathcal{O}$. Dès lors,

$$yz\mathcal{O} + a\mathcal{O} = d\mathcal{O} \subset y\mathcal{O}.$$

Ainsi il existe $u \in \mathcal{O}$ tel que $d = yu$. Comme $n(d) = n(y) = a$, on a $u \in \mathcal{O}^\times$ et donc $d\mathcal{O} = y\mathcal{O}$. ■

Avec ce lemme, on peut donner le nombre de représentations d'un entier quelconque par la forme norme de $\mathcal{O}$; comme précédemment, on note

$$R(m) = \{x \in \mathcal{O} \mid n(x) = m\}$$

et

$$r(m) = |R(m)|,$$

où m est un entier (positif) quelconque.

THÉORÈME 4.2. — 1. *La fonction* $r(m)/|\mathcal{O}^\times|$ *est multiplicative, c'est-à-dire que si* a *et* b *sont des entiers positifs premiers entre eux, alors :*

$$\frac{r(ab)}{|\mathcal{O}^\times|} = \frac{r(a)}{|\mathcal{O}^\times|} \cdot \frac{r(b)}{|\mathcal{O}^\times|}.$$

2. *Soit* m *un entier positif. On a*

$$r(m) = |\mathcal{O}^\times| \cdot \Bigg(\sum_{\substack{d|m \\ \text{pgcd}(d, \text{disc } H)=1}} d \Bigg).$$

Démonstration :

1. Soient a et b des entiers positifs premiers entre eux. La multiplication dans $\mathcal{O}$ définit une application : $R(a) \times R(b) \to R(ab)$, qui est surjective d'après le lemme 4.1. Pour démontrer la première partie de l'énoncé, il suffit de prouver que tout élément de $R(ab)$ est l'image de $|\mathcal{O}^\times|$ éléments de $R(a) \times R(b)$, puisqu'alors

$$r(ab) = \frac{|R(a) \times R(b)|}{|\mathcal{O}^\times|} = \frac{r(a)r(b)}{|\mathcal{O}^\times|}.$$

Fixons $x \in R(ab)$. Si (y, z) et (y', z') sont des éléments de $R(a) \times R(b)$ tels que

$$yz = x = y'z',$$

alors d'après la seconde partie du lemme 4.1 on a :

$$x\mathcal{O} + a\mathcal{O} = y\mathcal{O} = y'\mathcal{O}.$$

Dès lors, il existe un élément inversible $u \in \mathcal{O}^\times$ tel que $y' = yu$ (et donc $z' = u^{-1}z$). Cela prouve que les éléments de $R(a) \times R(b)$ qui ont x pour image sont les couples $(yu, u^{-1}z)$, où $u \in \mathcal{O}^\times$. Le nombre de ces couples est bien égal au nombre d'éléments de $\mathcal{O}^\times$.

2. On a déjà calculé, dans le corollaire 3.4, le nombre de représentations des puissances d'un nombre premier p : si p ne divise pas disc H, alors :

$$\frac{r(p^n)}{|\mathcal{O}^\times|} = \frac{p^{n+1} - 1}{p - 1} = \sum_{d|p^n} d$$

et si p divise disc H,

$$\frac{r(p^n)}{|\mathcal{O}^\times|} = 1.$$

On voit ainsi que les fonctions $\frac{r(m)}{|\mathcal{O}^\times|}$ et $(\sum_{\substack{d|m \\ \mathrm{pgcd}(d,\mathrm{disc}\, H)=1}} d)$ prennent la même valeur lorsque m est une puissance d'un nombre premier. Comme ces deux fonctions sont multiplicatives, elles doivent prendre la même valeur pour tout m. ■

Le théorème 4.2 s'applique aux ordres principaux dans les algèbres de quaternions rationnelles définies positives, qui sont au nombre de cinq [16, p.155]. L'ordre principal est alors unique (à conjugaison près) [16, p.26, Cor. 4.11]. Voici la liste des cinq formes quadratiques (à $\mathbb{Z}$-équivalence près) auxquelles le résultat s'applique ainsi que, pour chacune, la formule pour le nombre de représentations d'un entier quelconque.

– Le nombre de représentations d'un entier positif n par la forme

$$X_1^2 + X_2^2 + X_3^2 + X_4^2 + X_1X_4 + X_2X_4 + X_3X_4$$

est

$$24 \sum_{\substack{d|n \\ 2\nmid d}} d\,,$$

– Le nombre de représentations d'un entier positif n par la forme

$$X_1^2 + X_2^2 + X_3^2 + X_4^2 + X_1X_4 + X_2X_3$$

est

$$12 \sum_{\substack{d|n \\ 3\nmid d}} d\,,$$

– Le nombre de représentations d'un entier positif n par la forme

$$X_1^2 + 2X_2^2 + 5X_3^2 + X_4^2 + X_1X_2 + X_1X_4 + X_2X_4 + 5X_2X_3$$

est

$$6 \sum_{\substack{d|n \\ 5\nmid d}} d\,,$$

– Le nombre de représentations d'un entier positif n par la forme

$$X_1^2 + X_2^2 + 2X_3^2 + 2X_4^2 + X_1X_4 + X_2X_3$$

est

$$4 \sum_{\substack{d|n \\ 7\nmid d}} d\,,$$

– Le nombre de représentations d'un entier positif n par la forme

$$X_1^2 + 4X_2^2 + 13X_3^2 + 2X_4^2 + X_1X_2 + X_1X_4 + X_2X_4 + 13X_2X_3$$

est

$$2\sum_{\substack{d\mid n\\ 13\nmid d}} d\,.$$

Nous allons maintenant déduire la formule de Jacobi pour la somme de quatre carrés à partir de la première de ces formules.

La forme quadratique $X_1^2+X_2^2+X_3^2+X_4^2$ est la forme norme de l'ordre $\mathcal{O}'$ de base $(1,i,j,k)$ dans l'algèbre de quaternions $H=(-1,-1)_{\mathbb{Q}}$. Cet ordre n'est pas maximal : il est strictement contenu dans l'ordre maximal $\mathcal{O}$ de base $(1,i,j,(1+i+j+k)/2)$. On ne peut donc pas lui appliquer la technique développée ci-dessus. Cependant, les relations entre $\mathcal{O}$ et $\mathcal{O}'$ sont telles qu'il est quand même possible de déduire le nombre de représentations d'un entier en somme de quatre carrés à partir du nombre de représentations d'un entier par la forme norme de $\mathcal{O}$.

Pour éviter la confusion, la notation $r(m)$ désigne, dans la fin de ce paragraphe, le nombre de représentations de m par la forme norme de $\mathcal{O}$ tandis que $r_4(m)$ désigne le nombre de représentations de m en somme de quatre carrés (qui est la forme norme de $\mathcal{O}'$).

Commençons par indiquer la relation entre $\mathcal{O}$ et $\mathcal{O}'$.

Proposition 4.3. — *Soit x un élément non nul de $\mathcal{O}$. Si $n(x)$ est paire, alors $x\in\mathcal{O}'$; si $n(x)$ est impaire, alors x est associé (à droite) à 8 éléments de $\mathcal{O}'$ (et à 16 éléments de $\mathcal{O}\smallsetminus\mathcal{O}'$).*

Démonstration : par rapport à la base $(1,i,j,a)$ de $\mathcal{O}$, la forme norme s'exprime de la manière suivante :

$$n(x_1+ix_2+jx_3+ax_4) = x_1^2+x_2^2+x_3^2+x_4^2+x_1x_4+x_2x_4+x_3x_4.$$

Dès lors, pour $x=x_1+ix_2+jx_3+ax_4$,

$$n(x)\equiv(x_1+x_2+x_3)^2+(x_1+x_2+x_3)x_4+x_4^2 \bmod 2.$$

Comme la forme quadratique X^2+XY+Y^2 est anisotrope sur le corps à deux éléments, on a donc $n(x)\in 2\mathbb{Z}$ si et seulement si $x_1+x_2+x_3\equiv x_4\equiv 0 \bmod 2$. En particulier, si $n(x)$ est paire, alors x_4 est pair, ce qui entraîne : $x\in\mathcal{O}'$.

Si $n(x)$ est impaire, alors $(x+2\mathcal{O})(\overline{x}+2\mathcal{O})=1+2\mathcal{O}$, donc $\overline{x}+2\mathcal{O}$ est l'inverse de $x+2\mathcal{O}$ dans l'anneau quotient $\mathcal{O}/2\mathcal{O}$. Par ailleurs, on montre aisément (voir par exemple [13, p.100]) que x est associé à droite à un élément de $\mathcal{O}'$; pour terminer la démonstration, on peut donc supposer $x \in \mathcal{O}'$. Si $u \in \mathcal{O}^\times$ est tel que $xu \in \mathcal{O}'$, alors dans $\mathcal{O}/2\mathcal{O}$, on a

$$u + 2\mathcal{O} = \overline{x}(xu) + 2\mathcal{O},$$

ce qui prouve que $u \in \mathcal{O}'$, puisque $\overline{x}(xu) \in \mathcal{O}'$ et $2\mathcal{O} \subset \mathcal{O}'$. On a donc $xu \in \mathcal{O}'$ si et seulement si $u \in \mathcal{O}'^\times$, ce qui prouve que les associés à droite de x qui sont dans $\mathcal{O}'$ sont en bijection avec $\mathcal{O}'^\times$, d'où la proposition, car $|\mathcal{O}'^\times| = 8$. ■

La formule de Jacobi se déduit alors aisément de la formule pour le nombre de représentations par la forme norme de $\mathcal{O}$:

THÉORÈME 4.4. (Jacobi). — *Le nombre de représentations d'un entier positif m en somme de quatre carrés est donné par*

$$r_4(m) = 8(\sum_{\substack{d|m \\ 4\nmid d}} d).$$

Démonstration : soit, comme précédemment :

$$R(m) = \{x \in \mathcal{O} \mid n(x) = m\}.$$

Si m est impair, on déduit de la proposition précédente que chacune des classes d'éléments associés à droite qui constituent $R(m)$ contient 8 éléments de $\mathcal{O}'$ et 16 éléments de $\mathcal{O} \setminus \mathcal{O}'$; donc

$$r_4(m) = \tfrac{1}{3}r(m) = 8\sum_{d|m} d.$$

Si m est pair, alors d'après la proposition précédente $R(m) \subset \mathcal{O}'$, donc

$$r_4(m) = r(m) = 24\sum_{\substack{d|m \\ 2\nmid d}} d.$$

On peut encore exprimer ce résultat comme suit :

$$r_4(m) = 8(\sum_{\substack{d|m \\ 2\nmid d}} d + \sum_{\substack{d|m \\ 2\nmid d}} 2d) = 8\sum_{\substack{d|m \\ 4\nmid d}} d.$$

Dans les deux cas, on a donc bien le résultat annoncé. ■

manuscrit reçu le 22 février 1994

Bibliographie

[1] A. Blanchard. — *Les corps non commutatifs*, Presses Universitaires de France, Paris, 1970.

[2] J.W.S. Cassels. — *Rational Quadratic Forms*, Academic Press, London, 1978.

[3] L.E. Dickson. — *History of the Theory of Numbers*, vol. II, Chelsea Publishing Co., New York, 1952.

[4] B. Gordon. — *An Application of Modular Forms to Quadratic Forms*, BA-thesis, 1975.

[5] H. Gross. — *Darstellungsanzahlen von quaternären quadratischen Stammformen mit quadratischer Diskriminante*, Comment. Math. Helv **34**, (1960) 198-221.

[6] E. Grosswald. — *Representations of Intergers as Sums of Squares*, Springer-Verlag, New York Berlin Heidelberg Tokyo, 1985.

[7] G.H. Hardy and E.M. Wright. — *An Introduction to the Theory of Numbers*, 5th ed., Oxford University Press, 1979.

[8] E. Landau. — *Elementary Number Theory*, Chelsea Publishing Co., New York, 1966.

[9] G. Orzech ed. — *Conference on Quadratic Forms 1976*, Queen's Paper in Pure and Applied Math. **46**, Kingston, Ontario, Canada 1977.

[10] I. Reiner. — *Maximal Orders*, Academic Press, London 1975.

[11] A. Robert. — *Introduction to Modular Forms*, Queen's Paper in Pure and Applied Math. **45**, Kingston, Ontario, Canada 1976.

[12] G. Rousseau. — *On a construction for the representation of a positive integer as the sum of four squares*, L'Enseignement Math **33**, (1987) 301-306.

[13] P. Samuel. — *Théorie algébrique des nombres*, Hermann, Paris 1967.

[14] J.-P. Serre. — *Cours d'Arithmétique*, Coll. SUP, Presses Univ. France, Paris, 1970.

[15] J.V. Uspensky and M.A. Heaslet, *Elementary Number Theory*, McGraw-Hill, New York and London, 1939.

[16] M.–F. VIGNÉRAS. — *Arithmétique des algèbres de quaternions*, Lecture Notes in Math **800**, Springer-Verlag, Berlin Heidelberg New York, 1980.

[17] M.–F. VIGNÉRAS. — *Simplification pour les ordres des corps de quaternions totalement définis*, J. Reine Angew. Math. **286/287**, (1976) 257-287.

[18] A. WEIL. — *Sur les sommes de trois et quatre carrés*, L'Enseignement Math **20** (1974) 215-222.

Isabelle PAYS
Université de Mons–Hainaut
Avenue Maistriau, 15
B–7000 MONS
BELGIQUE

Number Theory
Paris 1992–93

On a conjecture that a product of k consecutive positive integers is never equal to a product of mk consecutive positive integers except for $8.9.10 = 6!$ and related questions

T.N. SHOREY

For an integer $m \geq 2$, we consider the equation

(1) $(x+1)\cdots(x+k) = (y+1)\cdots(y+mk)$ in integers $x \geq 0, y \geq 0, k \geq 2$.

We replace $x+1$ by x and $y+1$ by y in (1) for observing that it is identical to considering equation

$x(x+1)\cdots(x+k-1) = y(y+1)\cdots(y+mk-1)$ in integers $x>0$, $y>0$, $k\geq 2$.

If $m = 2$, equation (1) has a solution given by

$$8.9.10 = 6!$$

i.e.

(2) $$x = 7, y = 0, k = 3.$$

MacLeod and Barrodale [8] observed that this is the only solution of (1) with $m = 2$ and $k \leq 5$. I give their proof for $k = 2$. We write equation (1) with $m = 2$ and $k = 2$

(3) $$(x+1)(x+2) = (y+1)(y+2)(y+3)(y+4).$$

By putting $u = y^2 + 5y$, we have

$$(x+1)(x+2) = (u+4)(u+6).$$

Notice that

$$\left(x+\frac{3}{2}-\frac{1}{4}\right)^2 < (x+1)(x+2) < \left(x+\frac{3}{2}\right)^2$$

and

$$\left(u+5-\frac{1}{4}\right)^2 < (u+4)(u+6) < (u+5)^2.$$

We have

$$\left[\sqrt{(x+1)(x+2)}\right] = \left[\sqrt{(u+4)(u+6)}\right]$$

which implies that

$$x+1 = u+4$$

i.e.

(4) $$x = u+3.$$

By substituting (4) in (3), we have

$$(u+4)(u+5) = (u+4)(u+6)$$

and this is a contradiction.

Now, I give a few comments on the proof. We have written the right hand side of equation (1) as a product of translates of u and the translates are independent of y. This is typical of the case $m = 2$. In the general case, we shall be extracting k-th roots in place of square roots. For this, it is necessary for the above argument that x and y are large as compared with k.

Saradha and Shorey [11] proved that (2) is the only solution of equation (1) with $m = 2$. Further, Saradha and Shorey [12] showed that equation (1) with $m \in \{3, 4\}$ has no solution. Recently, Mignotte and Shorey showed that this is also the case when $m \in \{5, 6\}$. For $m \geq 2$, it is proved in [13] that equation (1) implies that

$$\max\,(x, y, k) \leq C$$

where C is an effectively computable[(1)] number depending only on m. We have not been able to replace C by an absolute constant. It is likely that equation (1) with $m > 2$ has no solution.

(5) $\left\{\text{Now, we give a sketch of the proof that equation (1) with } m \geq 2 \text{ implies that } \max(x, y, k) \text{ is bounded by a number depending only on } m.\right.$

As pointed out earlier, we secure that x and y are large as compared with k. We re-write equation (1) as

$$\frac{(x+1)\cdots(x+k)}{k!} = \frac{(y+1)\cdots(y+mk)}{(mk)!}\frac{(mk)!}{k!}.$$

(1) All the constants appearing in this paper are effectively computable.

We count the powers of 2 on both the sides to obtain

$$k \leq \operatorname{ord}_2\left(\frac{(mk)!}{k!}\right) \leq \operatorname{ord}_2\left(\frac{(x+1)\cdots(x+k)}{k!}\right)$$
$$\leq \max_{1\leq i\leq k} \operatorname{ord}_2(x+i) \leq \frac{\log\,(x+k)}{\log\,2},$$

which implies that :

$$x \geq 2^k - k. \tag{6}$$

By equation (1), we have

$$x^k \leq (y+mk)^{mk}$$

i.e.

$$x \leq (y+mk)^m. \tag{7}$$

By (6) and (7), we observe that x and y are large as compared with k. Further, we combine (6) and (7) to write

$$2^k - k \leq (y+mk)^m. \tag{8}$$

If y is bounded, we observe from (8) that k is bounded and equation (1) implies that x is bounded. Thus, we may always assume that y exceeds a sufficiently large number y_0 depending only on m. Further, by equation (1), we observe that $x > y > y_0$.

For extracting k-th roots on both the sides of equation (1), we need to introduce some notation. We write

$$(z+1)\cdots(z+mk) = \sum_{j=0}^{mk} A_j(m,k)z^{mk-j}. \tag{9}$$

Further, we determine rational numbers

$$B_j = B_j(m,k) \quad \text{with } 1 \leq j \leq m$$

such that :

$$(z^m + B_1 z^{m-1} + \cdots + B_m)^k = \sum_{j=0}^{mk} H_j(m,k)z^{mk-j} \tag{10}$$

satisfies

$$H_j(m,k) = A_j(m,k) \quad \text{for } 0 \le j \le m. \tag{11}$$

By (10) and (11),

$$kB_1 = A_1(m,k)$$
$$kB_2 + \binom{k}{2} B_1^2 = A_2(m,k)$$
$$\cdots\cdots\cdots\cdots\cdots\cdots\cdots\cdots\cdots\cdots$$
$$\cdots\cdots\cdots\cdots\cdots\cdots\cdots\cdots\cdots\cdots$$
$$\cdots\cdots\cdots\cdots\cdots\cdots\cdots\cdots\cdots\cdots$$

Therefore $B_1, \cdots, B_m$ are determined recursively.

Let z be a sufficiently large positive number as compared with k and m. The relations (11) imply that the left hand side of (9) is close to the left hand side of (10). Therefore, the k-th root of the left hand side of (9) is close to the k-th root of the left hand side of (10). We can use this observation with z replaced by x or y, since x and y are large as compared with k and m. Therefore, the k-th root of the left hand side of (1) is close to $x + \frac{k+1}{2}$ and the k-th root of the right hand side of (1) is close to $y^m + B_1 y^{m-1} + \cdots + B_m$. Consequently, by equation (1), we derive that $x + \frac{k+1}{2}$ is close to $y^m + B_1 y^{m-1} + \cdots + B_m$. In fact, we show that

$$\mid x - (y^m + B_1 y^{m-1} + \cdots + B_m - \frac{k+1}{2}) \mid < \tau^{-1} \tag{12}$$

where

$$\tau = (2\text{lcm }(\text{den}(B_1), \cdots, \text{den}(B_m))).$$

On the other hand, the left hand side of (12) is at least τ^{-1} whenever it is not equal to zero. Hence, we conclude that

$$x = y^m + B_1 y^{m-1} + \cdots + B_m - \frac{k+1}{2}. \tag{13}$$

We substitute (13) in (1) to derive that

$$H_j(m,k) = A_j(m,k) \quad \text{for } 0 \le j < 2m \tag{14}$$

and

$$H_{2m}(m,k) - A_{2m}(m,k) = \frac{k(k+1)(k-1)}{24}. \tag{15}$$

Thus, we have added $m-1$ relations to the relations (11) with which we started. This is the basic idea of the proof.

We calculate

$$B_1(2,k) = 2k+1\ ,\ B_2(2,k) = (k+1)(2k+1)/3$$

and

$$H_4(2,k) - A_4(2,k) = (4k^5 - 5k^3 + k)/90. \tag{16}$$

By (15) and (16), we derive that $m > 2$. Then, we apply a result of Balasubramanian [13, Appendix] to conclude from (14) that k is bounded by a number depending only on m. Thus, there are only finitely many possibilities for k and we fix k. We put

$$L(X,Y) = (X+1)\cdots(X+k) - (Y+1)\cdots(Y+mk)$$

and

$$\ell(Y) = L(\phi(Y), Y)$$

where

$$\phi(Y) = Y^m + B_1 Y^{m-1} + \cdots + B_m - \frac{k+1}{2}.$$

By equation (1) and (13),

$$\ell(y) = 0\,,$$

which implies that either $\ell(Y)$ is a zero polynomial or y is bounded by a number y_0' depending only on m and k. By taking $y_0 > y_0'$, we conclude that

$$\ell(Y) \equiv 0. \tag{17}$$

Now, I give two proofs to exclude the possibility (17).

We assume (17). Then

$$L(X,Y) = (X - \phi(Y))\,(X^{k-1} + R_1(Y)X^{k-2} + \cdots + R_{k-1}(Y)) \tag{18}$$

where $R_i(Y) \in \mathbb{Q}[Y]$ for $1 \le i < k$. By equating the terms independent of X in the factorisation (18), we observe that the polynomial

$$(Y+1)\cdots(Y+mk) - k!$$

is reducible over the field of rational numbers. Now, we apply a result of Brauer and Ehrlich [5] to conclude that

$$k! \geq \frac{(mk-1)!}{2^{(mk-1)}[(mk-2)/2]!}.$$

The right hand side is an increasing function of m and the inequality is not valid for $m = 4$. Therefore, we derive that $m = 3$ which implies that $k = 2$. Now, by looking at the constant term of $\ell(Y)$, we observe that

$$B_3^2(3,2) - \frac{1}{4} = 6!.$$

This is not possible, since $B_3(3,2)$ is a rational number.

Next, we turn to the second proof to exclude the possibility (17). We assume (17). Then, there exist pairwise distinct integers $i_1, \cdots, i_m, j_1, \cdots, j_m$ such that

$$\phi(Y) + 1 \equiv (Y + i_1) \cdots (Y + i_m)$$

and

$$\phi(Y) + 2 \equiv (Y + j_1) \cdots (Y + j_m).$$

Thus

$$(Y + j_1) \cdots (Y + j_m) - (Y + i_1) \cdots (Y + i_m) \equiv 1. \tag{19}$$

By putting $Y = -i_1$ in (19), we have

$$(j_1 - i_1) \cdots (j_m - i_1) = 1$$

which implies that $m = 2$. Then, we observe from (19) that

$$j_1 + j_2 = i_1 + i_2 \ , \ j_1 j_2 = i_1 i_2 + 1.$$

Consequently

$$(j_1 - j_2)^2 = (i_1 - i_2)^2 - 4$$

which is not possible.

This completes our sketch of the proof of (5). Now, I mention some extensions of (5). Let $f(X)$ be a monic polynomial of positive degree with rational coefficients. For an integer $m \geq 2$, we consider the equation

$$f(x+1) \cdots f(x+k) = f(y+1) \cdots f(y+mk) \quad \text{in integers } x, y, k \geq 2. \tag{20}$$

If f is a power of an irreducible polynomial, Balasubramanian and Shorey [3] proved that equation (20) with

$$f(x+j) \neq 0 \text{ for } 1 \leq j \leq k \tag{21}$$

implies that $\max(|x|, |y|, k)$ is bounded by a number depending only on m and f. This is an extension of (5), since (21) is satisfied whenever $f(X) = X$ and x is a non-negative integer. Further, it is easy to observe that the assumption (21) is necessary. The proof utilises an extension of the second proof for excluding the possibility (17). The second proof depends on the fact that $f(X) = X$ is irreducible. For extending this proof, we need that f is a power of an irreducible polynomial. Furthermore, this is the only place in the proof where the hypothesis that f is a power of an irreducible polynomial is used.

For an integer $d \geq 1$, Saradha and Shorey [14] obtained another extension of (5) by proving that if $x > 0, y > 0, k \geq 2$ are integers satisfying

$$x(x+d)\cdots(x+(k-1)d) = y(y+d)\cdots(y+(mk-1)d), \tag{22}$$

then max (x, y, k) is bounded by a number depending only on d and m. In fact, this is an immediate consequence of the following result ([14, Theorems 1,2]) : For $\epsilon > 0$, there exists a number C_1 depending only on m and ϵ such that equation (22) with max $(x, y, k) \geq C_1$ implies that

$$d \geq y^{(1-\epsilon)/(m+1)}, \ \log d \geq (\frac{M}{m(m+1)} - \epsilon)K \tag{23}$$

where K and M are positive numbers given by

$$K^2 = k \log k \ , \ M^2 = m(m-1)/2. \tag{24}$$

We observe from (23), (24), (25) and (22) that max (x, y, k) is bounded by a number depending only on d and m. Further, for positive integers d_1, d_2 and $m \geq 2$, Saradha and Shorey [15] considered a more general equation than (22) :

$$\begin{aligned} x(x+d_1)\cdots(x+(k-1)d_1) =& y(y+d_2)\cdots(y+(mk-1)d_2) \\ & \text{in integers } x > 0, y > 0, k \geq 2. \end{aligned} \tag{25}$$

It was shown in [15] that equation (25) with $m = 2$ implies that either $\max(x, y, k)$ is bounded by a number depending only on d_1, d_2 or $k = 2, d_1 = 2d_2^2, x = y^2 + 3d_2 y$. On the other hand, equation (25) with $m = 2$ is satisfied whenever the latter possiblities hold. If $m > 2$, it was proved in [15] that there exist numbers C_2 and C_3 depending only on d_1, d_2, m such that equation (25) implies that $k \leq C_2$ and moreover max $(x, y) \leq C_3$ unless (*) d_1/d_2^m is a product of m distinct positive integers composed of primes not exceeding m and $m \geq \alpha(k)$ where

$$\alpha(k) = \begin{cases} 14 \text{ for } 2 \leq k \leq 7 \\ 50 \text{ for } k = 8 \\ \exp(k\log k - (1.25475)k - \log k + 1.56577) \text{ for } k \geq 9. \end{cases}$$

This includes a result on the case $d_1 = d_2 = d$ mentioned above, since (*) is never satisfied. Further, we derive that equation (25) with $3 \leq m \leq 14$ or $3 \leq m \leq 2568, k \geq 9$ implies that max (x, y, k) is bounded by a number depending only on d_1, d_2, m. Finally, Saradha, Shorey and Tijdeman [17] showed that condition (*) is not necessary. Consequently, we conclude that equation (25) with $m > 2$ implies that max (x, y, k) is bounded by a number depending only on d_1, d_2, m. This is a consequence of a more general result which we describe now. For distinct positive integers ℓ and m with $\gcd(\ell, m) = 1$ and $\ell < m$, Saradha, Shorey and Tijdeman [17] proved that there exists a number C_4 depending only on d_1, d_2, m such that if $x > 0$, $y > 0$ and $k \geq 2$ are integers satisfying

$$x(x + d_1) \cdots (x + (\ell k - 1)d_1) = y(y + d_2) \cdots (y + (mk - 1)d_2),$$

then

$$\max(x, y, k) \leq C_4$$

unless $\ell = 1, m = 2$ which corresponds to equation (25) with $m = 2$ and we refer to the result already stated in this case. By applying the theory of linear forms in logarithms, it is shown in [17] that the preceding assertion is also valid for $k = 1$ provided that $\ell \in \{2, 4\}$ and $(\ell, m) = (3, 4)$. Now, we turn to equation (25) with $m = 1$. In this case, there is no loss of generality in assuming that $x > y$ and $\gcd(x, y, d_1, d_2) = 1$. Then Saradha, Shorey and Tijdeman [16] proved that equation (25) with $m = 1$ implies that there exists a number C_5 depending only on d_2 such that either

$$x = k + 1, y = 2, d_1 = 1, d_2 = 4$$

or

$$\max(x, y, k) \leq C_5.$$

We observe that

$$(k + 1) \cdots (2k) = 2.6 \cdots (4k - 1) \text{ for } k = 2, 3, \ldots$$

since the right hand side is equal to

$$2^k(2k)!/(2.4\ldots(2k)) = (2k)!/k!.$$

Therefore, the above possibilities for the case $d_1 = 1, d_2 = 4$ cannot be excluded. Further Saradha, Shorey and Tijdeman [18] showed that equation (25) with $m = d_1 = 1$ implies that $y \le k^2d_2^2/12$ and furthermore $k \le d_2 - 2$ unless $y \equiv 2 (\text{mod } 4)$ and $d_2 = 2^\ell$ for some integer $\ell \ge 2$. In the case $m = 2, d_1 = 1$, Saradha, Shorey and Tijdeman [18] proved that equation (25) implies that $y(y + (2k-1)d_2) \le (0.44)k^4d_2^4$ and furthermore $k \le d_2 - 2$ unless $k \le 35$ and $d_2 = 2^\ell$ for some integer $\ell \ge 2$. These results are applied in [18] to determine all the solutions of equation (25) with $m = d_1 = 1, d_2 \in \{2,3,5,6,7,9,10\}$ and $m = 2, d_1 = 1, d_2 \in \{5,6\}$.

Let a and b be positive integers. We consider

$$\begin{aligned} &a(x+1)\cdots(x+k) = b(y+1)\cdots(y+k+\ell) \\ &\quad \text{in integers } x \ge 0, y \ge 0, k \ge 2, \ell \ge 0. \end{aligned} \tag{26}$$

Equation (1) is a particular case of (26), namely, $a = b = 1$ and $k + \ell$ is an integral multiple of k. Erdős [7] conjectured that there are only finitely many integers $x \ge 0, y \ge 0, k \ge 2, \ell \ge 0$ with $k+\ell \ge 3$ and $x \ge y+k+\ell$ satisfying (26). This is a difficult problem. The assumption $k + \ell \ge 3$ is to exclude Pell's equations and the assumption $x \ge y + k + \ell$ is to guarantee that the two blocks of consecutive integers in equation (26) are non-overlapping.

Mordell [9] proved that equation (26) with $a = b = 1, k = 2, \ell = 1$ implies that $x = 1, y = 0$ and $x = 13, y = 4$. Mordell's result initiated much of research in this direction. Avanesov [1] confirmed a conjecture of Sierpinski by proving that $x = 0, y = 0; x = 3, y = 2; x = 14, y = 7; x = 54, y = 19$ and $x = 118, y = 33$ are the only solutions of equation (26) with $a = 3, b = 1, k = 2, \ell = 1$. Tzanakis and de Weger [20] determined all the solutions of equation (26) with $a = 1, b = 2, k = 2, \ell = 1$. Boyd and Kisilevsky [4] showed that $x = 1, y = 0; x = 3, y = 1$ and $x = 54, y = 18$ are the only solutions of equation (26) with $a = b = 1, k = 3, \ell = 1$. Cohn [6] proved that equation (26) with $a = 1, b = 2, k = 4, \ell = 0$ is satisfied only if $x = 4, y = 3$. Further, Ponnudurai [10] showed that $x = 2, y = 1$ and $x = 6, y = 4$ are the only solutions of equation (26) with $a = 1, b = 3, k = 4, \ell = 0$.

Let us consider equation (26) with $\ell = 0$. We re-write equation (26) with $\ell = 0$ as

$$(ax^k - by^k) + A_1(ax^{k-1} - by^{k-1}) + \cdots + A_{k-1}(ax - by) + A_k(a-b) = 0 \tag{27}$$

where $A_1, \ldots, A_k$ are given by

$$F(z) = (z+1)\cdots(z+k) = z^k + A_1z^{k-1} + \cdots + A_k. \tag{28}$$

If $a = b$, we observe from (27) that

$$(x^k - y^k) + A_1(x^{k-1} - y^{k-1}) + \cdots + A_{k-1}(x - y) = 0 \quad (29)$$

which implies that $x = y$, since all the summmands in (29) are of the same sign. Now, we assume that $a \neq b$. Shorey [19] showed that there exists a number C_6 depending only on a and b such that equation (26) with $\ell = 0, x \geq y+k$ and $k \geq C_6$ implies that the first summand in (27) is positive and the second summand in (27) is negative (then all the summands in (27) following the second one will be negative). This is equivalent to saying that equation (26) with $\ell = 0$ and $x \geq y + k$ implies that either $k \leq C_6$ or $k = [\alpha + 1]$ where

$$\alpha = \log\left(\frac{b}{a}\right) / \log\left(\frac{x}{y}\right).$$

We have not been able to exclude the latter possibility. This is the case if we allow C_6 to depend also on $P(x)$ and $P(y)$.(2) In fact, Shorey [19] showed that equation (26) with $\ell = 0$ and $x \geq y + k$ implies that $\max(x, y, k)$ is bounded by a number depending only on $a, b, P(x)$ and $P(y)$. Saradha and Shorey [11] extended these results to equation (26) where ℓ is not necessarily equal to zero.

Now, we give a sketch of the proof that equation (26) with $x \geq y + \ell + k$ implies that $\max(x, y, k, \ell)$ is bounded by a number depending only on $a, b, P(x)$ and $P(y)$.

The proof depends on Gel'fond - Baker theory of linear forms in logarithms. We assume (26). By (26) and (28), we observe that

$$0 < aF(x) - bF(y)y^\ell = U_k + A_1U_{k-1} + \cdots + A_kU_0 \quad (30)$$

where

$$U_i = ax^i - by^{i+\ell} \quad \text{for } 0 \leq i \leq k. \quad (31)$$

If $U_k \leq 0$, we observe from $x > y$ that $U_i \leq 0$ for $0 \leq i \leq k$ which contradicts (30). Thus

$$U_k > 0. \quad (32)$$

We write $c_1, c_2, \ldots, c_{12}$ for positive numbers depending only on $a, b, P(x)$ and $P(y)$. We may assume that $x \geq c_1$ with c_1 sufficiently large. In view

(2) For an integer $\nu > 1$, we write $P(\nu)$ for the greatest prime factor of ν and we put $P(0) = P(1) = 1$.

of the result of Shorey [19] mentioned above, we may suppose that $\ell > 0$ which implies that $k + \ell \geq 3$. Then, it is proved in [11, Corollary 2] that

$$x - y \leq c_2 \ell x (\log x)/k. \tag{33}$$

As in the proof of (5), we count the power of 2 on both the sides of (26) for deriving that

$$\ell \leq c_3 \log x. \tag{34}$$

By (33) and (34),

$$x - y \leq c_4 x (\log x)^2/k. \tag{35}$$

On the other hand, we apply an estimate of Fel'dman (see Baker [2]) on linear forms in logarithms to obtain

$$x - y \geq x (\log x)^{-c_5}. \tag{36}$$

By (35) and (36),

$$k \leq (\log x)^{c_6}. \tag{37}$$

Now, we show that

$$y \leq (\log x)^{c_7}. \tag{38}$$

We may assume that

$$y > (k + \ell)^4, \tag{39}$$

otherwise (38) follows from (34) and (37). Further, we derive from (26), (31), (32), (37) and (39) that

$$0 < U_k < c_8((k + \ell)^2 y^{k+\ell-1} + k^2 x^{-1}). \tag{40}$$

On the other hand, we apply again the estimate of Fel'dman on linear forms in logarithms for deriving that

$$U_k \geq \max(x^k, y^{k+\ell}) \; ((k + \ell) \log x)^{-c_9}$$

which, together with (34) and (37), implies that

$$U_k \geq \max(x^k, y^{k+\ell}) \; (\log x)^{-c_{10}}. \tag{41}$$

Finally, we combine (40), (41), (34) and (37) to conclude (38).

By (37), we have

$$x > k^{P(x)} > k^{\omega(x)}$$

where $\omega(x)$ denotes the number of distinct prime divisors of x. Therefore, there exists a prime p dividing x such that

$$p^{\mathrm{ord}_p(x)} > k. \tag{42}$$

Now, we count the power of p on both the sides of (26) which we re-write as

$$a\frac{(x+1)\cdots(x+k)}{k!} = b\frac{(y+1)\cdots(y+k+\ell)}{(k+\ell)!}\,\frac{(k+\ell)!}{k!}.$$

By (42), we obtain

$$\left[\frac{\ell}{p}\right] - 1 \leq \mathrm{ord}_p\left(a\frac{(x+1)\cdots(x+k)}{k!}\right) \leq \mathrm{ord}_p(a),$$

which sharpens (34) as :

$$\ell \leq c_{11}. \tag{43}$$

By (26),

$$x \leq (y+k+\ell)^{1+(\ell/k)}$$

which, together with (37), (38) and (43), implies that $x \leq c_{12}$. This completes the proof.

We refer to [11] for more results on equation (26). For example, it is proved in [11] that equation (26) with $x \geq y+k+\ell$ implies that :

$$x \geq C_7 k^3 (\log k)^{-4}$$

and

$$x - y \geq C_8 x^{2/3}$$

where C_7 and C_8 are positive numbers depending only on a and b.

Manuscrit reçu le 10 décembre 1993

References

[1] E.T. AVANESOV. — *Solution of a problem on polygonal numbers* (Russian), Acta Arith. **12** (1967), 409–419.

[2] A. BAKER. — *The theory of linear forms in logarithms*, Transcendence Theory : Advances and Applications, Academic Press (1977), 1–27.

[3] R. BALASUBRAMANIAN and T.N. SHOREY. — *On the equation* $f(x+1)\cdots f(x+k) = f(y+1)\cdots f(y+mk)$, Indag. Math. N.S. **4** (1993), 257–267.

[4] D.W. BOYD and H.H. KISILEVSKY. — *The diophantine equation* $u(u+1)(u+2)(u+3) = v(v+1)(v+2)$, Pacific Jour. Math. **40** (1972), 23–32.

[5] A. BRAUER and G. EHRLICH. — *On the irreducibility of certain polynomials*, Bull. Amer. Math. Soc. **52** (1946), 844–856.

[6] J.H.E. COHN. — *The diophantine equation* $Y(Y+1)(Y+2)(Y+3) = 2X(X+1)(X+2)(X+3)$, Pacific Jour. Math. **37** (1971), 331–335.

[7] P. ERDŐS. — *Problems and results on number theoretic properties of consecutive integers and related questions*, Proc. Fifth Manitoba Conf. Numerical Math. (Univ. Manitoba Winnipeg) (1975), 25–44.

[8] R.A. Mac LEOD and I. BARRODALE. — *On equal products of consecutive integers*, Canadian Math. Bull. **13** (1970), 255–259.

[9] L.J. MORDELL. — *On the integer solutions of* $y(y+1) = x(x+1)(x+2)$, Pacific Jour. Math. **13** (1963), 1347–1351.

[10] T. PONNUDURAI. — *The diophantine equation* $Y(Y+1)(Y+2)(Y+3) = 3X(X+1)(X+2)(X+3)$, Jour. London Math. Soc. **10** (1975), 232–240.

[11] N. SARADHA and T.N. SHOREY. — *On the ratio of two blocks of consecutive integers*, Proc. Indian Acad. Sci. (Math. Sci.) **100** (1990), 107-132.

[12] N. SARADHA and T.N. SHOREY. — *On the equation* $(x+1)\cdots(x+k) = (y+1)\cdots(y+mk)$ with $m = 3, 4$, Indag. Math., N.S. **2** (1991), 489–510.

[13] N. SARADHA and T.N. SHOREY. — *On the equation* $(x+1)\cdots(x+k) = (y+1)\cdots(y+mk)$, Indag. Math., N.S. **3** (1992), 79–90.

[14] N. SARADHA and T.N. SHOREY. — *On the equation* $x(x+d)\cdots(x+(k-1)d) = y(y+d)\cdots(y+(mk-1)d)$, Indag. Math., N.S. **3** (1992), 237–242.

[15] N. SARADHA and T.N. SHOREY. — *On the equation* $x(x+d_1)\cdots(x+(k-1)d_1) = y(y+d_2)\cdots(y+(mk-1)d_2)$, Proc. Indian Acad. Sci. (Math. Sci.), **104** (1994).

[16] N. SARADHA, T.N. SHOREY and R. TIJDEMAN. — *On arithmetic progressions of equal lengths with equal products*, Math. Proc. Camb. Phil. Soc. (1994), to appear.

[17] N. SARADHA, T.N. SHOREY and R. TIJDEMAN. — *On arithmetic progressions with equal products*, Acta Arith., to appear.

[18] N. SARADHA, T.N. SHOREY and R. TIJDEMAN. — *On the equation* $x(x+1)\cdots(x+k-1)=y(y+d)\cdots(y+(mk-1)d), m=1,2$, Acta Arith., to appear.

[19] T.N. SHOREY. — *On the ratio of values of a polynomial*, Proc. Indian Acad. Sci. (Math. Sci.) **93** (1984), 109–116.

[20] N. TZANAKIS and B.M.M. de WEGER. — *On the practical solution of the Thue equation*, Jour. Number Theory **31** (1989), 99–132.

T.N. SHOREY
School of Mathematics
Tata Institute of
Fundamental Research
Homi Bhabha Road
Bombay 400 005, India

Number Theory
Paris 1992–93

Rédei-matrices and applications

Peter Stevenhagen

1. — Introduction

In this paper we describe an algebraic method to study the structure of (parts of) class groups of abelian number fields. The method goes back to the Hungarian mathematician L. Rédei, who used it to study the 2-primary part of class groups of quadratic number fields in a series of papers [[18]–[24]] that appeared between 1934 and 1953. The case of the l-primary part of the class group of an arbitrary cyclic extension of prime degree l was studied by Inaba [[12], 1940], who realized that one should look at the class group as a module over the group ring. The matter was then taken up by Fröhlich [[6], 1954], who generalized Inaba's results by extending Rédei's quadratic method to the case of a cyclic field of prime power degree. In the seventies, generalizations in the line of Inaba were given by G. Gras [[10]]. In all cases, one studies l-primary parts of the class group of an abelian extension for primes l that divide the degree.

Recently, completely different methods have been developed by Kolyvagin and Rubin, showing that the structure of any l-primary part of the class group of an abelian field of degree coprime to l can be described 'algebraically'. For primes dividing the degree it is not yet clear whether the approach works. The Kolyvagin-Rubin methods can be seen as refinements of the analytic class number formula, and they are more general than the Rédei-Fröhlich method as they work for most l. On the other hand, they depend on the existence of infinite collections of auxiliary prime numbers, so effective versions of the Čebotarev density theorem are needed to yield deterministic algorithms. Because of this somewhat involved nature they can only be used in practice for abelian fields of very small degree. Moreover, the method does not give any clue as to the average behaviour that is to be expected when it is applied to a family of fields. For instance, it cannot be used to compute the class number h_n^+ of the maximal real subfield of the n-th cyclotomic field for any n that is not very small. Also, it does not tell us whether the fact that h_n^+ is either 1 or very small, which can be shown to be the case for all $n < 200$ having at least two prime divisors if one assumes the generalized Riemann hypothesis [[31]], should be seen as a common occurrence.

The Rédei–Fröhlich method is based only on class field theory and therefore of a rather different nature. It can be used only to describe p-primary parts of the class group when p does divide the degree of the extension, which is exactly the case that had to be excluded before. We will see that it gives in this case rise to density statements telling us how many fields in a given infinite family will have some prescribed part in their class group. An example : the class number h_{13p}^+ is even for some explicit collection of primes p of Dirichlet density $1/16$ and divisible by 3 for a collection of density $1/18$.

An application of Rédei's quadratic method that goes back to Rédei himself concerns criteria for the solvability of the negative Pell equation $x^2 - Dy^2 = -1$. This is a question that is closely related to the behaviour of 2-class fields, as was made clear by Scholz [[25]]. We will discuss it in the two final sections of this paper.

2. — Rédei-matrices

In this section, K will denote a number field that is cyclic of prime degree l with Galois group $G = \mathrm{Gal}(K/\mathbb{Q})$. It is our intention to study the l-part C of the class group of K. If $l = 2$, our convention will be that C is the 2-part of the *narrow* class group of K. Correspondingly, we call an extension unramified if all *finite* primes are unramified. The difference between the narrow and the ordinary class group that may exist is of interest only when $l = 2$. It will be discussed in detail the following two sections.

As C is a finite abelian l-group with a natural G-action, it is a module over the group ring over the l-adic integers $\mathbb{Z}_l[G]$. The norm $N = \sum_{g \in G} g$ annihilates the class group, so we can study C as a module over $\mathbb{Z}_l[G]/N$. If ζ_l denotes a primitive l-th root of unity, we have an isomorphism

$$\begin{aligned} \mathbb{Z}_l[G]/N &\xrightarrow{\sim} \mathbb{Z}_l[\zeta_l] \\ \sigma &\longmapsto \zeta_l \end{aligned}$$

showing that $A = \mathbb{Z}_l[G]/N$ is a discrete valuation ring whose maximal ideal is generated by $\sigma - 1$, with σ a generator of G. The residue class field of A is the finite field of l elements $\mathbb{F}_l$. Every finite A-module M is isomorphic to a module of the form

$$\prod_{i=1}^{s} A/A^{(\sigma-1)^{n_i}}$$

with $n_i \in \mathbb{Z}_{>1}$ for $i = 1, 2, \ldots, s$. Thus, we can specify the isomorphism class of the A-module M by giving the sequence of integers

$$r_k = \#\{i : n_i \geq k\} = \dim_{\mathbb{F}_l}(M^{(\sigma-1)^{k-1}}/M^{(\sigma-1)^k}).$$

Note that $r_1(M) = s$ and that $\{r_k(M)\}_k$ is a decreasing sequence with $r_k(M) = 0$ for k sufficiently large.

The evaluation of $r_1(C)$ amounts to doing genus theory for the field K. More precisely, class field theory associates to C an unramified extension H of K, called the l-class field of K, for which the Galois group $\mathrm{Gal}(H/K)$ is canonically isomorphic to C. The quotient $C/C^{\sigma-1}$ corresponds to an unramified extension H_1 of K that is known as the genus field of K. It is the maximal unramified extension of K that is abelian over $\mathbb{Q}$, and $\mathrm{Gal}(H_1/\mathbb{Q})$ is isomorphic to the elementary abelian l-group $G \times C/C^{\sigma-1}$. If χ denotes a Dirichlet character generating the character group X of G that corresponds to K, we can write χ as a product $\chi = \prod_{i=1}^{t} \chi_i$, where t is the number of primes that ramifies in the extension $K/\mathbb{Q}$ and χ_i is a character of conductor a power of some ramifying prime p_i and of order l. The conductor of χ_i is equal to p_i if and only if $p_i \neq l$. The field H_1 corresponds to the group of Dirichlet characters

$$X_1 = \prod_{i=1}^{t} \langle \chi_i \rangle.$$

It follows that $\mathrm{Gal}(H_1/K)$ has order l^{t-1}, i.e. the $(\sigma - 1)$-rank $r_1(C)$ is equal to $t - 1$, where t is the number of ramifying primes in $K/\mathbb{Q}$.

The subgroup $C^G = C[\sigma - 1]$ of G-invariant ideal classes in C is known as the subgroup of ambiguous ideal classes. As G is cyclic and C is finite, the order of $C^G = \widehat{H}^0(G, C)$ equals the order of $\widehat{H}^1(G, C) = C/C^{\sigma-1}$. It is not difficult to check that C^G is generated by the t classes $[\mathfrak{p}_i]$ of the ramified primes $\mathfrak{p}_i$ of K. The order of C^G is l^{t-1}, so there is exactly one additional relation between these classes that is independent of the obvious relations $[\mathfrak{p}_i^l] = 0$.

The Rédei–Fröhlich theorem gives a description of $r_2(C)$ by combining the two descriptions of $r_1(C)$. Note that as abelian groups, we have :

$$C/C^{(\sigma-1)^2} \cong \begin{cases} C/C^4 & \text{if } l = 2; \\ C/C^{\sigma-1} \times C^{\sigma-1}/C^{(\sigma-1)^2} & \text{if } l > 2, \end{cases}$$

and that the l-rank of C/C^l is equal to the sum $\sum_{k=1}^{l-1} r_k$. The theorem is based on the observation that $r_2(C)$ can be obtained from an explicit description of the natural map

$$\phi : C[\sigma - 1] \longrightarrow C/C^{\sigma-1}.$$

This is a homomorphism between elementary abelian l-groups, so it can be viewed as a linear map between vector spaces over $\mathbb{F}_l$. With this terminology, the $(\sigma-1)^2$-rank of C is nothing but the $\mathbb{F}_l$-dimension of the kernel of ϕ. This dimension can be given in terms of the rank of a certain matrix over $\mathbb{F}_l$, called the Rédei matrix of K, as follows.

1. THEOREM (Rédei–Fröhlich). — *Let $K/\mathbb{Q}$ be a cyclic extension of prime degree l with group $\langle\sigma\rangle$ and conductor f, and let $\chi : (\mathbb{Z}/f\mathbb{Z})^* \to \mathbb{F}_l$ be a generator of its character group, with values taken in the additive group $\mathbb{F}_l$. Let $p_1, p_2, \ldots, p_t$ be the prime factors of f, and $\chi = \sum_{i=1}^t \chi_i$ the corresponding decomposition of χ. Then the l-primary part C of the narrow class group of K has $(\sigma-1)^2$-rank*

$$r_2(C) = t - 1 - \mathrm{rank}_{\mathbb{F}_l} R,$$

where the entries $a_{ij} \in \mathbb{F}_l$ of the Rédei matrix $R = (a_{ij})_{i,j=1}^t$ are defined by :

$$a_{ij} = \chi_i(p_j) \qquad \text{if } i \neq j;$$
$$\sum_{i=1}^t a_{ij} = 0.$$

Proof : as $C[\sigma-1]$ is generated by the classes of the ramified primes $\mathfrak{p}_i$, we have a natural surjection $\rho : \mathbb{F}_l^t \to C[\sigma-1]$ that maps the j-th basis vector e_j to the class of $\mathfrak{p}_j$. The group $C/C^{\sigma-1}$ is canonically isomorphic to the subgroup $\mathrm{Gal}(H_1/K)$ of $\mathrm{Gal}(H_1/\mathbb{Q})$ under the Artin map. We know H_1 explicitly from genus theory : it is the compositum of the cyclic fields $\mathbb{Q}(\chi_i)$ of conductor a power of p_i corresponding to the characters χ_i. Each character χ_i furnishes an isomorphism $\mathrm{Gal}(\mathbb{Q}(\chi_i)/\mathbb{Q}) \xrightarrow{\sim} \mathbb{F}_l$, and they can be combined into an isomorphism $\oplus_{i=1}^t \chi_i : \mathrm{Gal}(H_1/\mathbb{Q}) \xrightarrow{\sim} \mathbb{F}_l^t$. The Rédei map $R : \mathbb{F}_l^t \to \mathbb{F}_l^t$ is defined as the composed map

$$R : \mathbb{F}_l^t \xrightarrow{\rho} C[\sigma-1] \xrightarrow{\phi} C/C^{\sigma-1} \xrightarrow{\sim} \mathrm{Gal}(H_1/K) \subset \mathrm{Gal}(H_1/\mathbb{Q}) \xrightarrow{\oplus\chi_i} \mathbb{F}_l^t$$

of vector spaces over $\mathbb{F}_l$. As the kernel of ρ is of dimension 1, one has

$$r_2(C) = \dim_{\mathbb{F}_l}[\ker\phi] = \dim_{\mathbb{F}_l}[\ker R] - 1 = t - 1 - \mathrm{rank}_{\mathbb{F}_l} R, \tag{2}$$

as desired. The image of a basis vector e_j is the Artin symbol of $\mathfrak{p}_j$ in $\mathrm{Gal}(H_1/K)$. If $i \neq j$, the restriction of this symbol to $\mathrm{Gal}(\mathbb{Q}(\chi_i)/\mathbb{Q})$ is the Artin symbol of p_j, and this is mapped to $a_{ij} = \chi_i(p_j)$ by χ_i. For the diagonal entry a_{ii} the Artin symbol and $\chi_i(p_i)$ are not defined, but we can use the fact that $\oplus\chi_i$ maps $\mathrm{Gal}(H_1/K)$ to the hyperplane $\{(a_i)_i \in \mathbb{F}_l^t : \sum_{i=1}^t a_i = 0\}$. The desired identity $\sum_{i=1}^t a_{ij} = 0$ follows immediately. ∎

The Rédei matrix R is by definition a singular $(t \times t)$-matrix since the sum of its rows is zero. It is said to have *maximal rank* if the rank equals $t-1$. Obviously, the rank is maximal if and only if $r_2(C) = 0$. We will meet this condition in the next section when investigating the solvability of the negative Pell equation.

The field H_2 corresponding to the quotient $C/C^{(\sigma-1)^2}$ is the central l-class field of K, i.e. the largest unramified extension E of H_1 that is normal over $\mathbb{Q}$ and for which the group extension

$$0 \longrightarrow \mathrm{Gal}(E/H_1) \longrightarrow \mathrm{Gal}(E/\mathbb{Q}) \longrightarrow \mathrm{Gal}(H_1/\mathbb{Q}) \longrightarrow 0$$

is a central extension. In Fröhlich's terminology [[7]], the central class field H_2 is a field of class two : its Galois group $\Omega = \mathrm{Gal}(H_2/\mathbb{Q})$ is not in general abelian but its lower central series has length at most two. This is equivalent to saying that the commutator subgroup $[\Omega, \Omega]$ is contained in the center of Ω, or that $[\Omega, [\Omega, \Omega]] = 0$. The Rédei-Fröhlich method enables us to obtain $\mathrm{Gal}(H_2/K)$ from very simple rational data. More precisely, we can determine $r_i = r_i(C)$ for $i = 1, 2$ in terms of the prime factors of the discriminant, and this leads to

$$\begin{aligned}\mathrm{Gal}(H_2/K) &\cong_A (A/A^{\sigma-1})^{r_1-r_2} \times (A/A^{(\sigma-1)^2})^{r_2} \\ &\cong \begin{cases} (\mathbb{Z}/2\mathbb{Z})^{r_1-r_2} \times (\mathbb{Z}/4\mathbb{Z})^{r_2} & \text{if } l = 2; \\ (\mathbb{Z}/l\mathbb{Z})^{r_1+r_2} & \text{if } l > 2. \end{cases}\end{aligned}$$

The first isomorphism is an isomorphism of modules over the ring $A = \mathbb{Z}_l[G]/N$, the second is an isomorphism of abelian groups.

3. — Applications

As a first application, we will obtain divisibility results for the real cyclotomic class numbers h_n^+ of the type discussed in the introduction. Recall that h_n^+ is the class number of the maximal real subfield $F_n = \mathbb{Q}(\zeta_n + \zeta_n^{-1})$ of the cyclotomic field of conductor n.

3. LEMMA. — *Suppose that $l > 2$ and that the l-class group C of the field K in theorem 1 has $(\sigma-1)^2$-rank $r_2(C) = r$. Then l^r divides h_n^+, with n the conductor of K.*

Proof : as K is real of conductor n, it is contained in F_n. The genus field H_1 of K is equal to $H \cap F_n$, so H_2F_n/F_n is an unramified abelian extension of F_n of degree $[H_2 : H_1] = l^r$. This degree divides h_n^+ by class field theory, so we are done. ∎

This lemma provides us with an easy method of constructing infinitely many n for which h_n^+ is divisible by an arbitrarily high power of a prime number

l. One simply takes those n for which F_n contains a subfield K of degree l over $\mathbb{Q}$ for which the Rédei-matrix is of rank much smaller than $t-1$. By taking it equal to the zero matrix, the following result is obtained.

4. THEOREM. — *Let n be divisible by t distinct primes congruent to* 1 mod l *such that each of these primes is an l-th power modulo all others. Then l^{t-1} divides h_n^+.* ∎

For fixed t, there are infinitely many pairwise coprime n satisfying the hypothesis of the theorem. This follows easily from Dirichlet's theorem on primes in arithmetic progressions. If $t-1$ primes congruent to 1 mod l have been chosen such that each is an l-th power modulo the others, the t-th prime that makes the hypothesis of the theorem hold true can be chosen from an infinite collection of Dirichlet density $[(l-1)l^{2(t-1)}]^{-1}$. This follows from the Čebotarev density theorem, as the condition on the t-th prime is that for each of the $t-1$ previous primes p, it splits completely in the field E_p that is obtained by adjoining an l-th root of unity ζ_l and $\sqrt[l]{p}$ to the subfield of degree l in the p-th cyclotomic field. The fields E_p are all of degree $(l-1)l^2$ over $\mathbb{Q}$, and they are linearly disjoint over $\mathbb{Q}(\zeta_l)$. As a very special case, we obtain the claim made in the introduction that h_{13p}^+ is divisible by 3 for a set of primes p of Dirichlet density $1/18$.

For odd l, results similar to those in the preceding theorem have been proved by Cornell and Rosen [[3], 1984] using cohomological methods that go back to Furuta [[9]]. For $t=2$ and $t=3$ their results are identical to those following from lemma 3, for large t their method is better. Neither method gives any result for the prime conductor case $t=1$.

For $l=2$, the lemma and the arguments given above have to be adapted for several reasons. First of all one has to take care of the ramification of real primes. This leads one to consider only those quadratic fields K that have real genus fields, i.e. real quadratic fields for which all odd prime divisors of the discriminant are congruent to 1 modulo 4. One only obtains a divisibility result $2^{r-1}|h_n^+$, which is weaker than lemma 3. In order to find $2^r|h_n^+$ one needs to show that H_2 is real. This can sometimes be done using a method of Scholz [[25]]. Secondly, one has to adapt the density computation given above as the fields E_p have smaller degree, a statement equivalent to the quadratic reciprocity law. Rather than working out all details here, we give a characteristic example. It is stronger than the cohomological result in [[3]].

5. THEOREM. — *Let p and q be primes congruent to* 1 *modulo* 4 *that are mutual quadratic residues, and suppose that the fourth power residue symbols $\left(\frac{p}{q}\right)_4$ and $\left(\frac{q}{p}\right)_4$ are equal. Then h_{pq}^+ is even.* ∎

It is not difficult to see that we obtain the claim in the introduction for $q=13$. For $n=pq=5\cdot 29=145$ it follows that h_{145}^+ is even. This is one of the two values $n < 200$ with n not a prime power for which $h_n^+ > 1$. In fact, one can use Odlyzko's discriminant minorations to show [[31]] that $h_{145}^+ = 2$.

There are no results of a similar algebraic nature in the prime conductor case $t = 1$, and we cannot produce infinite families of primes p for which h_p^+ is even. See [[28]] for a more complete discussion.

A second application of the technique of Rédei matrices arises in the study of the solvability in integers of the negative Pell equation $x^2 - Dy^2 = -1$, where $D > 1$ is a squarefree integer. With ϵ_D a fundamental unit in $\mathbb{Q}(\sqrt{D})$ and N the norm to $\mathbb{Q}$ one has

$$x^2 - Dy^2 = -1 \text{ is solvable in integers} \iff N\epsilon_D = -1.$$

Indeed, if the equation is solvable there are units of norm -1, so the fundamental unit cannot have norm $+1$. Conversely, if $N\epsilon_D = -1$ it may be that ϵ_D is not in $\mathbb{Z}[\sqrt{D}]$, but as its cube ϵ_D^3 always is we still get an integral solution to the equation. As it is more natural to work with discriminants than radicands, we will further take D to be a quadratic discriminant and say that the negative Pell equation is solvable for D if the equation $x^2 - Dy^2 = -4$ has integral solutions. If the equation is solvable for D, then D is positive and -1 is a quadratic residue modulo every prime divisor of D, so D must be in the set $\mathcal{D}$ of real quadratic discriminants that are not divisible by any prime congruent to 3 mod 4. A question that has been studied by many people but that is still completely open is the following.

6. PROBLEM. — *Let $\mathcal{D}(-1) \subset \mathcal{D}$ be the set of real quadratic discriminants for which the negative Pell equation is solvable. Decide whether the limit*

$$\lim_{X\to\infty} \frac{\#\{D \in \mathcal{D}(-1) : D \leq X\}}{\#\{D \in \mathcal{D} : D \leq X\}}$$

exists and if so, determine it.

This is a very hard problem, and to my knowledge is is not even known whether the liminf and the limsup of this expression are in the open interval $(0, 1)$.

The relation between the solvability of the negative Pell equation and the previous section is given by the following immediate consequence of class field theory.

7. LEMMA. — *The negative Pell equation is solvable for a quadratic discriminant D if and only if the narrow 2-Hilbert class field of $\mathbb{Q}(\sqrt{D})$ is real.*

Proof : both statements are equivalent to the fact that $\mathbb{Q}(\sqrt{D})$ is a real field for which the narrow Hilbert class field coincides with the ordinary Hilbert class field. ∎

Let H be the narrow 2–Hilbert class field of $K = \mathbb{Q}(\sqrt{D})$. This is the situation of the preceding section, with l equal to 2. From the lemma, we see that the negative Pell equation is solvable for D if and only H_k is real for all $k \geq 1$. The condition that the genus field H_1 is real is equivalent to the requirement that D is in $\mathcal{D}$, since H_1 is obtained from K by adjoining a square root of $(-1)^{(p-1)/2}$ for each odd prime divisor p of D. If $H = H_1$ this condition is also sufficient for solvability of the negative Pell equation.

8. Lemma. — *The negative Pell equation is solvable for $D \in \mathcal{D}$ if the Rédei matrix of $\mathbb{Q}(\sqrt{D})$ has maximal rank.*

Proof : the condition implies an equality $H_1 = H_2$, so $H = H_1$ is real and we are done by the previous lemma. ∎

If $D \in \mathcal{D}$ has t distinct prime divisors, the corresponding Rédei matrix R is by the quadratic reciprocity law a symmetric $(t \times t)$-matrix over $\mathbb{F}_2$ whose rows and columns add up to zero. Let R' be the $(t-1) \times (t-1)$-minor obtained by leaving out the last row and column from R. If D ranges over the subset $\mathcal{D}_t$ of $\mathcal{D}$ consisting of those discriminants that have exactly t distinct prime divisors, it is intuitively clear that the corresponding Rédei minor R'_D behaves like a random symmetric $(t-1) \times (t-1)$-matrix over $\mathbb{F}_2$, i.e. that

$$\lim_{X\to\infty} \frac{\#\{D \in \mathcal{D}_t : D \leq X \quad \text{and} \quad R'_D = S\}}{\#\{D \in \mathcal{D}_t : D \leq X\}}$$

exists and does not depend on the choice of the symmetric matrix S. The statement is a reformulation of the fact that the vector consisting of the $\binom{t}{2}$ Legendre symbols $\left(\frac{p_i}{p_j}\right)$ of an element $D = p_1 p_2 \ldots p_t$ is randomly distributed as a function on $\mathcal{D}_t$. The details of a correct proof are not trivial. Rédei's original proof [[22]] proceeds by induction on t, and so does the proof of the rediscovery of the result in [[5]]. There is also an easy way out by adapting the notion of density [[17]].

Once one knows that a discriminant $D \in \mathcal{D}_t$ gives rise to a random symmetric matrix, one can determine how likely it is that such a matrix is non–singular. We give a slightly more general result for future reference.

9. Proposition. — *Let $n \geq 1$ be an integer and q a prime power. Then there are*

$$A_n(q) = q^{\binom{n+1}{2}} \prod_{\substack{1 \leq k \leq n \\ k \text{ odd}}} (1 - q^{-k})$$

symmetric $(n \times n)$-matrices over the field of q elements $\mathbb{F}_q$ that are non-singular. The number of matrices of arbitrary rank $r \in \{0, 1, \ldots n\}$ equals $\left[{n \atop r}\right]_q A_r(q)$, where $\left[{n \atop r}\right]_q$ denotes the number of r-dimensional subspaces of a vector space of dimension n over $\mathbb{F}_q$.

Proof : the result for $q = 2$ occurs in rather cumbersome terminology and with a lengthy proof in [[22]]. A completely elementary proof by induction on n can be found in [[13]]. In order to see that the statement given there is identical to ours one needs the explicit value

$$\begin{bmatrix} n \\ r \end{bmatrix}_q = \frac{\prod_{i=1}^{n}(q^i - 1)}{\prod_{i=1}^{r}(q^i - 1)\prod_{i=1}^{n-r}(q^i - 1)}.$$

The first half of the proposition immediately implies the second half, as symmetric matrices correspond bijectively to symmetric bilinear forms and giving a symmetric bilinear form of rank r on $V = \mathbb{F}_q^n$ is equivalent to giving a subspace $W \subset V$ of dimension $n - r$ and a non-degenerate symmetric bilinear form of the factor space V/W. This remark also shows that the numbers $A_n(q)$ can be computed inductively from the relation

$$\sum_{r=0}^{n} \begin{bmatrix} n \\ r \end{bmatrix}_q A_r(q) = q^{\binom{n+1}{2}},$$

so it suffices to check that the given expression satisfies this relation. An elegant way of doing this is given in [[5]]. ∎

We will only use the preceding proposition for $q = 2$ and $n = t - 1$, so we write $A_{t-1}(2) = 2^{\binom{t}{2}}\alpha_t$ with

$$\alpha_t = \prod_{j=1}^{[t/2]}(1 - 2^{1-2j}).$$

Set $\mathcal{D}_t(-1) = \mathcal{D}_t \cap \mathcal{D}(-1)$. We now know that for *fixed* t, we have a lower density for $\mathcal{D}_t(-1)$ in $\mathcal{D}_t$, since the two preceding lemmas imply

$$\liminf_{X \to \infty} \frac{\#\{D \in \mathcal{D}_t(-1) : D \le X\}}{\#\{D \in \mathcal{D}_t : D \le X\}} \ge \alpha_t > \alpha_\infty = \prod_{j=1}^{\infty}(1 - 2^{1-2j}).$$

The numerical value $\alpha_\infty = .4194224\ldots$ is already in [[22]]. The density result has been reproved in [[17]], [[11]] and [[5]]. The formulation given by these authors is different from Rédei's, as they interpret the Rédei matrix as an incidence matrix of a graph on t points.

The equations $\mathcal{D} = \cup_{t\geq 1}\mathcal{D}_t$ and $\mathcal{D}(-1) = \cup_{t\geq 1}\mathcal{D}_t(-1)$ make it very plausible that the lower density of $\mathcal{D}(-1)$ in $\mathcal{D}$ is not smaller than α_∞. However, I do not know how to prove this. The problem is that each $\mathcal{D}_t$ is a subset of zero density in $\mathcal{D}$, and it seems non-trivial to prove a density result for $\mathcal{D}$ from a density result for each of the subsets $\mathcal{D}_t$.

The preceding argument can be further refined in order to obtain a still higher value of the lower density of $\mathcal{D}_t(-1)$ in $\mathcal{D}_t$. In particular, we will push the limit value for $t \to \infty$ over the value $1/2$ that has been suggested as a possible value [[17]]. However, we need to pass to the next higher level, i.e. the field H_3, in order to do this.

4. — Higher levels

In principle, the Rédei-Fröhlich method for determining $r_2(C)$ can be extended to determine inductively all values $r_k(C)$. Having defined the first Rédei map

$$R = R_1 : \mathbb{F}_l^t \longrightarrow C[\sigma - 1] \xrightarrow{\phi} C/C^{\sigma-1} \rightarrowtail \mathbb{F}_l^t,$$

one can repeat the procedure and consider the higher Rédei maps

$$R_k : \quad \ker R_{k-1} \longrightarrow C[\sigma - 1] \cap C^{(\sigma-1)^{k-1}} \xrightarrow{\text{can}} C^{(\sigma-1)^{k-1}}/C^{(\sigma-1)^k}.$$

Just as in the case $k = 1$, we obtain the $(\sigma - 1)^{k+1}$-rank from this map by

$$r_{k+1}(C) = \dim_{\mathbb{F}_l}[\ker R_k] - 1 = r_k(C) - \operatorname{rank}_{\mathbb{F}_l} R_k,$$

which is the analogue of (2). Despite the close analogy, a serious complication arises for these higher levels. For $k = 1$, we were able to embed $C^{(\sigma-1)^{k-1}}/C^{(\sigma-1)^k}$ in a canonical way in a vector space of dimension t over $\mathbb{F}_l$. This was due to the fact that we could describe the genus field H_1 very explicitly in terms of Dirichlet characters. The fields H_k for $k \geq 2$ are no longer abelian over $\mathbb{Q}$, and no general method is known to describe them explicitly. This is a serious drawback that accounts for the fact that there is no generalisation of theorem 1 to higher levels that is of a comparable simplicity. For the same reason, we do not have general density results for these levels that resemble those in the preceding section.

Only in the special case where $k = l = 2$, there is a more explicit version of the theory that goes back to Rédei [[22]] and was further developed by Fröhlich [[7]]. We can formulate it in modern terms as follows.

Let D be a quadratic discriminant, and $D = \prod_{i=1}^t d_i$ its factorization into prime power discriminants. The set V of discriminantal divisors of D is defined as the set of divisors d of D of the form $d = \prod_{i=1}^t d_i^{e_i}$ with $e_i \in \{0, 1\}$. This is in a natural way a vector space of dimension t over $\mathbb{F}_2$ with a canonical basis consisting of the divisors d_i. The natural surjection

$V \longrightarrow C[2]$ maps d_i to the class of the ramified prime $\mathfrak{d}_i$ of K that divides d_i. As the genus field H_1 of $K = \mathbb{Q}(\sqrt{D})$ is generated over $\mathbb{Q}$ by the square roots $\sqrt{d_i}$, Kummer theory tells us that the Galois group $\mathrm{Gal}(H_1/\mathbb{Q})$ can be seen as the dual space $V^* = \mathrm{Hom}(V, \mathbb{F}_2)$ of V. The kernel of the Rédei map $R_1 : V \to V^*$ consists of divisors $d \in V$ for which the associated Artin symbol $\sigma_{\mathfrak{d}} \in \mathrm{Gal}(H/K)$ is the identity on H_1. The kernel of the dual map $R_1^* : V = V^{**} \to V^*$ consists of those $d \in V$ for which $\sqrt{d}$ is left invariant by the Artin symbols of all ideals that have order 2 in the class group. Note that D itself is always in this kernel.

A decomposition $D = D_1 \cdot D_2$ with $D_1 \in \ker R_1^*$ is called a *decomposition of the second kind*. These decompositions are characterized by the fact D_1 is a square modulo all prime divisors of D_2 and vice versa. The prime 2 needs special attention here. Given a decomposition $D = D_1 \cdot D_2$ that is of the second kind, Rédei explicitly constructs a quadratic extension of $\mathbb{Q}(\sqrt{D_1}, \sqrt{D_2})$ that is cyclic of degree 4 and unramified over K. This is possible since the equation $x^2 - D_1y^2 - D_2z^2 = 0$ has non-zero rational solutions by Legendre's theorem and the assumption on the decomposition. For a primitive integral solution (x, y, z) with well-chosen 2-adic behavior, the extension E that is generated over K by a square root γ_{D_1} of $x + y\sqrt{D_1}$ has the desired properties. The extension E/K depends on the choice of the solution, but the quadratic extension $EH_1 = H_1(\gamma_{D_1})$ of H_1 does not. Every element $\sigma \in \mathrm{Gal}(H_2/H_1)$ is determined by its action on the elements γ_{D_1} for $D_1 \in \ker R_1^*$, so we can view this Galois group as a subspace of $\mathrm{Hom}(\ker R_1^*, \mathbb{F}_2) = (\ker R_1^*)^*$. With these identifications, we can describe the second Rédei map :

$$R_2 : \ \ker R_1 \longrightarrow C[2] \cap C^2 \longrightarrow C^2/C^4 \xrightarrow{\sim} \mathrm{Gal}(H_2/H_1) \subset (\ker R_1^*)^*$$

explicitly as an $\mathbb{F}_2$-linear map between vector spaces of dimension $1 + r_2(C)$. The 8-rank $r_3(C)$ of the narrow class group of K is given by the formula

$$r_3(C) = r_2(C) - \mathrm{rank}_{\mathbb{F}_l} R_2,$$

which is non-negative as R_2 is always singular.

As soon as one chooses a basis for $\ker R_1$ and for the space $\ker R_1^*$ of decompositions of the second kind, R_2 is given by a matrix whose entries describe the action of the Artin symbols $\sigma_{\mathfrak{d}}$ coming from $d \in \ker R_1$ on explicit elements $\gamma_{d'}$ with $d' \in \ker R_1^*$. Note the equality $R_1 = R_1^*$ in case D is in the set $\mathcal{D}$ of discriminants that are of interest for the negative Pell equation. The entries of R_2 are quadratic symbols of quadratic irrationals and can be computed rather easily. Rédei's paper [[22]] has numerous identities that express these 'new number theoretic symbols', as he calls them, in rational terms, and Fröhlich [[8]] does the same in a more systematic way. However, these expressions are usually given in terms of the chosen solution of $x^2 - D_1y^2 - D_2z^2 = 0$, and this makes it difficult to obtain density results in terms of the prime divisors of D. Special cases have been dealt with by Morton [[14]–[16]], and density statements for the behavior of C/C^8 have been proved by the author [[26]] in the case that $t-1$ prime divisors of the discriminant are fixed and the last one varies. For $t = 2$ this yields results that had been known for some time.

In the previous section, we showed that the negative Pell equation is solvable for all D in a subset of density α_t of $\mathcal{D}_t$ since these D have $r_2(C) = 0$. Following an idea that goes back to Rédei [[21]] and Scholz [[25]], we can use the 8-rank theory to enlarge this set even further by looking at those D that have $r_2(C) = 1$. We will indicate briefly how this is done.

The density β_t of the set of $D \in \mathcal{D}_t$ having $r_2(C) = 1$ follows easily from proposition 9. One has $\beta_t = \alpha_t$ if t is even and $\beta_t = (1 - 2^{1-t})\alpha_t$ if t is odd. Note that $\lim_{t\to\infty} \beta_t = \lim_{t\to\infty} \alpha_t = \alpha_\infty$. For D as above, there is exactly one non-trivial decomposition $D = D_1D_2$, and one can show that the higher Rédei matrix R_2 equals

$$R_2 = \begin{pmatrix} \left(\frac{D_1}{D_2}\right)_4 & \left(\frac{D_1}{D_2}\right)_4 \\ \left(\frac{D_1}{D_2}\right)_4 & \left(\frac{D_2}{D_1}\right)_4 \end{pmatrix},$$

which has to be interpreted in the obvious way as a matrix over $\mathbb{F}_2$. As the biquadratic residue symbols have value ± 1, there are 4 possible values for this matrix, and they each occur for a set of D that has density $\frac{1}{4}\beta_t$ in $\mathcal{D}_t$. If $\left(\frac{D_1}{D_2}\right)_4$ and $\left(\frac{D_1}{D_2}\right)_4$ are both equal to 1, the matrix R_2 is the zero matrix and H_2 is strictly smaller than the 2-Hilbert class field H. In all other cases, its rank is one and $H = H_2$. We can determine whether H_2 is real by a generalization of the argument used in proving theorem 5. One has :

$$H_2 \text{ is real} \iff \left(\frac{D_1}{D_2}\right)_4 \left(\frac{D_1}{D_2}\right)_4 = 1.$$

It follows that $H = H_2$ is totally complex and the Pell equation is not solvable if the biquadratic residue symbols are not equal, which happens for a collection of discriminants of density $\beta_t/2$ in $\mathcal{D}_t$. If both symbols equal -1, the Pell equation is solvable, and this happens for a set of density $\beta_t/4$. Taking together the two collections of D for which the Pell equation is solvable, we conclude that for each fixed t, the set $\mathcal{D}_t(-1)$ has lower density $\alpha_t + \frac{1}{4}\beta_t$ and upper density $1 - \frac{1}{2}\cdot\beta_t$ inside $\mathcal{D}_t$. For increasing t these values rapidly converge to :

$$\frac{5}{4}\cdot\alpha_\infty = .52428\ldots$$
$$1 - \frac{1}{2}\cdot\alpha_\infty = .79029\ldots$$

It remains a challenging problem to deduce any non-trivial density result for $\mathcal{D}(-1)$ in $\mathcal{D}$.

Hardly anything is known about the distribution of the ranks $r_k(C)$ when $k \geq 4$. Some numerical data are available in the quadratic case [[2], [27]], mainly for cyclic C. The best known example is probably that of the quadratic field $\mathbb{Q}(\sqrt{-p})$, with p a prime congruent to 1 mod 4. In this case the discriminant $D = -4p$ has two prime divisors and C is a non-trivial cyclic 2-group. It follows from theorem 1 that the order of C is divisible by 4 exactly when $p \equiv \bmod 8$, and the 8-rank results quoted above imply that the order is divisible by 8 if and only if p splits completely in $\mathbb{Q}(\zeta_8, \sqrt{1+i})$, which is a non-abelian field of degree 8 over $\mathbb{Q}$. All numerical evidence suggests strongly that the order of C is divisible by 16 for a set of primes of density $1/16$, but the existing techniques do not even suffice to show that this happens infinitely often. The question is closely related to the 2-adic behavior of the fundamental unit ϵ_p in the field $\mathbb{Q}(\sqrt{p})$, see [[27]].

Note added in proof

It is now conjectured that the limit value in problem 6 exists and equals $1 - \alpha_\infty = .5805775582\ldots$, with α_∞ as in section 3, see [[29]]. The heuristics can be extended to the case of quadratic orders [[30]]. They have been confirmed by extensive computer calculations [[1]].

Manuscrit reçu le 7 septembre 1994

References

[1] W. Bosma, P. Stevenhagen. — *Density computations for real quadratic units*, preprint (1994).

[2] H. Cohn, J.C. Lagarias. — *On the existence of fields governing the 2-invariants of the class group of* $\mathbb{Q}(\sqrt{dp})$ *as* p *varies*, Math. Comp. **41**, 711–730 (1983).

[3] G. Cornell, M.I. Rosen. — *The* ℓ*-rank of the real class group of cyclotomic fields*, Compositio Math. **53**, 133–141 (1984).

[5] J. E. Cremona, R.W.K. Odoni. — *Some density results for negative Pell equations; an application of graph theory*, J. London Math. Soc. (2) **39**, 16–28 (1989).

[6] A. Fröhlich. — *The generalization of a theorem of L. Rédei's*, Quart. J. Math. Oxford (2) **5**, 130–140 (1954).

[7] A. Fröhlich. — *On fields of class two*, Proc. Lond. Math. Soc. (3) **4**, 235–256 (1954).

[8] A. Fröhlich. — *A prime decomposition symbol for certain non Abelian number fields*, Acta Sci. Math. **21**, 229–246 (1960).

[9] Y. Furuta. — *On class field towers and the rank of ideal class groups*, Nagoya Math. J. **48**, 147–157 (1972).

[10] G. Gras. — *Sur les* l*-classes d'idéaux dans les extensions cycliques relatives de degré premier* l, Ann. Inst. Fourier, Grenoble **23**,3, 1–48 (1973).

[11] J. Hurrelbrink. — *On the norm of the fundamental unit*, preprint, Louisiana State University (1990).

[12] E. Inaba. — *Über die Struktur der* l*-Klassengruppe zyklischer Zahlkörper vom Primzahlgrad* l, J. Fac. Sci. Imp. Univ. Tokyo, section I, vol. IV **2**, 61–115 (1940).

[13] J. MacWilliams. — *Orthogonal matrices over finite fields*, Amer. Math. Monthly **76**, 152–164 (1969).

[14] P. Morton. — *Density results for the 2–classgroups of imaginary quadratic fields*, J. reine angew. Math. **332**, 156–187 (1982).

[15] P. Morton. — *Density results for the 2-classgroups and fundamental units of real quadratic fields*, Studia Scientiarum Math. Hungarica **17**, 21–43 (1982).

[16] P. Morton. — *The quadratic number fields with cyclic 2-class groups*, Pac. J. Math. **108**, 165–175 (1983).

[17] R. V. PERLIS. — *On the density of fields with* $N(\epsilon) = -1$, preprint, Louisiana State University (1990).

[18] L. RÉDEI, H. REICHARDT. — *Die Anzahl der durch 4 teilbaren Invarianten der Klassengruppe eines beliebigen quadratischen Zahlkörpers*, J. reine angew. Math. **170**, 69–74 (1934).

[19] L. RÉDEI. — *Arithmetischer Beweis des Satzes über die Anzahl der durch vier teilbaren Invarianten der absoluten Klassengruppe im quadratischen Zahlkörper*, J. reine angew. Math. **171**, 55–60 (1935).

[20] L. RÉDEI. — *Über die Grundeinheit und die durch 8 teilbaren Invarianten der absoluten Klassengruppe im quadratischen Zahlkörper*, J. reine angew. Math. **171**, 131–148 (1935).

[21] L. RÉDEI. — *Über einige Mittelwertfragen im quadratischen Zahlkörper*, J. reine angew. Math. **174**, 131–148 (1936).

[22] L. RÉDEI. — *Ein neues zahlentheoretisches Symbol mit Anwendungen auf die Theorie der quadratischen Zahlkörper*, J. reine angew. Math. **180**, 1–43 (1939).

[23] L. RÉDEI. — *Bedingtes Artinsches Symbol mit Anwendungen in der Klassenkörpertheorie*, Acta Math. Acad. Sci. Hung. **4**, 1–29 (1953).

[24] L. RÉDEI. — *Die 2-Ringklassengruppe des quadratischen Zahlkörpers und die Theorie der Pellschen Gleichung*, Acta Math. Acad. Sci. Hung. **4**, 31–87 (1953).

[25] A. SCHOLZ. — *Über die Lösbarkeit der Gleichung* $t^2 - Du^2 = -4$, Math. Zeitschrift **39**

[26] P. STEVENHAGEN. — *Class groups and governing fields*, Publ. Math. Fac. Sci. Besançon, année 1989/90, 1–94 (1990).

[27] P. STEVENHAGEN. — *On the 2-power divisibility of certain quadratic class numbers*, J. Number Theory **43** (1), 1–19 (1993).

[28] P. STEVENHAGEN. — *Class number parity for the p-th cyclotomic field*, Math. Comp. **63** no. 208 (to appear, 1994).

[29] P. STEVENHAGEN. — *The number of real quadratic fields having units of negative norm*, Exp. Math. **2** (2), 121–136 (1993).

[30] P. STEVENHAGEN. — *Frobenius distributions for real quadratic orders*, J. Théorie des Nombres Bordeaux (to appear, 1995).

[31] F.VAN DER LINDEN. — *Class number computations of real abelian number fields*, Math. Comp. **39**, 693–707 (1982).

Peter Stevenhagen
Faculteit Wiskunde en Informatica
Plantage Muidergracht 24
1018 TV Amsterdam, Netherlands
e-mail : `psh@fwi.uva.nl`

Number Theory
Paris 1992–93

Decomposition of the integers as a direct sum of two subsets

R. Tijdeman

1. — Introduction

Two subsets A and B of a set C induce a decomposition of C if every element of C has a unique representation $a + b$ with $a \in A$, $b \in B$. Notation : $C = A \oplus B$. We call A and B complementing C–pairs. A first study of such pairs arose in the forties from Hajós' proof of Minkowski's conjecture on systems of linear inequalities. Hajós reduced this conjecture to an equivalent statement on decompositions of finite abelian groups, which he was able to prove. A survey of the work on decompositions of finite abelian groups is given in Section 2.

The question of characterising all complementing $\mathbb{Z}$–pairs seems first to have been stated by de Bruijn in 1950. De Bruijn came to the problem while he studied bases for the integers. Let A be a finite set of integers including 0. A set of integers $\{b_1, b_2, \ldots\}$ is called an A–base whenever any integer x can be expressed uniquely in the form

$$x = \sum_{i=1}^{\infty} \varepsilon_i b_i \qquad (\varepsilon_i \in A, \ \sum_{i=1}^{\infty} |\varepsilon_i| < \infty).$$

An A–base is called simple if it can be rearranged in the form $\{d_1, hd_2, h^2 d_3, \ldots\}$ where h denotes the cardinality of A and $d_1, d_2, d_3, \ldots$ are integers. De Bruijn [2] considered the special case where the elements of A have no common factor and where h is a prime. He conjectured that under these assumptions $A \oplus B = \mathbb{Z}$ implies that B is the set of multiples of h. He remarked that a proof of his conjecture would imply that every A-base is simple. For later work on A-bases we refer to de Bruijn [6], Long and Woo [16], Swenson and Long [28].

In 1974, Swenson [27] showed that there is no effective characterisation of all complementing $\mathbb{Z}$–pairs. More precisely, he showed that any two finite sets of integers A, B with the property that all sums $a + b$ ($a \in A$, $b \in B$)

are distinct, can be extended to two infinite complementing $\mathbb{Z}$-pairs. For a similar construction, see Post [20].

In contrast, there is a particularly nice characterisation of all complementing $\mathbb{Z}_{\geq 0}$-pairs. The result, which was implicit in the work of de Bruijn [5], was rediscovered by Vaidya [30]. It is obvious that $A \cap B = \{0\}$ and $1 \in A \cup B$. Suppose $1 \in A$. Then A and B are infinite complementing $\mathbb{Z}_{\geq 0}$-pairs if and only if there exists an infinite sequence of integers $\{m_i\}_{i \geq 1}$ with $m_i \geq 2$ for all i, such that A and B are the sets of all finite sums of the form

$$a = \sum_{i=0}^{\infty} x_{2i} M_{2i}, \quad b = \sum_{i=0}^{\infty} x_{2i+1} M_{2i+1},$$

where $0 \leq x_i < m_{i+1}$ for $i \geq 0$ and $M_0 = 1$ and $M_i = \prod_{j=1}^{i} m_j$ for $i \geq 1$. If A or B is a finite set, a similar characterisation holds with the change that the sequence $\{m_i\}$ will be of finite length r and the only restriction on x_r is that it be nonnegative. C.T. Long [15] gave a corresponding characterisation of all complementing C-pairs in case $C = \{0, 1, \ldots, n-1\}$. He also showed that in this case the number $C(n)$ of complementing C-pairs is the same as the number of ordered nontrivial factorisations of n. The number $C(n)$ is determined by

$$2C(n) = \sum_{d|n} C(d) \qquad (n \in \mathbb{Z}_{>1}).$$

Hansen [14] and Niven [18] generalised these results to a characterisation of the complementing pairs of the set $\mathbb{Z}_{\geq 0} \times \mathbb{Z}_{\geq 0}$. Long [15] made the interesting observation that it follows from the above characterisation that if A and B are infinite sets such that $A \oplus B = \mathbb{Z}_{\geq 0}$ then $A \oplus (-B) = \mathbb{Z}$ (see also Brown [1]). In particular, we can take for A the set of finite sums of odd powers of 2 and for B the set of finite sums of even powers of 2. Eigen and Hajian [31] showed that if A and B are infinite sets such that $A \oplus B = \mathbb{Z}_{\geq 0}$, then there exists a continuum number of sets $\tilde{B}$ such that $A \oplus \tilde{B} = \mathbb{Z}$.

In Section 3 we formulate a conjecture which, if true, provides an inductive characterisation of all complementing sets A, B for which the cardinality of A is fixed integer n. It was already observed by Hajós [12] and de Bruijn [5] p. 240 that B is periodic if A is finite. In this way the problem is reduced to a finite problem which can be stated in terms of finite cyclic groups. If n is a prime number, then our conjecture coincides with de Bruijn's one stated above. This conjecture was proved by Sands [24] in 1957. We shall show how a proof of our conjecture can be derived from Sands' results if n is a prime power. The general case remains open.

We shall further show by a combinatorial argument that if m is coprime to n and $A \oplus B = \mathbb{Z}$, then $mA \oplus B = \mathbb{Z}$ (we define $mA = \{ma | a \in A\}$). On using this result we give an alternative proof of de Bruijn's conjecture.

The problem of characterising all sets B such that $A \oplus B = \mathbb{Z}$ in the special case where A consists of the finite sums of odd powers of 2 was posed to me by Yu. Ito. He was interested in the problem because of his joint research with S. Eigen and A. Hajian on exhaustive weakly wandering sequences for ergodic measure preserving transformations [7], [9], [8]. Such a characterisation will be presented in Section 4.

2. — Decomposition of finite abelian groups

About one century ago Minkowski [17] proved the following fundamental result on the geometry of numbers :

Let $\xi_1, \ldots, \xi_n$ be homogeneous linear forms in the variables $x_1, \ldots, x_n$ with real coefficients and determinant 1. Then there exist integers $x_1, \ldots, x_n$, not all zero, such that

$$|\xi_1| \leq 1, \ldots, |\xi_n| \leq 1. \tag{1}$$

Since $\xi_1, \ldots, \xi_n$ may have integer coefficients, the equality signs cannot be deleted. Minkowski conjectured that (1) can be replaced by

$$|\xi_1| < 1, \ldots, |\xi_n| < 1 \tag{2}$$

unless at least one of the linear forms has integer coefficients. Minkowski proved the statement for $n \leq 3$. Several mathematicians worked on it and in 1940 it was known to be true for $n \leq 9$. In 1941 Hajós [11] established Minkowski's conjecture in the affirmative. His proof consists of three parts :

(i) reduction to some equivalent geometric statement on k–multiple lattice tiling of the unit cube,

(ii) further reduction to the equivalent group theoretic statement given below.

(iii) proof of this group theoretic statement.

Hajós' theorem is very fundamental and has various aspects. Fáry [10] reformulated it as a result on the structure of commutative compact topological groups.

Now we state Hajós' result in terms of group theory. Let G be a finite abelian group with unit element 1. A subset of G is called a simplex if it is of the form $\{1, \alpha, \alpha^2, \ldots, \alpha^{e-1}\}$ where $\alpha \in G$ has order $\geq e$. Notation $[\alpha]_e$, or briefly $[\alpha]$. It is clear that $[\alpha]$ is a subgroup of G if and only if α has order e. We say that G is the free product of the sets $A_1, \ldots, A_n$ if every element of G has a unique representation $a_1 \cdots a_n$ with $a_j \in A_j$ for $j = 1, \ldots, n$. Hajós' theorem reads as follows.

If G is the free product of n simplices, then at least one of the simplices is a subgroup of G.

Hajós' proof has been simplified by Rédei [22] and Szele [29]. Szele [29] p. 57 conjectured that Hajós' theorem would hold true for any decomposition of the finite abelian group G. A simple example (cf. [13] p. 185) shows that this is false. Let G be the cyclic group defined by $a^8 = 1$ and let $A = \{1, a^2\}$, $B = \{1, a, a^4, a^5\}$. Then none of A and B are subgroups of G whereas G is the free product of A and B. Note, however, that $a^4B = B$. A subset A of G is said to be periodic whenever there exists an element $g \in G$, $g \neq 1$, such that $gA = A$. De Bruijn [2] conjectured that if G is a finite abelian group of order > 1 and G is the free product of the sets A and B, then A or B is periodic. He observed that the assertion is not true if G is the infinite cyclic group generated by g. Szele (cf. [13] p. 185) made the same observation. He took for A the product of the subsets $\{1, g^2\}$, $\{1, g^8\}$, $\{1, g^{32}\}, \ldots$ and for B the product of the subsets $\{1, g^{-1}\}$, $\{1, g^{-4}\}$, $\{1, g^{-16}\}, \ldots$. Here G is the free product of A and B and none of A and B is periodic.

Some years earlier, however, Rédei [21] had published two examples of Hajós which show that de Bruijn's conjecture is false. The simplest example refers to the abelian group generated by the elements a, b, c of orders 4, 4, 2 respectively. This group is the free product of the nonperiodic sets

$$\{1, a\}, \ \{1, b\} \text{ and } \{1, a^2, ab^2, a^3b^2, c, a^2bc, a^2b^3c, b^2c\}.$$

Later, Hajós [13] showed that any finite cyclic group the order of which is the product of three pairwise relatively prime numbers > 1, two of which are composite, can be represented as the free product of two nonperiodic subsets. He gave an explicit example of order $180 = 9 \times 4 \times 5$, which is the smallest number satisfying the conditions.

Let us follow de Bruijn in calling a group good if any factorisation of G as a free product of A and B implies that A or B is periodic and otherwise bad. De Bruijn [3] extended Hajós' result by showing that if $n = d_1d_2d_3$ with $(d_1, d_2) = 1$, $d_3 > 1$ and both d_1 and d_2 are composite numbers, then the cyclic group of order n is bad. The smallest order of this type is 72. De Bruijn [4] gave the explicit example ($g^{72} = 1$)

$$A: \ g^0, \ g^8, \ g^{16}, \ g^{18}, \ g^{26}, \ g^{34}$$
$$B: \ g^{12}, \ g^{17}, \ g^{18}, \ g^{21}, \ g^{24}, \ g^{41}, \ g^{45}, \ g^{48}, \ g^{54}, \ g^{60}, \ g^{65}, \ g^{69}.$$

On the other hand, cyclic groups of the following orders have been proved to be good (p, q, r, s are distinct primes) : p^λ ($\lambda \geq 1$) (Hajós [12]), pq, pqr (Rédei [23]), $p^\lambda q$ ($\lambda > 1$) (De Bruijn [4]), p^2q^2, p^2qr and $pqrs$ (Sands [24]).

This covers all cyclic groups. Already in 1947 Rédei [21] had shown that the non-cyclic group of order p^2 is good. The problem was completely solved by Sands [25, 26] who determined all good finite abelian groups. Sands [24] further proved that if the finite cyclic group G is the free product of the subsets A and B and the cardinality of A is a prime power, then either A or B is periodic. This had been conjectured by de Bruijn ([3] p. 371) for the case that the number of elements of A is prime.

Hajós [12] proposed the question whether every decomposition of a finite abelian group G is quasiperiodic. A factorisation of G as free product of A and B is called quasiperiodic if either A or B, B say, can be split into a number of parts $B_1, B_2, \ldots, B_m$ $(m > 1)$ such that $AB_i = g_i AB_1$ $(i = 1, \ldots, m)$ where the elements $g_1, \ldots, g_m$ form a subgroup of G. De Bruijn's example is quasiperiodic as we can take $B_1 = \{g^{12}, g^{17}, g^{18}, g^{24}, g^{41}, g^{65}\}$ and $g_1 = 1$, $g_2 = g^{36}$. Hajós' example is quasiperiodic as we can take $A = \{1, a, b, ab\}$,

$$B_1 = \{1, a^2, ab^2, a^3b^2\},\ B_2 = \{c, a^2bc, a^2b^3c, b^2c\} \text{ and } g_1 = 1,\ g_2 = c.$$

De Bruijn [4] obtained some partial result on Hajós' question.

3. — A is finite

Suppose $A \oplus B = \mathbb{Z}$ where A is finite. Let $A = \{a_0, a_1, \ldots, a_{n-1}\}$. Since, for any integer x, we have $(A - x) \oplus B = \mathbb{Z}$, we may assume without loss of generality that $0 = a_0 < a_1 < \ldots < a_{n-1}$. If $x = a + b$ with $a \in A$, $b \in B$, put $(x)_A = a$, $(x)_B = b$. The following result of Hajós [12] and de Bruijn [2] p. 240 reduces the problem of characterising all complementing $\mathbb{Z}$-pairs to a problem on finite sets which can be stated in terms of finite cyclic groups.

LEMMA 1. — *The sequence* $\{(x)_A\}_{x\in\mathbb{Z}}$ *is periodic. If the period length is* L *then* n *divides* L *and* B *is periodic* mod L.

Proof : put $M = a_{n-1}$. Consider the $n^M + 1$ vectors

$$((i)_A,\ (i+1)_A, \ldots, (i+M-1)_A) \quad \text{for } i = 0, 1, \ldots, n^M.$$

By the box principle at least two vectors are equal, for $i = s$ and $i = t$ with $s < t$, say. Hence $(x)_A = (x + t - s)_A$ for $x = s, s+1, \ldots, s+M-1$. Suppose k is the smallest integer with $k \geq s + M$ and $(k)_A \neq (k + t - s)_A$. If $(k)_A \neq 0$, then put $k = a + b$ with $a \in A$, $b \in B$. We infer that $(b)_A = 0$ and $s \leq k - M \leq b < k$. Hence $(b + t - s)_A = (b)_A = 0$ which implies $b + t - s \in B$. Since $k + t - s = a + (b + t - s)$, we obtain $(k + t - s)_A = a = (k)_A$, a contradiction. If $(k + t - s)_A \neq 0$, a similar argument yields a contradiction. Thus $(x)_A = (x + t - s)_A$ for all $x \geq s$. By symmetry we also have $(x)_A = (x+t-s)_A$ for all $x < s$. Let the period length of $\{(x)_A\}_{x\in\mathbb{Z}}$ be L. Since $\mathbb{Z} = \cup_{i=0}^{n-1}\{a_i + B\}$, all a_i have the same density in the sequence $\{(x)_A\}_{x\in\mathbb{Z}}$. Therefore, they occur with the same frequency in one period. This implies $n|L$. Since the elements of B are precisely the integers x with $(x)_A = 0$, we have that B is periodic $\bmod L$. □

By simple transformations each complementing $\mathbb{Z}$-pair A, B with A finite can be reduced to the standard situation that A is represented by $a_0 = 0 < a_1 < a_2 < \cdots < a_{n-1} < L$ with $\gcd(a_0, a_1, \ldots, a_{n-1}) = 1$ and B is represented by $\cup_{i=1}^{m-1}(b_i + \mathbb{Z}L)$ with $0 < b_1 < \cdots < b_{m-1} < L$. Here $L = nm$. Namely, $0 = a + b$ for some $a \in A$, $b \in B$. By taking $A - a$ in place of A and $B + a$ in place of B we have $0 \in A \cap B$. If $\gcd(a_0, a_1, \ldots, a_{n-1}) = d > 1$, then

$$\mathbb{Z} = \frac{A}{d} \oplus \frac{B_j - j}{d} \quad \text{for } j = 0, 1, \ldots, d-1 \tag{3}$$

where $B_j = \{b \in B | b \equiv j (\bmod d)\}$. The elements of A/d are coprime and we have obtained d complementing $\mathbb{Z}$-pairs with coprime a's. It is obvious that B can be represented as indicated. Without any trouble we can add or subtract multiples of L from the elements of A to obtain the required structure. The problem is now reduced to the decomposition problem for the cyclic group of residue classes $\bmod L$ (where we only know that L is a multiple of the cardinality of A).

It will be clear from the previous sections that a characterisation of all complementing $\mathbb{Z}$-pairs is not a simple matter, even if we assume one of the subsets to be finite. However, in the latter case, a kind of inductive characterisation would be possible, if we could prove the following statement.

CONJECTURE. — *If* $A \oplus B = \mathbb{Z}, 0 \in A \cap B, \gcd_{a\in A} a = 1$ *and* A *has exactly* n *elements, then there exists a prime factor* p *of* n *such that all elements of* B *are divisible by* p.

Suppose the statement is true. Then the elements of A are equally distributed among the residue classes $\bmod p$. We can make a splitting in p complementing $\mathbb{Z}$-pairs as indicated in (3), with $d = p$ and A and B interchanged. So the problem is reduced to the decomposition problem for the cyclic group of residue classes $\bmod(L/p)$ and the procedure can be repeated. The following examples show that p is not determined by L and n.

$$L = 12, \quad n = 6, \quad A = \{0, 1, 4, 5, 8, 9\}, \quad B = \{0, 2 (\bmod 12)\}, \quad p = 2;$$

$$L = 12, \quad n = 6, \quad A = \{0, 1, 2, 6, 7, 8\}, \quad B = \{0, 3 (\bmod 12)\}, \quad p = 3.$$

If the conjecture is true, then every such decomposition is quasiperiodic in accordance with Hajós' conjecture stated at the end of the previous section. For we can split A into residue classes mod p, $A_0, A_1, \ldots, A_{p-1}$, say, and $A_i + B = p\mathbb{Z} + i$ for $i = 0, 1, \ldots, p-1$. If the number n of elements of A is a prime, then the conjecture implies that A represents a complete residue system mod p and every element of B is divisible by p. This is precisely the conjecture of de Bruijn stated in the introduction. A proof of this conjecture can be obtained by combining results of de Bruijn and Sands. De Bruijn ([2]), p. 241) provided an argument which implies that his conjecture is true if the following statement is true : if the finite cyclic group G is the free product of the subsets A and B and the cardinality of A is prime, then either A or B is periodic. As remarked in the previous section, the latter statement was proved by Sands. I shall present a completely different proof of de Bruijn's conjecture (Theorem 2). Subsequently I shall extend de Bruijn's argument to the case where the cardinality of A is a prime power. By combining this with Sands' general result we obtain a proof of my conjecture, stated above, in case n is a prime power (Theorem 3). We start with a result without any restriction on n.

THEOREM 1. — *Let* $A \oplus B = \mathbb{Z}$ *with* $0 \in A \cap B$ *and cardinality* n *of* A *finite. Then, for any integer* h *with* $\gcd(h, n) = 1$, *we have* $hA \oplus B = \mathbb{Z}$.

We need some lemmas. Let $A = \{a_0, a_1, \ldots, a_{n-1}\}$ with $a_0 = 0$.

LEMMA 2. — *For any integer* x

$$\{(x + a_0)_A, (x + a_1)_A, \ldots, (x + a_{n-1})_A\} = A,$$
$$\{(x - a_0)_A, (x - a_1)_A, \ldots, (x - a_{n-1})_A\} = A.$$

Proof : suppose $(x + a_i)_A = (x + a_j)_A$. Then $x + a_i - \beta_1 = x + a_j - \beta_2$ for some $\beta_1, \beta_2 \in B$. Hence $a_i + \beta_2 = a_j + \beta_1$. Since such a representation is unique, we have $i = j$. The proof of the second statement is similar. $\square$

LEMMA 3. — *Let* q *be a prime power with* $\gcd(n, q) = 1$. *Then, for any integer* x,

$$\{(x + qa_0)_A,\ (x + qa_1)_A, \ldots, (x + qa_{n-1})_A\} = A.$$

Proof : let $q = p^k$, p prime. Put $D = \{(\alpha, \alpha, \ldots, \alpha) \in A^q | \alpha \in A\}$. Define $f : A^q \to A$ by $(\alpha_1, \alpha_2, \ldots, \alpha_q) \longmapsto (x + \alpha_1 + \alpha_2 + \cdots + \alpha_q)_A$. By Lemma 2 we have

$$\bigcup_{j=0}^{n-1} f(\alpha_1, \alpha_2, \ldots, \alpha_{q-1}, a_j) = A.$$

Hence $f(\alpha)$ $(\alpha \in A^q)$ assumes each element of A exactly n^{q-1} times. Note that $f(\alpha_1, \alpha_2, \ldots, \alpha_q)$ does not change value if we permute $\alpha_1, \alpha_2, \ldots, \alpha_q$. If $(\alpha_1, \alpha_2, \ldots, \alpha_q)$ contains entry a_j exactly l_j times $(j = 0, 1, \ldots, n-1)$, then it has precisely

$$\frac{q!}{l_0!\, l_1! \cdots l_{n-1}!}$$

permutations in A^q. This multinomial coefficient is divisible by p, unless all but one $l_0, l_1, \ldots, l_{n-1}$ equal zero, that is $(\alpha_1, \alpha_2, \ldots, \alpha_q) \in D$. It follows that f assumes on $A^q \backslash D$ each value of A a number of times which is divisible by p. Since p-n^{q-1}, we infer that f assumes each value of A on D. Thus $f(D) = A$. □

LEMMA 4. — *Let* $q = -1$ *or a prime power with* $\gcd(n, q) = 1$. *Then* $qA \oplus B = \mathbb{Z}$.

Proof : we first show that all numbers $\{qa + b\}_{a \in A,\, b \in B}$ are distinct. Suppose $\alpha_1, \alpha_2 \in A$, $\beta_1, \beta_2 \in B$ are such that $q\alpha_1 + \beta_1 = q\alpha_2 + \beta_2$. If $q = -1$, then the assertion follows from $\alpha_2 + \beta_1 = \alpha_1 + \beta_2$. Otherwise $q\alpha_1 - \beta_2 = q\alpha_2 - \beta_1$. This number has a unique representation $\alpha + \beta$ with $\alpha \in A$, $\beta \in B$. It follows that $-\beta + q\alpha_1 = \alpha + \beta_2$, $-\beta + q\alpha_2 = \alpha + \beta_1$. Hence $(-\beta + q\alpha_1)_A = (-\beta + q\alpha_2)_A$. By Lemma 3 we obtain $\alpha_1 = \alpha_2$ and therefore $\beta_1 = \beta_2$.

By Lemma 1, B is periodic $\bmod mn$ for some positive integer m. Hence B consists of m residue classes mod mn. Let $\{b_0, b_1, \ldots, b_{m-1}\}$ be a set of representatives. Since by the first paragraph all numbers $qa_i + b_j$ $(i = 0, 1, \ldots, n-1,\ j = 0, 1, \ldots, m-1)$ are in distinct residue classes $\bmod mn$, these mn numbers represent a complete residue system $\bmod mn$. Hence $qA \oplus B = \mathbb{Z}$. □

Proof of Theorem 1 : since h can be written as the product of prime powers and factors -1, each coprime to n, we reach the conclusion by repeated application of Lemma 4. □

COROLLARY. — *Let h be an integer with* $\gcd(h, n) = 1$. *Then* $h(\alpha_1 - \alpha_2) = \beta_1 - \beta_2$ *for some* $\alpha_1, \alpha_2 \in A$, $\beta_1, \beta_2 \in B$ *implies* $\alpha_1 = \alpha_2$, $\beta_1 = \beta_2$.

Proof : we have $h\alpha_1 = (h\alpha_1 + \beta_2)_{hA} = (h\alpha_2 + \beta_1)_{hA} = h\alpha_2$. □

Subsequently we show how a proof of de Bruijn's conjecture can be derived from Theorem 1.

THEOREM 2. — *Let $A \oplus B = \mathbb{Z}$ with $0 \in A \cap B$, the elements of A are coprime and the cardinality n of A is prime. Then every element of B is divisible by n.*

Proof : since B is periodic mod mn for some m, we assume without loss of generality that A consists of the nonnegative integers $a_0 = 0, a_1, \ldots, a_{n-1}$ and that $b_0, \ldots, b_{m-1}$ are integers with $0 \le b_0 < \cdots < b_{m-1} < mn$ such that $B = \cup_{j=0}^{m-1}(b_j + mn\mathbb{Z})$. Put $B_0(z) = 1 + z^{b_1} + z^{b_2} + \cdots + z^{b_{n-1}}$ and

$$(4) \qquad B(z) = \sum_{\substack{b \in B \\ b \ge 0}} z^b = B_0(z)\,(1 + z^{mn} + z^{2mn} + \cdots) = \frac{B_0(z)}{1 - z^{mn}}.$$

Note that every pole of B is an mn–th root of unity. We shall show that it is an n–th root of unity. Set $A_h(z) = 1 + z^{ha_1} + z^{ha_2} + \cdots + z^{ha_{n-1}}$. Then for every $h > 0$ with $\gcd(h, n) = 1$ we have, by Theorem 1,

$$(5) \qquad A_h(z)B(z) = \sum_{k=0}^{\infty} z^k - P_h(z) = \frac{1}{1-z} - P_h(z)$$

where $P_h(z) \in \mathbb{Z}[z]$. Hence every pole $\neq 1$ of B is a zero of A_h. Let ζ be a pole of B with $\zeta \neq 1$. Put $s_k = \sum_{j=0}^{n-1} \zeta^{ka_j}$ for $k \in \mathbb{Z}$. Since $A_h(\zeta) = 0$ whenever $\gcd(h, n) = 1$, we have $s_h = 0$ whenever $\gcd(h, n) = 1$. By the theorem on elementary symmetric functions (formulae of Newton–Girard) we obtain, since n is prime,

$$\Pi_{j=0}^{n-1}\,(z - \zeta^{a_j}) \;= z^n + c_n$$

where c_n is some constant. Since $a_0 = 0$, we have $c_n = -1$. Therefore, $\zeta^{a_0} = 1, \zeta^{a_1}, \ldots, \zeta^{a_{n-1}}$ is the complete set of n–th roots of unity. Since the a's are coprime, there exist integers $t_0, t_1, \ldots, t_{n-1}$ such that $1 = t_0a_0 + t_1a_1 + \cdots + t_{n-1}a_{n-1}$. Hence, putting $\zeta_1 = e^{2\pi i/n}$,

$$\zeta = (\zeta^{a_0})^{t_0}\,(\zeta^{a_1})^{t_1} \cdots (\zeta^{a_{n-1}})^{t_{n-1}} \;= \zeta_1^t \text{ for some } t \in \mathbb{Z}.$$

Thus ζ is an nth root of unity.

From (4) we see that every pole of B is simple and that $(z^{mn} - 1)/(z^n - 1)$ divides $B_0(z)$. This implies, by the choice of the b's,

$$B_0(z) = (1 + z^n + z^{2n} + \cdots + z^{(m-1)n})\;(1 + f_1z + \cdots + f_{n-1}z^{n-1})$$

for some coefficients $f_1, \ldots, f_{n-1}$. Since B_0 has only m nonzero coefficients, we see that $f_1 = \cdots = f_{n-1} = 0$. Thus $B_0(z) = (1 - z^{mn})/(1 - z^n)$ and $B(z) = (1 - z^n)^{-1} = 1 + z^n + z^{2n} + \cdots$, in other words, B consists of the multiples of n. □

We use Sands' result on finite cyclic groups [24] to obtain the following generalisation of Theorem 2.

THEOREM 3. — *Let* $A \oplus B = \mathbb{Z}$ *with* $0 \in A \cap B$, $\gcd_{a \in A} a = 1$ *and* A *has exactly* p^t *elements with* p *prime,* $t \in \mathbb{Z}_{\geq 1}$. *Then all elements of* B *are divisible by* p.

Proof : let $n = p^t$. By Lemma 1, B is periodic. Let L be its minimal period. If G denotes the group of residue classes mod L, then $\mathbb{Z} = A \oplus B$ furnishes a decomposition $G = A^* \oplus B^*$ where A^* and B^* consist of the residue classes mod L determined by the elements of A and B, respectively. Note that L is divisible by p^t.

For $t = 1$ the statement is true by Theorem 2. So suppose $t > 1$. We apply induction on t. It follows from Theorem 2 of Sands [24] that A^* or B^* is periodic. B^* cannot be periodic because of the minimality of L, so it has to be A^*. Note that the elements g with $g + A^* = A^*$ form a subgroup G_0 of G. We shall show that G_0 contains the residue class $L/p \pmod L$. If not, G_0 contains L/q for some prime $q \neq p$. Then $a \in A^*$ if and only if $a + vL/q \in A^*$ for all $v \in \mathbb{Z}$. Hence A^* splits into subsets of size q. Since A^* has p^t elements, this is impossible. Thus A^* is periodic mod L/p.

Let A^{**} and B^{**} consist of the residue classes mod L/p determined by the elements of A and B, respectively. By the previous paragraph A^{**} has p^{t-1} elements and $A^{**} \oplus B^{**} = \mathbb{Z}/(L/p)\mathbb{Z}$. Put $r = p^{t-1}$. Let $\tilde{A} = \{\tilde{a}_0 = 0, \tilde{a}_1, \ldots, \tilde{a}_{r-1}\}$ be a set of integers representing A^{**}. Since $\gcd_{a \in A} a = 1$ and every element of A is of the form $\tilde{a}_j + wL/p$, there exists integers $v_0, v_1, \ldots, v_{r-1}, v_r$ such that $v_0\tilde{a}_0 + v_1\tilde{a}_1 + \cdots + v_{r-1}\tilde{a}_{r-1} + v_r L/p = 1$. We infer that the greatest common divisor d of $\tilde{a}_0, \tilde{a}_1, \ldots, \tilde{a}_{r-1}$ is coprime to L/p. Let h be the inverse of $d \bmod (L/p)$. Then

$$h\tilde{a}_0,\ h\tilde{a}_1, \ldots, h\tilde{a}_{r-1} \equiv \frac{1}{d}\tilde{a}_0, \frac{1}{d}\tilde{a}_1, \ldots, \frac{1}{d}\tilde{a}_{r-1} \pmod{\frac{L}{p}}.$$

Therefore, by Theorem 1, $\frac{1}{d}\tilde{A} = \{\frac{1}{d}\tilde{a}_0 = 0, \frac{1}{d}\tilde{a}_1, \ldots, \frac{1}{d}\tilde{a}_{r-1}\}$ is a set of p^{t-1} relatively prime numbers with $\frac{1}{d}\tilde{A} \oplus B^{**} = h\tilde{A} \oplus B^{**} = \mathbb{Z}$. Hence, by the induction hypothesis, all elements of B^{**} are divisible by p. Since $B^{**} = B + (L/p)\mathbb{Z}$ and p is a divisor of L/p, all elements of B are divisible by p. □

4. — A is the set of finite sums of distinct odd powers of 2

We use the following notation. If $n = \pm \sum_{j=0}^{k} b_j 2^j$ with $b_j \in \{0, 1\}$ is the binary notation of n, then we write $n = \pm b_k b_{k-1} \cdots b_0$. We say that b_k is the first bit and b_0 the last.

The bit b_j is said to be at place j, for $j = 0, 1, \ldots, k$.

If j is even, then b_j is at an even place, otherwise at an odd place.

If b_j is the last bit 1 in the binary expansion of n, then $\mathrm{ord}_2(n) = j$.

Let A be the set of finite sums of distinct odd powers of 2 and $\overline{A}$ the set of the finite sums of distinct even powers of 2. Yu. Ito asked me to characterise all sets B such that $A \oplus B = \mathbb{Z}$. Obviously $A \oplus \overline{A} = \mathbb{Z}_{\geq 0}$, whence $A \oplus (-\overline{A}) = \mathbb{Z}$.

THEOREM 4. — *The above A satisfies $A \oplus B = \mathbb{Z}$ if and only if B is such that*

(i) *if $b, b' \in B$ with $b \neq b'$, then $\text{ord}_2(b - b')$ is even,*

(ii) *the set B is maximal with respect to* (i),

(iii) $-\overline{A} \subseteq A + B$.

The first condition says that there is an even number k such that $2^k \mid b - b'$, but $2^{k+1} \nmid b - b'$. The second means that B cannot be enlarged without affecting (i). The third condition is equivalent to saying that for every element $\overline{a} \in \overline{A}$ there is an $a \in A$ such that $\overline{a} + a \in -B$. Still another interpretation is that any finite collection of bits at even places can be completed to some nonpositive number in B by inserting suitable bits at odd places, zeros at even places and putting a minus sign in front of the number.

Recently it was proved by Eigen, Hajian and Kakutani [32] that if F is a finite set of integers, then F can be extended to a complementary set B of A if and only if (i) holds for F.

Proof : $\Longrightarrow$ (this proof was shown to me by Yu. Ito. A simpler proof of (i) can be found in S. Eigen, A. Hajian and S. Kakutani [32], Lemma 1).

(ii) Suppose $\tilde{b} \notin B$ and $\text{ord}_2(b - \tilde{b})$ is even for every $b \in B$. Since $\tilde{b} = a + b$ for some $a \in A$, $b \in B$, we have $\tilde{b} - b = a \in A$, whence $\text{ord}_2(\tilde{b} - b)$ is odd.

(iii) Obvious.

(i) Put $A_n = 2^{2n}A$ and $B_n = A_n \oplus B$. Then the 2^n sets

$$B_n + \sum_{j=0}^{n-1} \varepsilon_j 2^{2j+1}, \quad \varepsilon_0, \varepsilon_1, \ldots, \varepsilon_{n-1} \in \{0, 1\},$$

are disjoint and their union is $\mathbb{Z}$. We claim that $B_n + k \cdot 2^{2n} = B_n$ for $k \in \mathbb{Z}$. For $n = 0$ it is clear. Suppose the claim is valid for $n = m$. Then

$$B_{m+1} + k \cdot 2^{2m} \subseteq B_m + k \cdot 2^{2m} = B_m = B_{m+1} \cup (B_{m+1} + 2^{2m+1}).$$

Since B_{m+1} and $B_{m+1} + 2^{2m+1}$ are disjoint sets of the same cardinality, we conclude that adding or subtracting 2^{2m+1} to an element from one set yields an element from the other set. It follows by induction on $|l|$ that $B_{m+1} + l \cdot 2^{2m+1} = B_{m+1} + 2^{2m+1}$ for odd l and $B_{m+1} + l \cdot 2^{2m+1} = B_{m+1}$ for even l. This proves the claim.

Suppose $B - B$ contains an element z with $\text{ord}_2(z)$ is odd. Then

$$b - b' = z = (2k+1)2^{2l+1} = k \cdot 2^{2l+2} + 2^{2l+1}$$

for some $b, b' \in B$ and $k, l \in \mathbb{Z}$. Since $B \subseteq B_n$ for all n, we have

$$b = b' + k \cdot 2^{2l+2} + 2^{2l+1} \in B_{l+1} + 2^{2l+1}.$$

Thus $b \in B_{l+1} \cap (B_{l+1} + 2^{2l+1})$, but these sets are disjoint.

The proof of the sufficiency part of Theorem 4 requires some lemmas.

LEMMA 5. — *If* (i) *holds, then every integer is represented at most once as* $a + b$ *with* $a \in A$, $b \in B$.

Proof : suppose $a_1 + b_1 = a_2 + b_2$ for some $a_1, a_2 \in A$ and $b_1, b_2 \in B$.
Then $a_1 - a_2 = b_2 - b_1$. However, $\text{ord}_2(a_1 - a_2)$ is odd and $\text{ord}_2(b_2 - b_1)$ is even, unless $a_1 = a_2, b_1 = b_2$. □

LEMMA 6. — *Let* (i) *hold. If* b *and* b' *are elements of* B *with* $bb' \geq 0$ *such that* $b = \pm b_{2k-1}b_{2k-2}\cdots b_0$, $b' = \pm b'_{2k-1}b'_{2k-2}\cdots b'_0$ *and* $b_{2j} = b'_{2j}$ *for* $j = 0, 1, \ldots, m-1$. *Then* $b_j = b'_j$ *for* $j = 0, 1, \ldots, 2m-1$.

Proof : clear.

LEMMA 7. — *If* (i) *and* (iii) *hold, then every non-positive integer has a representation* $a + b$ *with* $a \in A$ *and* $b \in B$.

Proof : we have $0 \in \overline{A}$, whence $0 \in A \oplus B$. Consider $n \in \mathbb{Z}_{<0}$ with $-n = n_{2k-1}n_{2k-2}\cdots n_1 n_0$ where the first bit may be 0. We shall construct $a \in A$ and $b \in B$ such that $a + b = n$. Then $-b = -n + a \geq 0$.

Put $b_0 = n_0$. Since $b_0 \in \overline{A}$, there exists some nonpositive element of B ending with b_0 by (iii). Let b_1 be the bit at place 1 of this element. Define $a \in \{0, 1\}$ such that, in binary notation, $b_1 b_0 \equiv n_1 n_0 + a_1 0 \pmod 4$. Next consider $n_2 n_1 n_0 + a_1 0$. Let b_2 be the bit at place 2 of this sum. Then, by (iii), there is a nonpositive element in B such that the last two bits at even places are b_2 and b_0. Let b_3 be the bit at place 3 of this element. Define $a_3 \in \{0, 1\}$ such that $b_3 b_2 b_1 b_0 \equiv n_3 n_2 n_1 n_0 + a_3 0 a_1 0 \pmod{2^4}$. By considering $n_4 n_3 n_2 n_1 n_0 + a_3 0 a_1 0$ and continuing the procedure, we eventually construct bits $b_{2k+1}, b_{2k}, \ldots, b_1, b_0$ and $a_{2k+1}, a_{2k-1}, \ldots, a_3, a_1$ such that

$$b_{2k+1}b_{2k}\cdots b_1 b_0 \equiv -n + a_{2k+1}0a_{2k-1}0\cdots 0a_1 0 \pmod{2^{2k+2}} \tag{6}$$

and b_{2k+1} is the bit at place $2k+1$ of some nonpositive element $\tilde{b}$ of B with $b_{2k}, b_{2k-2}, \ldots, b_0$ at the last $k+1$ even places and zeros at all other even places.

Note that, by Lemma 6, the bit at place 1 of $\tilde{b}$ is b_1 (apply it for $m = 1$), the bit at place 3 of $\tilde{b}$ is b_3 (apply Lemma 6 for $m = 2$), and so on. Thus $\tilde{b}$ ends with the $2k+2$ bits $b_{2k+1}b_{2k}\cdots b_1b_0$, whence $-\tilde{b}$ can be written as $t\cdot 2^{2k+2}+b_{2k+1}b_{2k}\cdots b_1b_0$ with $t \in A$. Further, observe that on both sides of (6) the numbers are nonnegative and less than 2^{2k+2}, by $-n < 2^{2k}$, so that actually in (6) both sides are equal. Put $\tilde{a} = n - \tilde{b}$. Then, for some $t \in A$,

$$\tilde{a} = t\cdot 2^{2k+2} + b_{2k+1}b_{2k}\cdots b_1b_0 + n = t\cdot 2^{2k+2} + a_{2k+1}0\cdots 0a_10.$$

Hence $\tilde{a} \in A$. Thus $n = \tilde{a} + \tilde{b}$ with $\tilde{a} \in A$ and $\tilde{b} \in B$. □

LEMMA 8. — *Let $B \in \mathbb{Z}$ be a set satisfying* (i) *and* (iii) *and such that $A \oplus B$ represents every nonpositive integer, but not* 1. *Then every element of $B-1$ has its last nonzero bit at an even place.*

Proof : suppose b is a negative element of B such that the last nonzero bit of $b-1$ is at an odd place, at place $2m-1$, say. By (iii) there is a nonpositive element $b^* = -b_l^*b_{l-1}^*\cdots b_0^*$ in B with $b_0^* = b_2^* = \cdots = b_{2m-2}^* = 1$ and $b_{2k}^* = 0$ for $k \geq m$. Since the last bit 1 of $b-1$ is at place $2m-1$ we have

$$b = \cdots 0\underbrace{1\,1\cdots 1}_{2m-1}.$$

Hence, by Lemma 6, $b^* \equiv b (\mathrm{mod}\ 2^{2m+1})$. Thus

$$b^* - 1 = -b_l^*b_{l-1}^*\cdots b_{2m}^*1\ \underbrace{0\,0\ \cdots\ 0}_{2m-1}.$$

It follows that b^*-1 has zeros at all even places. Hence $a := -(b^*-1) \in A$ and $1 = a + b^* \in A \oplus B$, a contradiction.

Suppose b is a positive element of B such that the last nonzero bit of $b-1$ is at an odd place, at place $2m-1$ say. By (iii) there exists a negative element $b' = -b_l'b_{l-1}'\cdots b_0'$ in B such that $b_0' = b_2' = \cdots = b_{2m-2}' = 1$ and $b_{2k} = 0$ for $k \geq m$. Since we have proved in the previous paragraph that the last nonzero bit of $b'-1$ is at an even place, we find that $b_0' = b_1' = \cdots = b_{2m-1}' = 1$. Thus $\mathrm{ord}_2(b-b') = 2m-1$ which contradicts (i). □

Proof of the sufficiency part of Theorem 4 : by Lemmas 5 and 7 it remains to prove that every positive integer has a representation $a+b$ with $a \in A$ and $b \in B$. Let n be the smallest positive integer without such a representation. Then $n \notin B$, since $0 \in A$. Put $\tilde{B} = B-(n-1)$. Then $A \oplus \tilde{B}$ represents every nonpositive integer, but not 1. By Lemma 8 every element of $\tilde{B}-1 = B-n$ has its last nonzero bit at an even place. Put $B^* = B\cup\{n\}$. Then B^* is larger than B and $\mathrm{ord}_2(b-b')$ is even for every $b, b' \in B^*$ with $b \neq b'$. Thus B is not maximal with respect to (i), in contradiction to (ii). Hence every positive integer is contained in $A \oplus B$. □

Theorem 4 induces a similar characterisation for $\overline{A}$.

COROLLARY. — *$\overline{A} \oplus B = \mathbb{Z}$ if and only if B satisfies*

(i') *if $b, b' \in B$ with $b \neq b'$, then $\operatorname{ord}_2(b - b')$ is odd,*

(ii') *the set B is maximal with respect to* (i'),

(iii') $-A \subseteq \overline{A} + B$.

Proof : note that $A = 2\overline{A}$ and $\overline{A} = 2A \cup (2A + 1)$.
$' \Longrightarrow '$. By $2\overline{A} \oplus 2B = 2\mathbb{Z}$, we have $A \oplus (2B \cup (2B + 1)) = \mathbb{Z}$.
Hence $2B \cup (2B + 1)$ satisfies the conditions (i), (ii), (iii) of Theorem 3.
It follows immediately, that B satisfies (i'), (ii'), (iii').

$' \Longleftarrow '$. By (i') all elements of B are even or all are odd. In the latter case we replace B by $B + 1$. This involves no loss of generality. Let $\tilde{B}$ be the set of numbers of B divided by 2. Then $\tilde{B}$ satisfies conditions (i), (ii), (iii) of Theorem 1.

Thus $A + \tilde{B} = \mathbb{Z}$. Hence $2A + 2\tilde{B} = 2\mathbb{Z}$ and

$$\overline{A} \oplus B = (2A \cup (2A+1)) \oplus 2\tilde{B} = 2A \oplus 2\tilde{B} \cup (2A+1) \oplus 2\tilde{B} = 2\mathbb{Z} \cup (2\mathbb{Z}+1) = \mathbb{Z}.$$

$\square$

It is obvious that conditions (i) and (ii) of Theorem 3 are not enough to guarantee $A \oplus B = \mathbb{Z}$. The set $\overline{A}$ satisfies (i) and (ii), but $A \oplus \overline{A} = \mathbb{Z}_{\geq 0}$. Yu. Ito asked for some set of type A for which the complementing sets are characterised by (i) and (ii) only. He wondered whether $A' = \{(-1)^{a/2} \cdot a | a \in A\}$ is such a set. P. ten Pas [19] showed that there exist sets B' and B'' which both satisfy (i) and (ii) such that $A' \oplus B' = \mathbb{Z}$ and $A' \oplus B'' \neq \mathbb{Z}$.

Acknowledgement. I am indebted to Yu. Ito and J. Urbanowicz for useful discussions and to S. Eigen and Yu. Ito for remarks on an earlier version of this paper.

Manuscrit reçu le 3 décembre 1993

References

[1] J.L. Brown. — *Generalized bases for the integers*, Amer. Math. Monthly **71** (1964), 973-980.

[2] N.G. de Bruijn. — *On bases for the set of integers*, Publ. Math. Debrecen **1** (1950), 232-242.

[3] N.G. de Bruijn. — *On the factorisation of finite abelian groups*, Indag. Math. **15** (1953), 258-264.

[4] N.G. de Bruijn. — *On the factorisation of cyclic groups*, Indag. Math. **17** (1955), 370-377.

[5] N.G. de Bruijn. — *On number systems*, Nieuw Arch. Wisk. (3) **4** (1956), 15-17.

[6] N.G. de Bruijn. — *Some direct decompositions of the set of integers*, Math. Comp. **18** (1964), 537-546.

[7] S. Eigen and A. Hajian. — *A characterisation of exhaustive weakly wandering sequences for nonsingular transformations*, Comment. Math. Univ. Sancti Pauli **36** (1987), 227-233.

[8] S. Eigen and A. Hajian. — *Sequences of integers and ergodic transformations*, Advances Math. **73** (1989), 256-262.

[9] S. Eigen, A. Hajian and Y. Ito. — *Ergodic measure preserving transformations of finite type*, Tokyo J. Math. **11** (1988), 459-470.

[10] I. Fary. — *Die Äquivalente des Minkowski-Hajósschen Satzes in der Theorie der topologischen Gruppen*, Comm. Math. Helv. **23** (1949), 283-287.

[11] G. Hajós. — *Über einfache und mehrfache Bedeckung des n-dimensionalen Raumes mit einem Würfelgitter*, Math. Z. **47** (1941), 427-467.

[12] G. Hajós. — *Sur la factorisation des groupes abéliens, Časopis* Pěst. Mat. Fys. **74** (1950), 157-162.

[13] G. Hajós. — *Sur la problème de factorisation des groupes cycliques*, Acta. Math. Acad. Sci. Hungar. **1** (1950), 189-195.

[14] R.T. Hansen. — *Complementing pairs of subsets in the plane*, Duke Math. J. **36** (1969), 441-449.

[15] C.T. Long. — *Addition theorems for sets of integers*, Pacific J. Math. **23** (1967), 107-112.

[16] C.T. Long and N. Woo. — *On bases for the set of integers*, Duke Math. J. **38** (1971), 583-590.

[17] H. Minkowski. — *Geometrie der Zahlen*, Leipzig, 1896.

[18] I. Niven. — *A characterization of complementing sets of pairs of integers*, Duke Math. J. **38** (1971), 193-203.

[19] P. ten Pas. — *Complementing sets for* $\mathbb{Z}$ (in Dutch), Leiden, 1990.

[20] K. Post. — *Problem 71*, Nieuw Arch. Wisk. (3) **14** (1966), 274-275.

[21] L. Rédei. — *Zwei Lückensätze über Polynome in endlichen Primkörpern mit Anwendung auf die endlichen Abelschen Gruppen und die Gaussischen Summen*, Acta Math. **79** (1947), 273-290.

[22] L. Rédei. — *Kurzer Beweis des gruppentheoretischen Satzes von Hajós*, Comm. Math. Helv. **23** (1949), 272-282.

[23] L. Rédei. — *Ein Beitrag zum Problem der Faktorisation von endlichen Abelschen Gruppen*, Acta Math. Acad. Sci. Hungar. **1** (1950), 197-207.

[24] A.D. Sands. — *On the factorisation of finite abelian groups*, Acta Math. Acad. Sci. Hungar. **8** (1957), 65-86.

[25] A.D. Sands. — *The factorisation of abelian groups*, Quart. J. Math. Oxford (2) **10** (1959), 81-91.

[26] A.D. Sands. — *On the factorisation of finite abelian groups II*, Acta Math. Acad. Sci. Hungar. **13** (1962), 153-159.

[27] C. Swenson. — *Direct sum subset decompositions of* $\mathbb{Z}$, Pacific J. Math. **53** (1974), 629-633.

[28] C. Swenson and C. Long. — *Necessary and sufficient conditions for simple A-bases*, Pacific J. Math. **126** (1987), 379-384.

[29] T. Szele. — *Neuer vereinfachter Beweis des gruppentheoretischen Satzes von Hajós*, Publ. Math. Debrecen **1** (1949), 56-62.

[30] A. M. Vaidya. — *On complementing sets of nonnegative integers*, Math. Mag. **39** (1966), 43-44.

[31] S. Eigen and A. Hajian. — *Sequences of integers and ergodic transformations*, Advances in Mathematics **73** (1989), 256-262.

[32] S. Eigen, A. Hajian and S. Kakutani. — *Complementing sets of integers – A result from ergodic theory*, Japan J. Math. **18** (1992), 205-210.

R. TIJDEMAN
Mathematisch Instituut R.U.
Postbus 9512
2300 RA Leiden
The Netherlands

Number Theory
Paris 1992–93

CM Abelian varieties with almost ordinary reduction

Yuri G. ZARHIN*

In this note we discuss the Hodge group $\mathrm{Hdg}(X)$ of a simple Abelian variety X of CM-type. It is well-known that $\dim_{\mathbb{Q}} \mathrm{Hdg}(X) \leq \dim(X)$. Assuming that X has somewhere good *almost ordinary* reduction, we prove that

$$\dim_{\mathbb{Q}} \mathrm{Hdg}(X) = \dim(X)$$

and give an explicit description of $\mathrm{Hdg}(X)$.

1. — Almost ordinary Abelian varieties

Let A be an Abelian variety defined over a finite field k of characteristic p. We call A *almost ordinary* if $\dim(A) > 1$ and it has the same Newton polygon as the product of $(\dim(A) - 1)$–dimensional *ordinary* Abelian variety and a supersingular elliptic curve. This means that its set of slopes is $\{0, 1/2, 1\}$ and slope $1/2$ has length 2. For example, an Abelian surface is almost ordinary if and only if it is neither ordinary nor supersingular. One may easily check that if $g = \dim(A) > 1$ then A is almost ordinary if and only if its p–rank equals $g - 1$, i.e., the group of "physical" points of order p is isomorphic to $(\mathbb{Z}/p\mathbb{Z})^{g-1}$.

Almost ordinary varieties were studied by Oort [13] in connection with the lifting problem of CM Abelian varieties to characteristic zero. In particular, he proved that each almost ordinary Abelian variety can be lifted to characteristic zero as CM Abelian variety (recall [26] that each Abelian variety over a finite field can be lifted to characteristic zero as CM Abelian variety *up to an isogeny*). Of course, if we start with an (absolutely) simple Abelian variety over a finite field, then its lifting will be also (absolutely) simple. It follows from ([5], Th.7; [12], Th.4.1) that polarized almost ordinary Abelian varieties of given dimension constitute subvarieties of codimension 1 in the moduli spaces of Abelian varieties. See also [14].

A special case of a theorem of Lenstra and Oort [6] asserts that, for each positive integer $g > 1$ and for each prime number p there exists an absolutely simple almost ordinary g–dimensional Abelian variety defined

* Supported by C.N.R.S.

over a certain finite field field of characteristic p. It was proven by Oort [13] that the endomorphism algebra of simple almost ordinary Abelian variety (over finite field) is a number field of degree $2\dim(A)$. Notice (see Sect. 6.6 below), that each simple almost ordinary Abelian variety is absolutely simple.

One may easily check that each non-simple almost ordinary Abelian variety is isogenous either to the product of ordinary Abelian variety and simple almost ordinary Abelian variety or to the product of ordinary Abelian variety and a supersingular elliptic curve.

Let A be an Abelian variety over a finite field k of characteristic p. We write Γ_A for the multiplicative subgroup of $\mathbb{C}^*$ generated by the eigenvalues of the Frobenius endomorphism of A [29, 30, 31]. It is known ([30], Sect. 2.1; [33], Sect. 4.1), that the rank $\mathrm{rk}(\Gamma_A)$ of Γ_A is a positive number which does not exceed $\dim(A)+1$. The non-negative integer $\mathrm{rk}(\Gamma_A)-1$ is called the rank of A and denoted by $\mathrm{rk}(A)$ [31]. One may easily check ([31], Sect. 2.0), that

$$0 \leq \mathrm{rk}(A) \leq \dim(A)$$

and $\mathrm{rk}(A)=0$ if and only if A is supersingular.

Now, assume that A is simple and almost ordinary. In that case it is known ([7], Th. 5.7) that either $\mathrm{rk}(A)=\dim(A)$ or $\mathrm{rk}(A)=\dim(A)-1$. In addition, if $\dim(A)$ is even then $\mathrm{rk}(A)=\dim(A)$, i.e., $\mathrm{rk}(\Gamma_A)=\dim(A)+1$. If $\mathrm{rk}(\Gamma_A)=\dim(A)$ then the endomorphism algebra of A must contain an imaginary quadratic field; see [7], Th. 3.6. H.W. Lenstra (see [31], pp. 286-288) has constructed an example of 3-dimensional simple almost ordinary Abelian variety A with $\mathrm{rk}(\Gamma_A)=\dim(A)$. His construction also gives an example of a 3-dimensional absolutely simple CM Abelian variety having an almost ordinary reduction.

2. — ℓ-adic Lie Algebras

Let X be an Abelian variety defined over a number field K. We assume that K is sufficiently large, i.e., all endomorphisms of X are defined over K. We will also fix an embedding of K into the field $\mathbb{C}$ of complex numbers and consider K as a certain subfield of $\mathbb{C}$. We write $K(s)$ for the algebraic closure of K in $\mathbb{C}$. We write $G(K)$ for the Galois group of K. We write g for the dimension of X. Let E be the endomorphism algebra of X; it is a finite-dimensional semisimple $\mathbb{Q}$-algebra.

For a positive integer m, we denote by X_m the group

$$\{x \in X(K(s)) \mid mx = 0\}.$$

It is well known that X_m is a free $\mathbb{Z}/m\mathbb{Z}$-module of rank $2g$. Let us fix a prime number ℓ. Then one may define the $\mathbb{Z}_\ell$-Tate module $T_\ell(X)$ as the projective limit of the groups X_m where m runs through the set of all powers ℓ^i and the transition maps are multiplication by ℓ. It is well known that $T_\ell(X)$ is a free $\mathbb{Z}_\ell$–module of rank $2g$. Clearly, all X_m are finite Galois submodules of $X(K(s))$, and the Galois actions for $m = \ell^i$ glue together to give rise to a continuous homomorphism

$$\rho_\ell = \rho_{\ell,X} : G(K) \to \operatorname{Aut}_{\mathbb{Z}_\ell} T_\ell(X).$$

The image

$$G_\ell = G_{\ell,X} = \operatorname{Im}(\rho_{\ell,X}) \subset \operatorname{Aut}_{\mathbb{Z}_\ell} T_\ell(X)$$

is a compact ℓ-adic Lie subgroup in $\operatorname{Aut}_{\mathbb{Z}_\ell} T_\ell(X)$. Let us put

$$V_\ell(X) := T_\ell(X) \otimes_{\mathbb{Z}_\ell} \mathbb{Q}_\ell.$$

Clearly, $V_\ell(X)$ is a $\mathbb{Q}_\ell$–vector space of dimension $2g$ and one may naturally identify $T_\ell(X)$ with a certain $\mathbb{Z}_\ell$–lattice of rank $2\dim(X)$ in $V_\ell(X)$. In particular, $\operatorname{Aut}_{\mathbb{Z}_\ell} T_\ell(X)$ becomes an open compact subgroup in $\operatorname{Aut}_{\mathbb{Q}_\ell} V_\ell(X)$. This allows us to regard ρ_ℓ as an ℓ-adic representation ([19]) :

$$\rho_\ell = \rho_{\ell,X} : G(K) \to \operatorname{Aut}_{\mathbb{Z}_\ell} T_\ell(X) \subset \operatorname{Aut}_{\mathbb{Q}_\ell} V_\ell(X).$$

We have

$$G_\ell \subset \operatorname{Aut}_{\mathbb{Z}_\ell} T_\ell(X) \subset \operatorname{Aut}_{\mathbb{Q}_\ell} V_\ell(X).$$

Clearly, G_ℓ is a compact (and therefore) closed subgroup of $\operatorname{Aut}_{\mathbb{Q}_\ell} V_\ell(X)$ and therefore is a closed ℓ–adic Lie subgroup. Let

$$\mathfrak{g}_\ell = \mathfrak{g}_{\ell,X} \subset \operatorname{End}_{\mathbb{Q}_\ell} V_\ell(X)$$

be the *Lie algebra* of G_ℓ [19]. A theorem of Faltings [4] asserts that $\mathfrak{g}_\ell$ is a reductive $\mathbb{Q}_\ell$–Lie algebra, its natural representation in $V_\ell(X)$ is completely reducible and the centralizer of this representation is $E \otimes_{\mathbb{Q}} \mathbb{Q}_\ell$. A theorem of Bogomolov [1] asserts that $\mathfrak{g}_\ell$ is an *algebraic* Lie algebra containing homotheties $\mathbb{Q}_\ell\mathrm{id}$. It follows that

$$\mathfrak{g}_{\ell,X} = \mathbb{Q}_\ell \mathrm{id} \oplus \mathfrak{g}^0_{\ell,X}.$$

Here

$$\mathfrak{g}^0_{\ell,X} := \mathfrak{sl}(V_\ell(X)) \bigcap \mathfrak{g}_{\ell,X}$$

is an algebraic reductive $\mathbb{Q}_\ell$–Lie algebra. Its natural representation in $V_\ell(X)$ is completely reducible and the centralizer of this representation is $E \otimes_{\mathbb{Q}} \mathbb{Q}_\ell$. It is known that the rank of $\mathfrak{g}^0_\ell$ is a non-negative number which does not exceed g. If the equality holds then the Lie algebra is "as large as possible" and one may give an "explicit" description of $\mathfrak{g}_\ell^0$ in terms of E; see [32], Th. 3.2; [33].

Let v be a non-Archimedean place of K such that X has a good reduction $X(v)$ at v. Then G_ℓ contains a *Frobenius element*

$$\mathrm{Fr}_v \in G_\ell \subset \mathrm{Aut}_{\mathbf{Q}_\ell} V_\ell(X)$$

canonically defined up to conjugation in G_ℓ [19]. If we view Fr_v as a linear operator in $V_\ell(X)$, then its eigenvalues are just eigenvalues of the Frobenius endomorphism of $X(v)$. In particular, if $\Gamma(\mathrm{Fr}_v)$ is the multiplicative group generated by the eigenvalues of Fr_v, then

$$\Gamma(\mathrm{Fr}_v) = \Gamma_{X(v)}.$$

Notice, that the rank of $\mathfrak{g}_\ell$ is greater or equal than $\mathrm{rk}(\Gamma(\mathrm{Fr}_v))$ (see [33], Corollary 2.4.1). This implies that the rank of $\mathfrak{g}_\ell^0$ is greater or equal than

$$\mathrm{rk}(\Gamma(\mathrm{Fr}_v)) - 1 = \mathrm{rk}(\Gamma_{X(v)}) - 1 = \mathrm{rk}(X(v)).$$

We have

$$0 \leq \mathrm{rk}(X(v)) \leq \mathrm{rk}\mathfrak{g}_\ell^0 \leq g = \dim(X(v)).$$

In particular, if $\mathrm{rk}(X(v)) = \dim(X(v))$ then

$$\mathrm{rk}(\mathfrak{g}_\ell^0) = g.$$

For example, if $X(v)$ is a simple almost ordinary Abelian variety and g is even then (see the end of Sect. 1)

$$\mathrm{rk}(\mathfrak{g}_\ell^0) = g$$

(recall that $g = \dim(X) = \dim(X(v))$).

The aim of the present paper is to prove that if $X(v)$ is an almost ordinary Abelian variety then

$$\mathrm{rk}(\mathfrak{g}_\ell^0) = g$$

under an additional assumption that X is an absolutely simple Abelian variety of CM-type. (Compare with the corresponding results for Abelian varieties having a reduction of K3 type ([32], Th. 3.0 and Sect. 7.1).

3. — Abelian varieties of CM-type

Let X be an absolutely simple Abelian variety of CM-type. Then its endomorphism algebra E is a CM-field of degree $2g$. We write $a \mapsto a'$ for the complex conjugation on E. We write T_E for the Weil restriction $\mathbf{R}_{E/\mathbb{Q}}\mathbb{G}_m$ of the multiplicative group $\mathbb{G}_m$. Clearly, T_E is a $2g$–dimensional algebraic torus. Let $\mathbf{U}_E$ be the g–dimensional algebraic subtorus of T_E defined by the condition

$$\mathbf{U}_E(\mathbb{Q}) = \{a \in T_E(\mathbb{Q}) = E^* \mid aa' = 1\}.$$

3.1. — The Hodge group

We write $V(X)$ for the first rational homology group $H_1(X(\mathbb{C}), \mathbb{Q})$ of $X(\mathbb{C})$: it is a $2g$–dimensional $\mathbb{Q}$–vector space. It also carries a natural structure of 1–dimensional E–vector space. The choice of a polarization on X gives rise to a certain non-degenerate skew-symmetric bilinear form

$$\varphi : V(X) \times V(X) \to \mathbb{Q}$$

such that

$$\varphi(ax, y) = \varphi(x, a'y)$$

for all $x, y \in V(X)$ and $a \in E$.

Let us choose a non-zero $\epsilon \in E$ with

$$\epsilon' = -\epsilon.$$

Then there exists a non-degenerate E–Hermitian sesquilinear form

$$\psi_\epsilon : V(X) \times V(X) \to E$$

such that

$$\varphi(x, y) = \mathrm{Tr}_{E/\mathbb{Q}}(\epsilon^{-1}\psi_\epsilon(x, y))$$

where $\mathrm{Tr}_{E/\mathbb{Q}} : E \to \mathbb{Q}$ is the trace map (see [21]; [2], Sect. 4; [17], p. 531). If we change ϵ by ϵ_1 then the form is multiplied by a non-zero totally real element ϵ_1/ϵ of E . The unitary group $\mathbf{U}(V(X), \psi_\epsilon)$ viewed as a $\mathbb{Q}$–algebraic group does not depend on the choice of ϵ and can be naturally identified with $\mathbf{U}_E$. In particular,

$$\mathbf{U}(V(X), \psi_\epsilon)(\mathbb{Q}) = \{a \in E^* \mid aa' = 1\}.$$

Here we identify E with its image in $\mathrm{End}_{\mathbb{Q}} V(X)$. Its Lie algebra

$$\mathfrak{u}_E := \mathrm{Lie}(\mathbf{U}(V(X), \psi_\epsilon)) = \mathrm{Lie}(\mathbf{U}_E) = \{a \in E \mid a + a' = 0\}.$$

Let $\mathrm{Hdg}(X)$ be the corresponding Hodge or as it sometimes called the special Mumford-Tate group of X (see [10, 15, 17, 18, 11]). It is a connected commutative reductive algebraic $\mathbb{Q}$-group. It is well-known ([17], p. 531) that

$$\mathrm{Hdg}(X) \subset \mathbf{U}_E.$$

Let $\mathfrak{hdg} = \mathfrak{hdg}_X$ be its Lie algebra. Clearly,

$$\mathfrak{hdg} \subset \mathfrak{u}_E = \{a \in E \mid a + a' = 0\}.$$

It is known that it is a commutative $\mathbb{Q}$-Lie algebra, i.e., its rank and dimension coincide, and

$$\mathrm{rk}(\mathfrak{hdg}) = \dim_{\mathbb{Q}} \mathfrak{hdg} \leq \dim_{\mathbb{Q}} \mathbf{U}_E = g;$$

the equality holds true if and only if

$$\mathrm{Hdg}(X) = \mathbf{U}_E.$$

For example, it is known that this equality holds true when g is a prime (a theorem of Tankeev–Ribet [17, 23]). For arbitrary dimensions there is a Ribet's inequality ([18], p.87)

$$\log_2(2g) \leq \dim_{\mathbb{Q}} \mathrm{Hdg}(X)$$

(see also [8]). For further properties and examples of the Hodge groups of CM-Abelian varieties see [18, 3, 8, 28].

There is a well-known natural isomorphism of $\mathbb{Q}_\ell$–vector spaces

$$V_\ell(X) = V(X) \otimes_{\mathbb{Q}} \mathbb{Q}_\ell.$$

It is known that for Abelian varieties of CM-type the $\mathbb{Q}_\ell$–Lie algebra $\mathfrak{g}_\ell^0$ is a commutative $\mathbb{Q}_\ell$–Lie algebra, i.e., its rank and dimension coincide. A theorem of Pohlman [16] asserts that the isomorphism of the $\mathbb{Q}_\ell$–vector spaces mentioned above gives us an identification

$$\mathfrak{hdg} \otimes_{\mathbb{Q}} \mathbb{Q}_\ell = \mathfrak{g}_\ell^0$$

of commutative $\mathbb{Q}_\ell$–Lie algebras.

Clearly, if $\mathrm{rk}(\mathfrak{g}_\ell^0) = g$, then it follows easily that $\dim_{\mathbb{Q}} \mathfrak{hdg} = g$ and, therefore,

$$\mathrm{Hdg}(X) = \mathbf{U}_E.$$

3.2. — Remark.

One may define the Hodge group $\mathrm{Hdg}(X)$ for any (complex) Abelian variety X not necessarily of CM-type [10]. It is a connected reductive algebraic $\mathbb{Q}$-group which is commutative if and only if X is of CM-type. The Mumford–Tate conjecture [20] asserts that the ℓ-adic Lie algebra $\mathfrak{g}^0_{\ell,X}$ can be obtained from the Lie algebra of $\mathrm{Hdg}(X)$ by extensions of scalars from $\mathbb{Q}$ to $\mathbb{Q}_\ell$. The theorem of Pohlman cited above proves the Mumford-Tate conjecture for Abelian varieties of CM-type.

4. — Main result.

The main result of the present paper is the following statement.

MAIN THEOREM. — *Let X be an absolutely simple g-dimensional Abelian variety of CM-type defined over a number field K and all endomorphisms of X are also defined over K. Let E be the endomorphism algebra of X.*

Assume that there exists a non-Archimedean place v of K such that X has a good reduction $X(v)$ at v and $X(v)$ is an almost ordinary Abelian variety $X(v)$. Then

$$\mathrm{Hdg}(X) = \mathbf{U}_E.$$

In other words,

$$\dim_{\mathbb{Q}} \mathrm{Hdg}(X) = \dim(X) = g.$$

4.2. — Remark.

Assume that g is *even* and $X(v)$ is a *simple* almost ordinary Abelian variety. Then $\dim(X(v)) = \dim(X) = g$ is also *even* and, as we have already seen,

$$g = \mathrm{rk}(X(v)) \le \mathrm{rk}(\mathfrak{g}^0_\ell) = \dim_{\mathbb{Q}} \mathfrak{hdg} \le \dim_{\mathbb{Q}} \mathbf{U}_E = g$$

and, therefore,

$$\dim_{\mathbb{Q}} \mathfrak{hdg} = \dim_{\mathbb{Q}} \mathbf{U}_E = g.$$

This proves the Theorem under our additional assumptions.

4.3. — Remark.

Assume that g is *odd* and $X(v)$ is a *simple* almost ordinary Abelian variety. Then $\dim(X(v)) = \dim(X) = g$ and, as we have already seen,

$$g - 1 \le \mathrm{rk}(X(v)) \le \mathrm{rk}(\mathfrak{g}^0_\ell) = \dim_{\mathbb{Q}} \mathfrak{hdg} \le \dim_{\mathbb{Q}} \mathbf{U}_E = g$$

and, therefore,

$$g - 1 \le \dim_{\mathbb{Q}} \mathfrak{hdg} = \dim_{\mathbb{Q}} \mathrm{Hdg}(X) \le \dim_{\mathbb{Q}} \mathbf{U}_E = g.$$

4.4. — Combining the last two Remarks, we obtain that the Theorem follows from the next two lemmas.

4.5. LEMMA. — *Let X be an absolutely simple g–dimensional Abelian variety of CM-type defined over a number field K and all endomorphisms of X are also defined over K. Let E be the endomorphism algebra of X.*

Assume that there exists a non-Archimedean place v of K such that X has a good reduction at v and this reduction is an almost ordinary Abelian variety $X(v)$. Then $X(v)$ is a simple Abelian variety.

4.6. LEMMA. — *Let Y be an absolutely simple g–dimensional Abelian variety of CM-type. Let E be the endomorphism algebra of Y.*

Assume that

$$\dim_{\mathbb{Q}} \mathrm{Hdg}(Y) = g - 1 = \dim_{\mathbb{Q}} \mathbf{U}_E - 1.$$

Then g is even.

4.7. — Remark.

It is well-known ([17], Th. 0, p. 524) that the equality

$$\mathrm{Hdg}(X) = \mathbf{U}_E$$

implies that all Hodge classes on all powers of X are linear combinations of the products of divisors classes. In particular, all these Hodge classes are algebraic, i.e., the Hodge conjecture holds true for all powers of X. Since the Mumford-Tate conjecture holds true for Abelian varieties of CM–type [16], we obtain that all Tate classes on all powers of X are linear combinations of the products of divisors classes. Indeed, by a theorem of Faltings [4], each 2–dimensional Tate class on an Abelian variety over a number field is a linear combination of divisor classes. In particular, all these Tate classes are algebraic, i.e., the Tate conjecture [24, 25] holds true for all powers of X.

5. — Proof of the Lemma 4.6.

We start this section with the explicit description of $\mathbb{Q}$–algebraic subtori in $\mathbf{U}_E$ of codimension 1. This description had tacitly appeared in [6] and, later, was explicitly formulated and proved in [9]. In our exposition we follow [9].

Suppose E contains an imaginary quadratic subfield k. Let us define the algebraic subtorus $\mathbf{SU}_{E/k}$ of $\mathbf{U}_E$ by the condition

$$\mathbf{SU}_{E/k}(\mathbb{Q}) = \{a \in \mathbf{U}_E(\mathbb{Q}) \mid \mathrm{Norm}_{E/k}(a) = 1\}.$$

One may easily check that $\mathbf{SU}_{E/k}$ has codimension 1 in $\mathbf{U}_E$. Clearly, its Lie algebra

$$\mathfrak{su}_{E/k} := \mathrm{Lie}(\mathbf{SU}_{E/k}) = \{a \in \mathfrak{u}_E \mid \mathrm{Tr}_{E/k}(a) = 0\}.$$

Here $\mathrm{Tr}_{E/k} : E \to k$ is the trace map. Notice, that this trace map commutes with the complex conjugation (if someone is unhappy with the definition of $\mathbf{SU}_{E/k}$ by its $\mathbb{Q}$–rational points then there is another description of $\mathbf{SU}_{E/k}$. Namely, it is a $\mathbb{Q}$–algebraic (connected) subtorus of $\mathbf{U}_E$ such that its Lie algebra coincides with $\mathfrak{su}_{E/k}$).

The following statement was proven in [9], Sect. 7.3.

5.1. Key Lemma. — *Let H be an algebraic subtorus of codimension 1 in $\mathbf{U}_E$. Then there exists an imaginary quadratic subfield k of E such that:*

$$H = \mathbf{SU}_{E/k}.$$

5.2. — Since $H := \mathrm{Hdg}(X)$ is an algebraic subtorus of codimension 1 in $\mathbf{U}_E$, we obtain, applying the Key Lemma , that there exists an imaginary quadratic subfield k of E such that

$$\mathrm{Hdg}(Y) = \mathbf{SU}_{E/k}.$$

This means that

$$\mathfrak{hdg} = \mathfrak{su}_{E/k}.$$

Now, let us choose a non-zero $\epsilon \in k \in E$ such that

$$\epsilon' = -\epsilon.$$

Now, if we consider $V(X)$ as a g–dimensional k–vector space, then the E–Hermitian form ψ_ϵ gives rise to the k–Hermitian form

$$\psi := \mathrm{Tr}_{E/k}\psi_\epsilon : V(X) \times V(X) \to k,$$

$$\psi(x, y) = \mathrm{Tr}_{E/k}(\psi_\epsilon(x, y)).$$

It follows easily that

$$\varphi(x, y) = \mathrm{Tr}_{E/k}(\epsilon^{-1}\psi(x, y))$$

for all $x, y \in V(X)$ and ψ is non-degenerate. Clearly,

$$\mathfrak{u}_E \subset \mathfrak{u}(V(X), \psi) :=$$

$$\{a \in \mathrm{End}_k V(X) \mid \psi(ax, y) + \psi(x, a'y) = 0 \; \forall x, y \in V(X)\}$$

and

$$\mathfrak{su}_{E/k} \subset \mathfrak{su}_k(V(X), \psi) := \{a \in \mathfrak{u}(V(X), \psi) \mid \mathrm{Tr}_{V(X)/k}(a) = 0\}.$$

Here

$$\mathrm{Tr}_{V(X)/k} : \mathrm{End}_k V(X) \to k$$

is the usual trace map on the algebra of k-linear operators of the k-vector space $V(X)$ (notice, that the maps $\mathrm{Tr}_{E/k}$ and $\mathrm{Tr}_{V(X)/k}$ coincide on E). So, we obtained that

$$\mathfrak{h}\mathfrak{d}\mathfrak{g} = \mathfrak{su}_{E/k} \subset \mathfrak{su}_k(V(X), \psi).$$

It turns out that the inclusion

$$\mathfrak{h}\mathfrak{d}\mathfrak{g} \subset \mathfrak{su}_k(V(X), \psi)$$

can be rewritten in terms of the action of k on the tangent space of X (see Weil [27]). Namely, if $\mathrm{Lie}(X(\mathbb{C}))$ is the tangent space of the *complex* Abelian variety X then the inclusion means that $\mathrm{Lie}(X(\mathbb{C}))$ is a *free* $k \otimes_{\mathbb{Q}} \mathbb{C}$-module ([9], Lemma 2.8; see also [17], p. 525). Since $\mathrm{Lie}(X(\mathbb{C}))$ is a g-dimensional complex vector space and $k \otimes_{\mathbb{Q}} \mathbb{C} = \mathbb{C} \oplus \mathbb{C}$, the dimension g must be *even*. This ends the proof.

6. — Proof of the Lemma 4.5.

By functoriality of Néron models, there is a natural embedding

$$E = \mathrm{End}(X) \otimes \mathbb{Q} \to \mathrm{End}(X(v)) \otimes \mathbb{Q}$$

and $1 \in E$ acts on $X(v)$ as the identity map. Notice, that E is a number field and

$$[E : \mathbb{Q}] = 2\dim(X) = 2\dim(X(v)).$$

The following proposition will be proved at the end of this Section.

6.1. Proposition. — *Let Y be an Abelian variety over an arbitrary field $\mathcal{K}$ and assume that the semisimple $\mathbb{Q}$-algebra $\mathrm{End}^0(Y) = \mathrm{End}(Y) \otimes \mathbb{Q}$ contains a number field F of degree $2\dim(Y)$ such that $1 \in E$ is the identity automorphism of Y. Then there exists a $\mathcal{K}$-simple Abelian variety Z over k such that Z is $\mathcal{K}$-isogenous to the power Z^r of Z with $r = \dim(Y)/\dim(Z)$.*

6.2. — Applying the Proposition 6.1, we obtain that there exists a $k(v)$-simple Abelian variety Z over $k(v)$ such that $X(v)$ is isogenous to Z^r for a certain positive integer r. In order to prove the lemma 4.5, we have only to check that

$$r = 1.$$

First, notice, that each slope of the Newton polygon of $X(v)$ has length divisible by r. Since the slope $1/2$ has length 2, either $r = 1$ or $r = 2$. If $r = 2$ then $X(v)$ is isogenous to Z^2 and, therefore, $1/2$ is the slope of the Newton polygon of Z with length 1. But it cannot be true, since the length of the slope $1/2$ must always be even [30, 7], due to the fact that all the break-points of the Newton polygon are integral. This rules out the case $r = 2$. So, $r = 1$ and we are done.

6.3. — Proof of the Proposition 6.1.

Assume that Y is not $\mathcal{K}$-isogenous to a power of a $\mathcal{K}$-simple Abelian variety. Then, using the Poincaré reducibility theorem, one may easily check that there exist Abelian $\mathcal{K}$-subvarieties $Y_1, Y_2 \subset Y$ of *positive* dimensions, enjoying the following properties :

a) the natural homomorphism

$$Y_1 \times Y_2 \to Y, \ (y_1, y_2) \to y_1 + y_2$$

is an isogeny;

b) $\mathrm{Hom}(Y_1, Y_2) = \{0\}, \ \mathrm{Hom}(Y_2, Y_1) = \{0\}$.

This implies that

$$0 < \dim(Y_1) < \dim(Y); \ 0 < \dim(Y_2) < \dim(Y);$$

$$\mathrm{End}^0(Y) = \mathrm{End}^0(Y_1) \oplus \mathrm{End}^0(Y_2).$$

Let $\mathrm{pr}_i : \mathrm{End}^0(Y) \to \mathrm{End}^0(Y_i)$ be the corresponding projection homomorphisms. Clearly, if $\mathrm{id}_Y \in \mathrm{End}^0(Y)$ is the identity automorphism of Y then $\mathrm{pr}_i(\mathrm{id}_Y) \in \mathrm{End}^0(Y_i)$ is the identity automorphism id_{Y_i} of Y_i. This implies that $F_i := \mathrm{pr}_i(F) \subset \mathrm{End}^0(Y_i)$ is a number field isomorphic to F; in particular, its degree equals $2\dim(Y) > 2\dim(Y_i)$ $(i = 1, 2.)$ Now, in order to get a contradiction let us recall the following well-known fact (see [22], Sect. 5.1, Proposition 2).

6.4. SUBLEMMA. — *If the endomorphism algebra of an m-dimensional Abelian variety contains a number field which, in turn, contains the identity automorphism, then the degree of this field divides $2m$. In particular, it does not exceed $2m$.*

6.5. — Now, in order to finish the proof by coming to the contradiction, one has only to apply the Sublemma to the Abelian variety Y_i of dimension $m = \dim(Y_i)$ and the number field F_i of degree $2\dim(Y) > \dim(Y_i)$.

6.6. — Remark.

Similar arguments prove that if k is a finite field and A is a g-dimensional k-simple almost ordinary Abelian variety over k then A is absolutely simple. Indeed, for each extension k' of k the Abelian variety $A' := A \times k'$ is an almost ordinary Abelian variety and $\mathrm{End}^0 A'$ contains a number field $\mathrm{End}^0 A$ of degree $2g = 2\dim(A')$, which, in turn, contains the identity automorphism. By the Sublemma, A' must be k'–isogenous to the power Z^r of k'–simple Abelian variety Z. Now, the same arguments with the Newton polygons as in Sect. 6.2, prove that $r = 1$, i.e., $A' = Z$ is $k\prime$–simple.

7. — Acknowledgements

I am deeply grateful to H.W. Lenstra and B. Moonen for helpful discussions. This paper is a result of my stay in Paris in June-July of 1993 and I am deeply grateful to the Groupe d'Etudes sur les Problèmes Diophantiens (Université de Paris VI) for the hospitality. The support of the Université Paris Nord is also gratefully acknowledged. I am grateful to Frans Oort who had read the manuscript and made many valuable remarks. My special thanks go to Daniel Bertrand and Larry Breen, whose efforts made my trip to France possible.

Manuscrit reçu le 21 janvier 1994

REFERENCES

[1] F.A. Bogomolov. — *Sur l'algébricité des représentations ℓ-adiques*, C.R. Acad. Sci. Paris Sér. I Math. **290**, 1980, 701–704.

[2] P. Deligne. — (notes by J.S. Milne). *Hodge cycles on abelian varieties*, Springer Lecture Notes in Math. **900**, 1982, 9–100.

[3] B. Dodson. — *On the Mumford-Tate group of an abelian variety with complex multiplication*, J. Algebra **111**,1987,49-73.

[4] G. Faltings. — *Endlichkeitssätze für abelsche Varietäten über Zahlkörpern*, Invent. Math. **73**,1983, 349–366.

[5] N. Koblitz. — *p-adic variation of the zeta-function over families of varieties defined over finite fields*, Compositio Math. **31**, 1975, 119–218.

[6] H.W. Lenstra, Jr. and F. Oort. — *Simple Abelian varieties having a prescribed formal isogeny type*, J. Pure Appl. Algebra **4**, 1974, 47–53.

[7] H.W. Lenstra Jr. and Yu.G. Zarhin. — *The Tate conjecture for almost ordinary abelian varieties over finite fields*, Advances in Number Theory, Proc. of the Third Conf. of the CNTA, 1991 (F. Gouvêa and N. Yui, eds.), 179–194. Clarendon Press, Oxford, 1993.

[8] L. Mai. — *Lower bounds for the rank of a CM-type*, J. Number Theory **32**, 1989, 192-202.

[9] B.J.J. Moonen and Yu.G. Zarhin. — *Hodge classes and Tate classes on simple abelian fourfolds*, Duke Math. J., to appear.

[10] D. Mumford. — *A note of Shimura's paper "Discontinous groups and abelian varieties*, Math. Ann. **181**, 1969, 345–351.

[11] V.K. Murty. — *Computing the Hodge group of an abelian variety*, Séminaire de Théorie des Nombres, Paris 1988–89, (C. Goldstein éd.), Progress in Math., Birkhäuser **91**, 1990, 141–158.

[12] P. Norman and F. Oort. — *Moduli of abelian varieties*, Ann. of Math. **112**, 1980, 413–439.

[13] F. Oort. — *CM-liftings of Abelian varieties*, J. Algebraic Geometry **1**, 1992, 131–146.

[14] F. OORT. — *Moduli of Abelian varieties and Newton polygons*, C.R. Acad. Sci. Paris Sér. I Math. **312**, 1991, 385–389.

[15] I.I. PIATETSKII–SHAPIRO. — *Interrelations between the Tate and Hodge conjectures for abelian varieties*, Math. USSR Sbornik **14**, 1971, 615–625.

[16] H. POHLMAN. — *Algebraic cycles on abelian varieties of complex multiplication type*, Ann. of Math. **88**, 1968, 161–180.

[17] K. RIBET. — *Hodge classes on certain types of abelian varieties*, Amer. J. of Math. **105**, 1983, 523-538.

[18] K. RIBET. — *Division fields of abelian varieties with complex multiplication*, Mémoires de la S.M.F., nouvelle série **2**, 1980, 75–94.

[19] J.-P. SERRE. — *Abelian l-adic representations and elliptic curves*, Addison Wesley, second edition, 1989.

[20] J.-P. SERRE. — *Représentations l-adiques*, Kyoto International Symposium on Algebraic Number Theory, Japan Society for the Promotion of Science, Tokyo (1977), 177–193 (= Œ 112).

[21] G. SHIMURA. — *On the field of definition for a field of automorphic functions*, Ann. of Math. (2) **80**, 1964, 160–189.

[22] G. SHIMURA and Y. TANIYAMA. — *Complex multiplication of abelian varieties and its applications to number theory*, Publ. Math. Soc. Japan **6**, 1961.

[23] S.G. TANKEEV. — *Cycles on simple Abelian varieties of prime dimension*, Izv. Akad. Nauk SSSR ser. matem.; English translation in Math. USSR Izvestija **46**, 1982, 155–170.

[24] J. TATE. — *Algebraic cycles and poles of zeta functions*, Arithmetical Algebraic Geometry, Harper and Row, New York, 1965, 93–110.

[25] J. TATE. — *Endomorphisms of Abelian varieties over finite fields*, Invent. Math. **2**, 1966, 134–144.

[26] J. TATE. — *Classes d'isogénie des variétés abéliennes sur un corps fini* (d'aprés T. Honda), Séminaire Bourbaki **352** (1968), Springer Lecture Notes in Mathematics **179** (1971), 95–110.

[27] A. WEIL. — *Abelian varieties and the Hodge ring*, Collected papers, Springer–Verlag **III**, 1980, 421–429.

[28] S.P. WHITE. — *Sporadic cycles on CM abelian varieties*, Compositio Math. **88**, 1993, 123–142.

[29] Yu.G. ZARHIN. — *Abelian varieties of K3 type and ℓ-adic representations.* Algebraic Geometry and Analytic Geometry Tokyo 1990, ICM-90 Satellite Conference Proceedings, Springer-Verlag, Tokyo (1991), 231–255.

[30] Yu.G. ZARHIN. — *Abelian varieties of K3 type*, Séminaire de Théorie des Nombres, Paris 1990–91, (S. David éd.), Progress in Math., Birkhäuser **108**, 1993, 263–279.

[31] Yu.G. ZARHIN. — *The Tate conjecture for non-simple Abelian varieties over finites fields.* Algebra and Number Theory, Proceedings of a Conference held at the Institute for Experimental Mathematics, University of Essen, Germany, December 2-4, 1992 (G. Frey and J. Ritter, eds.) de Gruyter, Berlin, (1994), 267–296.

[32] Yu.G. ZARHIN. — *Abelian varieties having a reduction of K3 type*, Duke Math. J. **65**, 1992, 511–527.

[33] Yu.G. ZARHIN. — *ℓ-adic representations and Lie algebras.* Elliptic Curves and Related Topics, (M. Ram Murty and H. Hisilevsky, eds.) CRM Proceedings & Lecture Notes **4** (1994), AMS, 183–195.

Yuri G. ZARHIN
The Pennsylvania State University,
Department of Mathematics
325 McAllister Building,
University Park, PA 16802,
USA
e-mail address : zarhin@math.psu.edu
and
Institute for Mathematical
Problems in Biology
Russian Academy of Sciences
Pushchino, Moscow Region,
142292 Russia

Printed in the United States
By Bookmasters